PHYSICS
STUDENTS' GUIDE 1
UNITS A to G

D1344525

Revised Nuffield Advanced Science

**General editor,
Revised Nuffield
Advanced Physics**
John Harris

Consultant editor
E. J. Wenham

Editors of Units in this _Guide_
Mark Ellse
David Grace
Roger Hackett
Peter Harvey
Paul Jordan
Charles Milward
Trevor Sandford
Maurice Tebbutt
Nigel Wallis

Contributors to this _Guide_
Michael Carrick
Bev Cox
Mark Ellse
John Gardner
David Grace
Tom Gregory
Roger Hackett
Peter Harvey
Paul Jordan
April Bueno de Mesquita
Charles Milward
John Minister
Susan Ross
Trevor Sandford
Richard Spiby
Maurice Tebbutt
Mark Tweedle
Nigel Wallis

The Nuffield–Chelsea Curriculum Trust is grateful to the authors and editors of the first edition:

Organizers: P. J. Black, Jon Ogborn; **Contributors:** W. Bolton, R. W. Fairbrother, G. E. Foxcroft, Martin Harrap, John Harris, A. L. Mansell, A. W. Trotter.

PHYSICS
STUDENTS' GUIDE 1
UNITS A to G

REVISED NUFFIELD ADVANCED SCIENCE
Published for the Nuffield–Chelsea Curriculum Trust
by Longman Group Limited

Addison Wesley Longman Limited
Edinburgh Gate, Harlow, Essex CM20 2JE, England
and Associated Companies throughout the world

First published 1971
Revised edition first published 1985
Nineth impression 1997
Copyright © 1971, 1985 The Nuffield-Chelsea Curriculum Trust

Design and art direction by Ivan Dodd
Illustrations by Oxford Illustrators Limited

Filmset in Times Roman and Univers
Produced by Longman Singapore Publishers Pte Ltd
Printed in Singapore

ISBN 0 582 35415 3

The Publisher's policy is to use paper manufactured from sustainable forests.

Cover

An electron gun, part of a television picture tube or cathode ray tube. When electrons from the gun strike the phosphor on the screen the screen emits light, so building up a visible image.

Paul Brierley

CONTENTS

A
B
C
D
E
F
G

FOREWORD

When the Nuffield Advanced Science series first appeared on the market in 1970, they were rapidly accepted as a notable contribution to the choices for the sixth-form science curriculum. These courses were devised by experienced teachers working in consultation with the universities and examination boards, and subjected to extensive trials in schools before publication and they introduced a new element of intellectual excitement into the work of A-level students. Though the period since publication has seen many debates on the sixth-form curriculum, it is now clear that the Advanced Level framework of education will be with us for some years in its established form. That period saw various proposals for change in structure which were not accepted but the debate to which we contributed encouraged us to start looking at the scope and aims of our A-level courses and at the ways they were being used in schools. Much of value was learned during those investigations and has been extremely useful in the planning of the present revision.

The revision of the physics course under the general editorship of John Harris has been conducted with the help of a committee under the chairmanship of K. F. Smith, Professor of Physics, University of Sussex. We are grateful to him and to the committee. We also owe a considerable debt to the Oxford and Cambridge Schools Examinations Board which for many years has been responsible for the special Nuffield examinations in physics and to the Assistant Secretary of the Board, Mrs B. G. Fraser, who has been an invaluable adviser.

The Nuffield–Chelsea Curriculum Trust is also grateful for the advice and recommendations received from its Advisory Committee, a body containing representatives from the teaching profession, the Association for Science Education, Her Majesty's Inspectorate, universities, and local authority advisers; the committee is under the chairmanship of Professor P. J. Black, educational consultant to the Trust.

Our appreciation also goes to the editors and authors of the first edition of Nuffield Advanced Physics, who worked with Jon Ogborn and P. J. Black, the project organizers. Their team of editors and writers included: W. Bolton, R. W. Fairbrother, G. E. Foxcroft, Martin Harrap, John Harris, A. L. Mansell, and A. W. Trotter. Much of their original work has been preserved in the new edition.

I particularly wish to record our gratitude to the General Editor of the revision, John Harris, Lecturer at the Centre for Science and Mathematics Education, Chelsea College, and a member of the team responsible for the first edition. To him, to E. J. Wenham, Consultant Editor of the revision, and to the editors of the Units in the revised course – all teachers with a wide experience of the needs of students and of the current state of physics education – Roger Hackett, Nigel Wallis, David Grace, Mark Ellse, Charles Milward, Trevor Sandford, Paul Jordan, Peter Harvey, Maurice Tebbutt, David Chaundy, Wilf Mace, Stephen Borthwick, Peter Bullett, and Jon Ogborn, we offer our most sincere thanks.

I would also like to acknowledge the work of William Anderson, publications manager to the Trust, his colleagues, and our publishers, the Longman Group, for their assistance in the publication of these books. The editorial and publishing skills they contribute are essential to effective curriculum development.

K. W. Keohane,
Chairman, Nuffield–Chelsea Curriculum Trust

ABOUT THE COURSE AND ABOUT THIS BOOK

At the beginning of a new course in physics we think you should know something about its special features, the ways in which it may be different from other physics courses at A-level, and the reasons for these differences.

Physics, and the world with it, are changing so fast that no one can tell what aspects of physics you will use in, say, ten years' time: however, one can be pretty sure that there are some basic ideas that will be relevant to the new problems of tomorrow. We have tried to build the course around these basic ideas.

A-level physics courses do not all cover exactly the same topics. So you will find some topics in this course that do not appear in other A-level courses. You will also find that this course does not include some items that are included in others. But you should know that there is a core syllabus which is common to all A-level physics courses, including this one. The core of physics which universities and other institutions regard as essential to further study is covered in this course.

You will study fewer physics topics in this course than you would in most other A-level courses because we believe that it is more important to have a good understanding of some basic ideas and to be able to use them confidently, than to know less about more. Those who understand their physics are the ones who can offer sensible, relevant ideas that would help towards clearing up a problem.

The course aims to show you what doing physics is like, and this is another reason for encouraging plenty of discussion of problems, for that is the way physicists work. To use physics successfully and to understand what it can, and what it cannot do, it is important to know about such things as how theories, models, experiments, and facts fit together. Physicists also guess, estimate, and speculate, and you will be asked to do these things too.

But physics is more than facts and theories: it has practical value and through its application in science and technology it has social and economic effects on the way we live. Some of the Units in the course emphasize this aspect of physics, and in most of the Units there are questions and reading about how physics can be used in engineering and technology.

This course regards students as active, creative people making their own sense out of the physical phenomena they meet. The alternative view of students as passive recipients of knowledge, dry sponges absorbing facts and formulae provided by a teacher or text book is *not* one we have adopted in designing this course.

You will find that this emphasis on activity takes many forms: you will do many experiments – and sometimes be asked to demonstrate them to others; you will read from a variety of sources including text books, magazines, and special articles – and you may be asked to give a written or verbal report on what you have found out; you will do two independent investigations, in which you plan and carry out your own work on a topic of your choice for a couple of weeks. And there should be plenty of opportunities to *talk* about the physics you are learning – because that is probably the best way to find out how well you understand it.

We think that to learn effectively, now and in the future, certain skills are very valuable, and in designing the course we have sought ways of helping you develop skills of various sorts: including experimental, communication, and mathematical skills.

This *Student's guide* is not the 'text book' of the course. It is important to realize that you must keep your own notes and records of the course as you go along. You will also need access to other sources: text books and reference books in the school or college library, data books, journals, magazines, and so on.

More about this book

The course is divided into twelve Units. Units A to G are covered in this book, Units H to L in *Students' guide 2*. For each Unit there are:

Summaries   These are very short and are not intended to do more than the name suggests. They only *summarize* the most important ideas of each part of the course. They do not show you how results are derived; they do not give full explanations of methods, or evidence for the facts stated. You can follow some of these derivations by working through questions in the book; some of the evidence is provided by the experiments and demonstrations suggested for the course. You will find references in the margin to these questions and to the practical work.

These summaries will be most useful to you *after* you have studied a particular part of the course in class, and for revision. We hope they will help you remember what some of the key ideas are, and what the experimental evidence and theoretical basis for them are.

Readings   As part of most Units we have included a few short passages which depend on the ideas developed in the Unit. We have included questions about them, to help you develop the skill of reading with a purpose, which is one of the aims of the course.

Laboratory notes   You will find brief notes on most of the experiments and demonstrations suggested for the course – notes, not instructions. Their purpose is to show what equipment is needed, how to set it up, and to suggest some questions which can be answered, or some observations which can be made with it.

Home experiments   Learning and doing science is not something to be confined to school or the laboratory. So in some of the Units we have included suggestions for practical activities that you can do at home.

Questions   With each Unit we have included a large number of questions – more than any one student will be able to do while studying the Unit. There are various kinds of questions. Each has a letter as well as a number: the letter tells you what kind of a question it is.

The code is:

I Introductory questions: These are questions which you should be able to do confidently *before* you start the Unit. These questions make use of the simpler ideas which the Unit takes as its starting point.

L Learning questions: Many of the derivations of standard results, formulae, relationships, etc., are presented in the form of questions. These questions are intended to help you learn new physics rather than test what you have learned before.

P Practice questions: You should find these relatively simple and straightforward. They are intended to give you practice in using the new ideas you have learned in fairly simple situations.

E Essay, estimation, and discussion questions: A more open category of question to encourage a different kind of thinking and writing. There are no right and wrong answers to these.

R Review and revision questions: As the name suggests, these are examples of the kind of questions that you should be able to answer at the end of the course. This kind of question may cover work from more than one Unit. Several of them are taken from past examination papers.

Formulae, relationships, data, and symbols are gathered together at the end of each *Students' guide*.

Experimental work   The short passage on pages xii to xvi of this book makes some general points about doing experiments, taking measurements, and interpreting results. You will find it useful to refer to it from time to time throughout the course.

So there are many things in this *Students' guide* that you should find helpful during the course. But it is important to realize that you will need to use other sources as well, and to keep your own notes of the work as you go along.

More about skills

Some of the useful skills we hope the course will help you acquire include:

Experimental skills: ability to use a range of modern tools used in physics; ability to design effective experiments; understanding of the factors which affect the reliability of measurements.

Communication skills: ability to extract and make use of information from technical articles; ability to present technical arguments; ability to convey technical ideas to a 'lay' reader; ability to discuss critically and constructively experimental or theoretical work, either your own or a colleague's. The course will give you opportunities to practise these communication skills both verbally and in writing.

Mathematical skills: algebraic manipulation; handling scientific notation quickly and efficiently; giving useful rough estimates and dealing sensibly with approximations; graphical work – use of graphs to display data, recognition of some important functions, interpretation of the physical significance of the slopes and areas under some graphs; using the idea of rate of change in many experimental situations; ability to discuss the exponential function in a number of different situations; numerical methods of solving equations, which lead into computer methods.

The course itself

This Revised Nuffield A-level physics course is the result of the combined efforts of many teachers with considerable experience of teaching the original course. One of the main reasons that we want to offer you some physics is that we like the subject and get excited about it. So we hope you enjoy it too.

The Editors

EXPERIMENTAL WORK

Wilf Mace
King Edward VII School, Sheffield

Physics is about how the world works, and the only way to find how it works is to observe it. This is why observation and measurement are vital to physics. As with all techniques, experimenting has its own set of particular skills, and these notes are to draw your attention to a number of things that need to be borne in mind in practical work.

Uncertainty

Nothing can ever be measured exactly. Even if an object apparently reads exactly 15 divisions against a millimetre scale, we know that if it were 14.98 mm or 15.02 mm we simply wouldn't be able to tell the difference: there is always an *uncertainty* in every measurement we make, however carefully we work, and the size of the uncertainty is governed by the instrument we are using. We therefore need to consider our instruments rather carefully, in order to make the best possible use of them.

Characteristics of measuring instruments

The variety of instruments and techniques at our disposal have some rather obvious properties, and also some less obvious ones that may be equally important. There is a vocabulary about this, and a glossary of the most important terms is given below. We don't suggest you should 'learn' these at the start, but as your experience grows you will find that the ideas they represent keep cropping up, and you will come to appreciate the kinds of situation in which they need to be taken into consideration.

Range: This term has the obvious meaning. Typical values would be 0–100 mA for a meter, −10 to +110°C for a thermometer, and so on.

Resolution: This refers to the smallest change (increment) that can be detected. For instruments with scales (micrometer, ammeter, thermometer, etc.) it is normally taken to mean the value indicated by one smallest scale division, though in practice, fractions of a scale division can often be judged quite reliably. For digital readout instruments it means the value of one unit in the last digit. Manufacturers often express resolution as a percentage of the range. Thus a typical school microammeter may have a range of 100 µA, with 50 divisions: this constitutes a resolution of 2 µA or 2 % (of the range). But note that resolution is by no means the only factor which may affect the uncertainty of a measurement: see, for example, *repeatability*, below.

Sensitivity: This refers particularly to instruments such as the light beam galvanometer, and the oscilloscope, where a scale of mm or cm is provided and may be made to represent different ranges of values. Statements of sensitivity take the form '16.2 mm μA^{-1}' or (in inverse form) '0.2 V cm^{-1}'.

Response time: A light beam galvanometer responds rather slowly to changes of voltage; an oscilloscope responds within a small fraction of a microsecond. The rapid response is obviously very necessary for following high frequency fluctuations, as in electronics, but there are situations when a long response time is advantageous. In a ratemeter for radioactivity measurements a long response time is deliberately built in, so that the pointer, instead of flickering at every count, tends to show a fairly steady averaged value of the count rate.

Response time can be an important property of transducers too. Temperature-measuring instruments tend to be very sluggish, so that rapid fluctuations of temperature are difficult to measure reliably. Some types of photocell, too, have response times of an appreciable fraction of a second, and may be unsuitable for fast photographic work.

Accuracy: Given an ammeter stated by the manufacturer to have a range of 5 mA, we tend to assume that this figure is reliable: similarly with metre rules, stopwatches, frequency generators, and indeed all scaled instruments. This is what we mean by accuracy. Sometimes instruments are stated to conform to some British Standard Specification (B.S.S.), and that is a guarantee that they are accurate to within some specified limit. In school laboratories, accuracy has sometimes to be taken on trust, though one learns to notice that some makes of meter, for example, tend to be consistent with one another and so can be hoped to be accurate, or perhaps that 10 g slotted masses can disagree by 0.4 g and therefore some of them must be inaccurate by at least 0.2 g.

Linearity: A meter may be extremely accurate at one point, but inaccurate over other parts of its scale: this is referred to as a fault in linearity, since a graph showing the true current against the scale reading would not be a straight line. In high accuracy work it is often necessary to draw a calibration curve to use in conjunction with an instrument to correct readings for non-linearity.

Repeatability: Experimental situations are very often not as simple as we would like to make them. Pointers may be sticking slightly, a bad contact may make a current change slightly, a clamp may not be sufficiently rigid, we may handle a micrometer clumsily: the list is endless. But effects like these are very easily missed unless we go out of our way to look for them, so the first thing we should do with *any* reading is to check that it is repeatable. Whatever the 'theoretical' resolution of an instrument may be, the actual uncertainty in a measurement must be judged in terms of repeatability.

Systematic error: Some factor in an experimental situation may cause a whole series of apparent values to be wrong by the same amount. If in radioactivity measurements we fail to take into account the background radiation which is always present, this will cause a systematic error in our readings. Another common cause of systematic error is the faulty zero setting of a meter. Often systematic error can be diagnosed by plotting a suitable graph: a line expected to show proportionality may not pass through the origin, for example.

Drift: This term speaks for itself: readings are not only not repeatable, but they move further and further in the same direction. Battery-powered instruments are prone to this, as their supply runs down, but there can be other causes: measurements of the length of a wire may drift because of changing temperature, for example. Sensitive electrical readings can sometimes drift because of temperature changes in components or at junctions between different conductors (causing thermoelectric e.m.f.s).

Loading effects: If we want to know the temperature of some water, we put a thermometer in it. But the thermometer takes some energy from the water, and thereby reduces the temperature it was put there to measure. In the same way, a voltmeter reduces the p.d. between the terminals of a battery to which it is applied, because it takes some current. In situations like this we say that the measuring instrument is *loading* the system to which we have applied it. (Load is a widely used term: the torch bulb is the load on the battery, the consumers' equipment is the load on a power station, and so on.) Loading does not always have to be taken into account, as it should be in the two examples above. The voltmeter built into a power supply is technically a load on the supply, but the reading it gives *is* the p.d. which that supply is providing, and as that is normally what we want to know, no correction is needed. Loading is relevant to more than measurement systems – see the article on 'Systems' in *Physics in engineering and technology.*

Validity: This is a term we apply to whole techniques rather than to instruments: is the apparatus measuring what we assume it is? To take a very obvious example, suppose we want to measure the rate of heating of some water by an immersion heater. If we put the thermometer high in the water it will show a rapid rise in temperature, but if we put it below the level of the heater its readings may change very slowly indeed. Neither set of readings is valid, because they are not values of the true average temperature of the water. Validity considerations can be much more subtle than this: does an ammeter read the correct value for an unsmoothed direct current? To answer that we have to think very hard about what we are going to use the reading for; in some cases the answer might be yes, in others it could well be no.

The effect of uncertainty upon conclusions

Once a set of readings has been taken, we have to assess what conclusions we can draw from them – *and* how uncertain the conclusions are. This involves a little calculation, and often makes use of graphs.

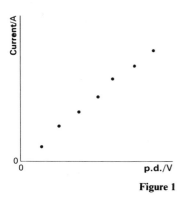

Figure 1

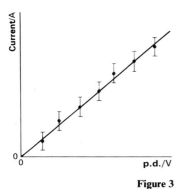

Figure 3

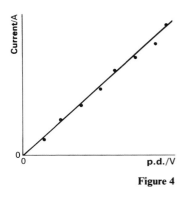

Figure 4

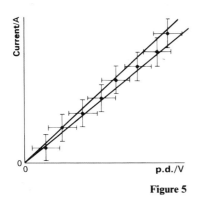

Figure 5

Graphs

Two examples will help you to appreciate how uncertainty can be dealt with in terms of graphs.

1   If we are trying to find out whether two quantities are proportional (current and p.d. for some new component, perhaps), we generally draw a graph like figure 1.

Do these points seem to be indicating a straight line or not? If we hold the graph almost up to eye level and look along it from the origin, it looks like figure 2. This seems to show a very definite curvature away to the right.

Figure 2

But suppose there were appreciable uncertainties in the readings plotted vertically. We can allow for this by plotting *error bars*, as in figure 3.

A straight line drawn through the origin and through all the error bars, now shows that a proportional relationship is *possible*. But 'possible' is the operative word. One inescapable fact in science is that no theory can ever be 'proved correct': we may believe very firmly in certain laws, but we have to recognize that one day a very careful experiment might disprove any one of them. Older text books often talked about 'verifying' laws. But no law can ever be positively verified: the most we can do is fail to contradict it – and to regard an experiment as an attempt to *disprove* something is often a very healthy attitude of mind.

2   Suppose we are satisfied that a conductor obeys Ohm's Law (as closely as we will ever need), and we want to find its resistance. We could take a number of readings and calculate V/I for each, getting a series of different values and taking the average.

Yet this might not be the best thing to do. Plotting the results might give something like figure 4.

We can look at the lie of these points using a ruler (or better, a length of cotton) to 'average' much more intelligently. One point stands out as further from the general line than the rest. If we had simply taken the average V/I value, instead of drawing the graph and calculating its slope, this point would have been distorting the average, and therefore distorting our conclusion as to the most likely value for the resistance.

Finally, we can obtain from our graph a firm value for the highest and lowest possible values of the resistance, by drawing in the most extreme lines that can be fitted within the error bars (figure 5).

In these graphs we have shown variations and uncertainties which are quite large. Graphs of your own results will generally (though perhaps not always) be much more precise. It will be up to you to decide, in the light of the above ideas, how best to interpret them intelligently.

Calculation of uncertainty

If a reading is 8.6 with an uncertainty of ± 0.1, we can consider the figure 0.1 itself: it is called the *absolute uncertainty*. We can also express it as a fraction of the measured value: 0.1/8.6. This is close to 0.1/10, or 1%, and we say that the *percentage uncertainty* is $\pm 1\%$. (Note that since uncertainties are themselves a matter of estimation, we rarely write them to more than one significant figure.)

There are two simple rules for calculating the uncertainty in the final result of an experiment:

i Where quantities are added or subtracted, *absolute* uncertainties must be added.

Thus if a flask containing water weighs 187 ± 2 g and when empty weighs 107 ± 2 g, then the mass of water is 80 ± 4 g. If a smaller quantity of water was used, and the readings were 117 ± 2 g and 107 ± 2 g, then the mass of water would be 10 ± 4 g. Note that the same absolute uncertainty (± 4 g) is now a much bigger percentage uncertainty.

ii Where quantities are multiplied or divided, *percentage* uncertainties must be added.

For example, in calculating the density of a sphere, measuring mass and radius, we write

$$\rho = \frac{m}{\frac{4}{3}\pi r^3}.$$

Suppose

$m = 90.2 \pm 0.1$ g

$\quad = 0.0902 \pm 0.000\,1$ kg (about 1 in 1000, *i.e.*, 0.1%)

and

$r = 4.8 \pm 0.1$ cm

$\quad = 0.048 \pm 0.001$ m (about 1 in 50, *i.e.*, 2%)

The percentage uncertainty in ρ will be the sum of the percentage uncertainty in m (0.1%), and *3 times* the percentage uncertainty in r (because the formula involves r^3, *i.e.*, $r \times r \times r$), that is, 6%.

This amounts to 6.1% in all, so we write

$$\rho = \frac{0.0902 \text{ kg}}{\frac{4}{3}\pi \times (0.048)^3 \text{ m}^3} \pm 6.1\%$$

$$= 195 \text{ kg m}^{-3} \pm 6.1\%$$

$$= 195 \pm 11.9 \text{ kg m}^{-3} \text{ by calculator}$$

Since we only know the uncertainty roughly we round this off as

$$\rho = 190 \pm 10 \text{ kg m}^{-3}$$

Note that just as the percentage uncertainty in r^3 is 3 times the percentage uncertainty in r, so the percentage uncertainty in $\sqrt{x}$ ($= x^{1/2}$) is *half* that in x, and so on.

ACKNOWLEDGEMENTS

One of the pleasantest aspects of the development of *Revised Nuffield Advanced Physics* has been the willing way in which so many people have contributed and become involved in the work. Above all, teachers have helped in many ways, and the very number who have done so makes it impossible to acknowledge the contribution of each individual. Many have offered suggestions at meetings or have written in with ideas for questions, demonstrations, and so on. We have tried to consider carefully all the suggestions put forward and, inevitably, it is impossible to give proper credit to the source or origin of every idea we have used. One who has made a particularly valuable contribution in this way is Colin Price. To him and the many others whose contributions go unacknowledged, we offer our sincere thanks.

Other teachers have helped by conducting trials of some of the more radically changed parts of the course, and of a major innovation – the 'Dynamic modelling system'. The trial schools are: Aylesbury Grammar School; Beechen Cliff School, Bath; Bexley-Erith Technical High School, Bexley; Bishop Hedley High School, Merthyr Tydfil; Cheltenham College; Esher College; Forest Hill School, London; Godolphin and Latymer School, London; The Grammar School, Batley; The Greenhill School, Tenby; Haverstock School, London; Heathland School, Hounslow; Henbury School, Bristol; Highfield School, Wolverhampton; Howell's School, Llandaff; King Edward VI College, Nuneaton; Kingsbridge School; Lady Margaret High School, Cardiff; Malvern College; Marlborough College; Netherhall School, Cambridge; North London Collegiate School; Northgate High School, Ipswich; Oulder Hill Community School, Rochdale; Richmond-upon-Thames College; Royal Grammar School, High Wycombe; Rugby School; Samuel Ward Upper School, Haverhill; and Sutton Manor High School.

We are grateful to the Inner London Education Authority for trying some of our material on electronics in their 1983 Summer School for sixth-form students at the North London Science Centre.

Mark Ellse has read and commented on much of the draft material, and has made particularly useful suggestions about the up-dating of some experiments and pieces of equipment.

Thanks are due to a group of teachers, convened by Bob Fairbrother, who met several times to discuss assessment. Their suggestions led to some changes in the structure of the examination.

Others, as well as teachers, have helped, of course. While he was working as a technician at the Centre for Science and Mathematics Education, Chelsea College, Phil Webb found time in a busy schedule to try out ideas for demonstrations and experiments, and to suggest ideas for new apparatus.

CLEAPSE School Science Service reviewed all the suggested experiments and demonstrations and made useful suggestions on the safety aspects of some of them.

Industry has helped too, and, among others, we are indebted to Rank Xerox, Amersham International P.L.C., and the CEGB for technical help and information.

Examination questions in the *Students' guide* are reprinted by permission of the Oxford and Cambridge Schools Examination Board. With one exception all are taken from Oxford and Cambridge Nuffield A-level Physics papers. The exception is one question taken from an Oxford and Cambridge O-level Nuffield Physics paper. Where guide lines for answers to examination questions are provided it must be understood that these are not the Examination Board's responsibility.

The Consultative Committee have, I believe, been asked to work harder and contribute more than is usually expected of such a group. As well as attending many meetings they have read and commented in detail on draft manuscripts – sometimes in a far from ideal state – and they have done all this most willingly.

It is a pleasure to acknowledge E. J. Wenham's help and sound advice. Much of what is written in these books has benefited from his knowledge and experience as teacher and author.

All of us who have contributed to these books owe a great debt of gratitude to Nina Konrad and her colleagues in the Publications office of the Nuffield–Chelsea Curriculum Trust for their thorough and painstaking work in preparing our manuscripts for the printers and our sometimes quite inadequate drawings for the artists.

Finally, I would like to express my sincere thanks to Paul Black and Jon Ogborn. Their help and support has been invaluable. During a period when both have been particularly busy, they have still found time to give advice both on general matters and on points of detail. They were, of course, the chief architects of the original Nuffield Advanced Physics course. Their willingness to be involved with what must at times have seemed like a severe distortion of their original plans, says much about their generosity of spirit.

John Harris

Unit A
MATERIALS
AND MECHANICS

Roger Hackett
Christ's Hospital, Horsham

A

SUMMARY OF THE UNIT

SUMMARY OF THE UNIT

INTRODUCTION

This Unit is about the science of the materials upon which a technologically based society depends for its buildings, bridges, tunnels, lorries, fabrics, telecommunications equipment, tubes, rails, and so on. The application of the science is the province of the engineer who must design and construct large and small structures, choose the most appropriate materials for the job, calculate the forces which will exist within the structure when in use, and achieve the best balance between cost and function.

The scientist is responsible for investigating the properties of the materials used by the engineer, whether these are metals, woods, glasses, plastics, etc. This involves the measurements of the effects of applied forces, studies of the behaviour of the materials under widely varying conditions, consideration of the energy stored within the material when in use, all of which determine the suitability of the material for its purpose.

To understand the information obtained, scientists need a theoretical model for materials in general. Such a model, based upon experience in the laboratory and in the field, will allow them to predict how the material under study will behave under various conditions. It may also allow them to develop new materials to meet specific needs. Such materials are often composite, for example, concrete, glass fibre, plywood.

Using simple ideas from mechanics, the kinetic theory (or model) of gases is developed to account for many properties of real gases.

Some of the topics and ideas within the Unit will be developed further later in the course. For example, the nature of the force between atoms will be taken up in Unit E, 'Field and potential', the oscillations of mechanical systems will be examined in Unit D, 'Oscillations and waves', and the important theme of randomness will be used in a different context in Unit K, 'Energy and entropy'.

Section A1 THE BEHAVIOUR OF MATERIALS

Descriptive terms

QUESTION 4 We need to study how materials behave under the action of forces so that we can choose the most suitable for a particular job. What, for example, are the best properties for the material with which to make the springs of an easy chair or the cables of a suspension bridge? The behaviour of constructional materials under the action of forces is described by words to which very specific meanings are attached. For example, steel which is strong under steady tension and compression deforms elastically and is stiff with soft yielding near the point of fracture. Brief explanations of some of these terms follow.

Tension, compression, and shear

QUESTIONS 1, 2

The size of the force needed to deform a material by a given amount, or to break it, depends on how the force is applied. Pulling on both ends of a specimen causes *tension*; pushing on both ends causes *compression*; twisting both ends in opposite directions causes *shear*. Most of the examples in the course are concerned with tension.

Stress and strain

$$\text{Tensile strain} = \frac{\text{extension}}{\text{original length}}$$

The size of a specimen must be taken into account when comparing the deformations produced by the applied forces. In the case of stretching under tension the ratio of the extension, x, to the original length, l, is called the *strain*.

$$\text{Tensile stress} = \frac{\text{applied force}}{\text{cross-sectional area}}$$

The ratio of the tension in the specimen, F, to the cross-sectional area, A, is known as the *stress*.

A

Stiffness

A *stiff* material is one in which a large stress is required to produce a small strain. Stiff materials have high Young moduli.

Strength

EXPERIMENTS A3, A4, A5
Strengths of aluminium, glass, and paper

HOME EXPERIMENT AH1
Saved by a hair!

QUESTION 9

A material is *strong* if a large stress is required to break it. The *breaking stress* (ultimate tensile stress) is the stress needed to break the material. A material may be stiff but not strong, such as a biscuit; quite strong but not very stiff, such as nylon; both strong and stiff, such as steel. Some materials are strong in compression but weak in tension, such as brick and concrete.

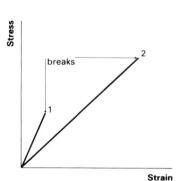

Figure A1
1 is twice as stiff as 2; 2 is twice as strong as 1.

Elastic; plastic

An *elastic* material returns to its original shape after the deforming forces are removed, with no loss of energy. Steel and many other metals are elastic for small strains (up to about one-tenth of one per cent). Materials which flow, slip, or slide internally long before they break are said to show *plastic* deformation. Plasticine is an obvious example, as are metals which can be hammered or pressed into new shapes.

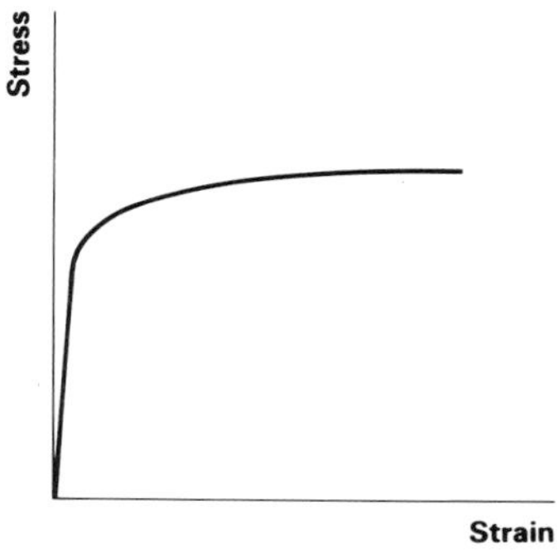

Figure A2
Stress–strain curve for a plastic material.

Tough

EXPERIMENT A2
Stretching rubber, nylon, polythene

A *tough* material will deform plastically before breaking. Examples include steel, which has a high breaking stress and nylon, which has a much lower breaking stress.

Brittle

EXPERIMENT A4
Breaking strength of glass fibre

Materials which do not deform plastically before breaking are *brittle*. They snap cleanly and the edges fit together after breaking. Glass, pottery, and concrete are brittle.

EXPERIMENT A7
Effect of cracks

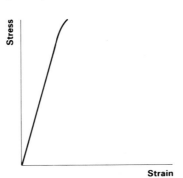

Figure A3
Stress–strain curve for a brittle material.

Ductile, malleable

EXPERIMENT A1
Stretching, bending, and breaking
materials; force–extension relationship

Ductile materials can be drawn into wires and *malleable* materials can be hammered into new shapes. Both processes involve plastic deformation.

Hard, soft

QUESTION 3

Hardness is a measure of the resistance of the material to scratching and indentation of its surface. Diamond is the hardest substance found in nature.

Hooke's Law and the Young modulus

What are the units of k?
k is called the *spring* or *force constant*

A wire or a spring, if not extended too far, will obey Hooke's Law. The extension, x, of the material is proportional to the applied force, F.

$$F = k\,x$$

QUESTIONS 5 to 8

Hooke's Law may be written: stress is proportional to strain, if the strain is not too large. Stress $= E \times$ strain, where E, the constant of proportionality, is called the *Young modulus*.

The Young modulus $= \dfrac{\text{tensile stress}}{\text{tensile strain}}$

What are the units of E?

$$E = \text{stress/strain} = \frac{F/A}{x/l}$$

EXPERIMENT A6
Behaviour of springs

Since

$$k = F/x$$

HOME EXPERIMENT AH2
An accurate newtonmeter

$$k = EA/l$$

QUESTION 10

EXPERIMENT A8
The Young modulus for the
material of a wire

QUESTION 9
READING
Mechanical testing of materials
(page 23)

The Young modulus is important because it depends only on the material, not the size and shape of the specimen. It can be found from the slope of the linear (elastic) region of the stress–strain curve. For a typical metal wire the limit of proportionality is reached for a strain of 0.001, and the Young modulus may be 10^{11} N m^{-2}.

An engineer may also need to know the *yield strength*, the stress at which plastic deformation becomes important.

The treatment which a wire specimen has undergone before it is tested often affects its properties. For example, a wire which has been flexed a number of times, or squashed at one point, will behave differently from a new, undamaged one. The stress–strain curve will be different and the wire will fail at a lower stress. Most wires found in the laboratory have been made by drawing through a circular die and have thus been subject to considerable stress. This partly explains why the stress–strain curves for new short rod specimens tested in automatic extensometers and reproduced in many textbooks exhibit features not observable in simple laboratory experiments with cold drawn wires. Automatic extensometers are hydraulic machines which can sense any reduction in stress in the specimen although the strain is still increasing. Such an effect cannot be observed with simple equipment.

A

Time and temperature effects

A material under a continuous stress which is not large enough to break it, may gradually extend more and more. This *creep* is shown by rubber bands, and by lead and materials such as pitch and tarmac. Metal wires creep if stressed beyond their yield point, and the effect becomes more pronounced near the breaking stress. Probably more materials show creep than we realize, but the time needed for an observable extension is very long. Raising the temperature usually increases the rate of creep and may affect other behaviour of the material. For example, a rubber ball becomes brittle and will shatter if dropped after being cooled to the temperature of liquid nitrogen.

EXPERIMENT A5
Strength of paper

The time scale of events can also alter the behaviour. Silicone putty ('potty putty') flows like a viscous liquid if left alone. It can also be moulded into a ball. If the ball is dropped, it bounces elastically. If hit with a hammer, it shatters.

Elastic strain energy

EXPERIMENT A9
Translational kinetic energy

EXPERIMENT A10
Energy stored in a spring

QUESTIONS 11 to 16

Energy is transformed when a specimen is stretched. Although the stretching force is not constant, the work required to produce a given extension is given by the area under the force–extension graph. For a material obeying Hooke's Law and not stretched beyond its elastic limit, the energy stored in the strained specimen, W, can be found from the area of the triangle under the force–extension curve (figure A4):

area = (average force) × (extension)

$$W = \tfrac{1}{2}Fx = \tfrac{1}{2}kx^2$$

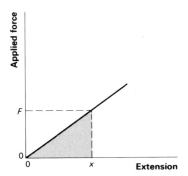

Figure A4

By expressing the applied force and extension in terms of tensile stress and strain, we can show that the elastic strain energy stored per unit volume = $\frac{1}{2}$(stress) × (strain).

QUESTION 14

If the stretch is elastic (that is, if the specimen returns to its original shape when the applied force is removed) all the stored energy is recoverable, for example as kinetic energy.

QUESTION 17

For a specimen not obeying Hooke's Law (figure A5), or one stretched beyond its elastic limit, the area under the force–extension graph is still a measure of the energy transformed, but $W=\frac{1}{2}kx^2$ cannot be used. When such a sample is released, the area between the loading (upper) curve and the unloading (lower) curve represents the energy transformed in deforming the specimen and so heating it. The closed curve shown for rubber is an example of elastic hysteresis and is important in the behaviour of tyres.

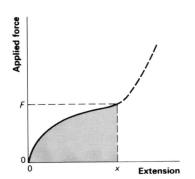

Figure A5

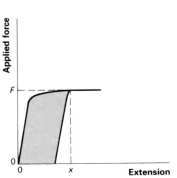

Figure A6
Force–extension curve for copper.

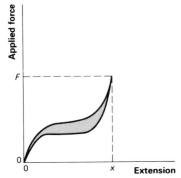

Figure A7
Force–extension curve for rubber.

EXPERIMENT A11
Changing elastic strain energy into gravitational P.E. or translational K.E.

Gravitational P.E. $= mgh$

Translational K.E. $= \frac{1}{2}\,mv^2$

DEMONSTRATION A12
Energy absorbed in deformation

Experimental attempts to check the expression for elastic strain energy assume that the Law of Conservation of Energy applies and arrange to convert the elastic strain energy into another measurable form. Typically this is the gravitational potential energy of a mass, or the kinetic energy of a vehicle. The success of such experiments depends on how completely the stored elastic energy is transformed into one single measurable form.

The transformation from kinetic to strain energy has to be considered in the design of safety belts and in the choice of materials for items such as climbing ropes, crash barriers, and car chassis.

Section A2　THE STRUCTURE OF SOLID MATERIALS

Atoms and molecules

All matter is made up of particles consisting of single atoms, molecules, or ions. Knowledge of their sizes and how they are arranged in various materials leads to a better understanding of why materials behave differently and exhibit different properties. The arrangement of atoms in a metal such as copper is quite simple. Such substances as rubber, polythene, wood, and concrete are rather complicated chemically, and have correspondingly complex structures.

As well as a physical model of the structure of materials, physicists need a mathematical model for the dependence of the forces between atoms or molecules and their separation.

The upper limit to the size of a molecule was estimated by Lord Rayleigh in 1899 by pouring a thin layer of oil onto water. He argued that the layer of oil could not be less than one molecule thick. Later (in 1912), von Laue passed a beam of X-rays through a crystal and observed different patterns of spots on a photographic plate depending on the crystal used. Sir Lawrence Bragg, using an X-ray spectrometer his father had invented, developed this technique so that it was possible to relate the systems of spots and rings to the arrangement and spacing of the atoms in the crystal. This same method has been used and refined to such an extent that the incredibly complex structures of such very large organic molecules as vitamin B12 and DNA have been unravelled. Several researchers have won the Nobel prize for this work.

Modern developments in electron and field-ion microscopes give more direct evidence for the arrangement and spacing of atoms in crystals. The photographs shown were taken with a 600 kV electron microscope at Cambridge.

READING
Models (page 31)
Theories – true or not? (page 32)

QUESTION 18

EXPERIMENT A13
Optical analogue for X-ray diffraction

Unit J, 'Electromagnetic waves'

READING
Looking at the structure of materials
(page 24)

Atom about 10^{-10} m in diameter

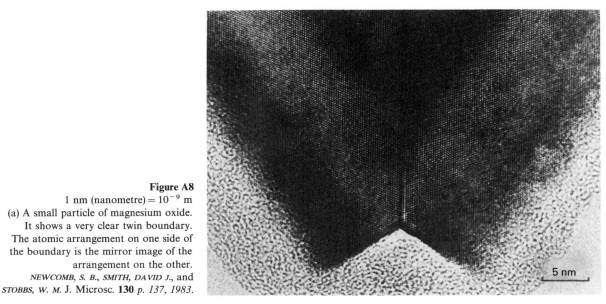

Figure A8
1 nm (nanometre) $= 10^{-9}$ m
(a) A small particle of magnesium oxide. It shows a very clear twin boundary. The atomic arrangement on one side of the boundary is the mirror image of the arrangement on the other.
NEWCOMB, S. B., SMITH, DAVID J., and *STOBBS, W. M. J. Microsc.* **130** *p. 137, 1983.*

5 nm

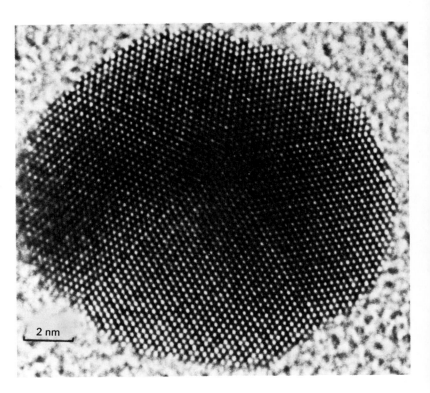

Figure A8
(b) A small particle of gold. The rows of atoms are viewed end-on with each row showing up as a white spot. Note the hexagonal arrangement of atoms and five twin boundaries.
MARKS, L. D. and *SMITH, DAVID J.* J. Microsc. **130** *page 249*, 1983.

2 nm

The mole and molar mass

'Mole' means heap, or pile

QUESTIONS 20, 24

The fact that the volumes of combining gases are always in simple ratio led Avogadro to suggest, early in the nineteenth century, that equal volumes of gases (at the same temperature and pressure) contain equal numbers of particles.

The Avogadro constant, L

It is useful to specify an amount of a substance that contains a specific number of individual particles. The appropriate SI unit is the mole: the amount of substance which contains as many elementary entities as there are atoms in 0.012 kg of carbon-12. One mole contains 6.02×10^{23} particles, so it is unusual to apply the idea to anything not of atomic or molecular size.

QUESTIONS 20 to 22

The molar mass of a substance is the mass of one mole of particles, and it is essential to say what the particles are – atoms, molecules, ... The molar mass of hydrogen atoms (H) is $0.001 \text{ kg mol}^{-1}$, but the molar mass of hydrogen molecules (H_2) is $0.002 \text{ kg mol}^{-1}$. Copper atoms have a molar mass of $0.0636 \text{ kg mol}^{-1}$.

The mass and size of an individual atom can be estimated by assuming that the atoms in a solid are in contact.

Forces between atoms and molecules

Ionic: NaCl
Covalent: H_2 CH_4
Metallic: metals

Atoms are held together in molecules by electrical forces. The different kinds of bonding which chemists call ionic, covalent, or metallic are all electrical. There are electrical forces between molecules too, and the forces depend on how far apart the molecules are. In a gas the average distance between two molecules is so large (about 10 molecular diameters) that the force between them is negligible.

The behaviour of solid materials under tension suggests that powerful attractive forces exist between molecules which are close together. Similarly, their behaviour under compression suggests that there are also repulsive forces which increase very rapidly with diminishing separation. At the equilibrium separation, r_0, these two forces are equal and opposite (zero resultant force).

QUESTIONS 25, 26

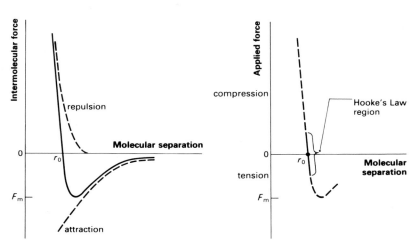

Figure A9   **Figure A10**

Hooke's Law
$$F = k\Delta r$$

The gradient of the force–distance curve indicates the stiffness of the solid, and if the curve is linear close to r_0, then the substance obeys Hooke's Law. Brittle fracture will occur when the force applied to the solid is greater than the maximum molecular attraction, F_m.

The Young modulus and the interatomic force constant

$$E = \frac{\text{tensile stress}}{\text{tensile strain}}$$

QUESTIONS 28, 29, 30

The origin of the elastic behaviour of a solid being stretched or compressed, and so of Hooke's Law, lies at atomic level. Imagine two sheets of atoms, equilibrium distance, r_0, apart, being pulled elastically a distance Δr, by a force F. The atoms can be pictured as being held together by small springs of force constant k. This interatomic force constant is related to the Young modulus, E, by the equation:

$$k = Er_0$$

For steel $E \approx 2 \times 10^{11} \, \text{N m}^{-2}$

and $\qquad r_0 \approx 3 \times 10^{-10} \, \text{m}$,

so $\qquad k \approx 60 \, \text{N m}^{-1}$.

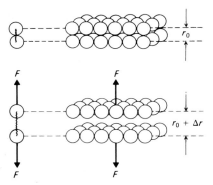

Figure A11

Intermolecular energies

DEMONSTRATION A15
Intermolecular force–distance model
using air track

Two molecules have zero potential energy when an infinite distance apart. As they move together they lose potential energy because of the attractive force between them: the potential energy becomes negative. At the equilibrium separation, r_0, the energy is a minimum. Work must be done to push them any closer together. The deeper this 'potential well', the greater the binding energy, the energy needed to separate the molecules completely from the equilibrium position.

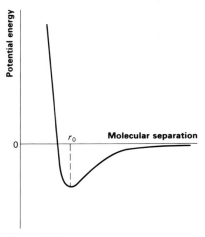

Figure A12

QUESTION 27

When a material is heated, its molecules gain energy: in a gas the molecules move faster; in a solid they depart from their equilibrium positions. The solid will sublime or the liquid vaporize when the kinetic energy of molecules is equal to the binding energy. The energy needed to sublime a mole of the solid or to vaporize a mole of the liquid (the molar latent heat) can be measured. From this measurement the binding energy per molecule can be estimated for solids and for liquids.

Structure of solids

EXPERIMENT A16b
Bubble raft model

Atoms linked by metallic bonds form simple structures, since this bond has equal attractions in all directions. A model of a metal structure in two dimensions can be made using a raft of soap bubbles. Each is surrounded by six neighbours. Areas of neatly ordered bubbles are formed, divided by discontinuities (*grain boundaries*), a useful model of a *polycrystalline solid*.

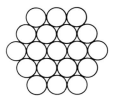

Figure A13

QUESTION A23

DEMONSTRATION A14
How atoms are arranged in solids

Many metal crystals (copper, aluminium, and magnesium, for example) are made up of layers of hexagonally arranged atoms packed together as closely as possible. Some metals (including sodium, and iron at temperatures below 900 °C) form crystals in which the atoms are in a slightly more open arrangement.

Sodium chloride is an example of an ionic crystal. The small sodium ions, Na^+, fit into the holes in the packing of the larger chloride ions, Cl^-.

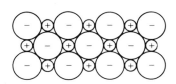

Figure A14

A

EXPERIMENTS A2
Stretching rubber, nylon, polythene

EXPERIMENT A16a
Splitting stretched rubber

QUESTIONS 31 to 33

Polythene, rubber, and nylon are examples of very long chains of organic molecules called *polymers*. Within each long chain, strong unidirectional covalent bonds hold carbon and hydrogen atoms together. Weak van der Waals forces cross-link between chains. The chains are tangled and tightly coiled. When the material is stressed the molecular chains unravel and line up (ordering of the structure), and this enables rubber and polythene to accept strains of more than 100 per cent. Once straightened the material becomes much stiffer, like pulling out a tangle of string. If the stress is released, the molecular chains return to their original tangled and coiled state.

Structures in which the same regular arrangement covers very many atoms or ions (long range order) are called *crystalline*. Completely irregular structures or those with only very short range order are called *glassy* or *amorphous*. In glass, the disordered liquid structure is maintained into the solid state. Polymeric solids, such as polythene, rubber, hair, and wool, can exhibit both amorphous and crystalline properties depending on temperature and method of formation. Crystalline substances have precise melting temperatures, glasses do not.

Metals, unlike rubber or glass, can be hammered flat or drawn through a die to make wires. Metallic bonds enable layers of atoms to *slip* over each other rather than suffer brittle fracture like the disordered but strongly bonded glasses (figure A15).

The presence of dislocations can make slip easier than it would be in a perfect structure.

Figure A15

Dislocations

EXPERIMENT A16b
Bubble raft model

A dislocation is a single defect in the otherwise perfect and regular arrangement of atoms in part of a crystal. There are several kinds. One of the simplest occurs where a row of atoms stops and meets two rows instead of continuing on as it would in a perfectly regular arrangement.

The presence of a few dislocations can make a material soft or plastic, because the dislocations can run through the crystal causing slip. *Creep* is an example of the slow movement of dislocations through the crystal. The dislocations move as far as the grain boundaries before stopping.

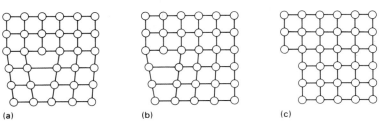

Figure A16 (a) (b) (c)

QUESTIONS 34 to 36

Large numbers of dislocations together reduce the ability of atoms to slip over each other, making the material stronger. *Work hardening* increases the number of dislocations through hammering, stretching, or bending, etc. In time, the dislocations can vanish again, especially if the material is kept hot, because dislocations of opposite types may come near each other and allow the atoms to 'snap back' into a regular arrangement, or they may travel to the edge of the crystal leaving only a step on the surface. This process is called *annealing*.

EXPERIMENT A16c
Heat treatment of steel

Introducing a certain number of foreign atoms into the material (*alloying*) can have the same effect as work hardening in reducing slip in a polycrystalline material, giving it greater strength. Heat treatment of the material can alter the arrangement of the foreign atoms in the material completely changing the properties of the alloy.

The presence of defects, such as dislocations, extra foreign atoms, or simply holes in the crystal pattern, means that the strength of real materials is usually much less than predicted for large perfect crystals.

Section A3 STATICS, STRUCTURES, AND COMPOSITE MATERIALS

Statics: structures and objects in equilibrium

Mechanics deals with the effect of forces on simple objects and complicated structures. Part of it, dynamics, is concerned with the effect of forces on bodies in motion. Another part, statics, deals with forces on bodies and in systems at rest. Forces in equilibrium don't cause change in rotation or acceleration – this is Newton's First Law. But as seen in Sections A1 and A2, they do deform and may break an object. Statics allows engineers to calculate the stresses at critical places in a complex structure such as a building or a bridge.

Static: 'standing, at rest'

Contact forces: the 'normal' force, and friction

Typically statics is concerned with gravitational force (the weight of a body), forces caused by deformation (tension, compression), and forces due to friction. When a book rests on a table, the two forces acting on it are in equilibrium: the upward force of the table just balances the weight of the book. This upward force is called the 'normal' force, because it is perpendicular to the surface of the table. It is due to the

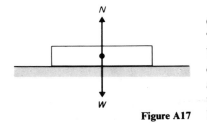

Figure A17

compression of the table under the weight of the book – the molecules are slightly closer together, the repulsive force between them is increased.

A mass suspended from a spring will be in equilibrium if the upward force on it due to the tension of the spring balances its weight. The tension is caused by the extension of the spring – its atoms have been pulled slightly further apart than in the unstressed spring, and the attractive force between them is increased. An object at rest on a slope is in equilibrium because the forces acting on it – its weight, the normal force, and friction – are in balance.

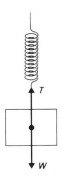

Figure A18

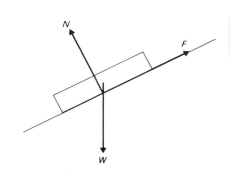

Figure A19

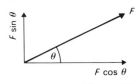

Figure A20

The components of a force

Force is a vector quantity, and any force can be resolved into two components. The vector sum of the two components is, of course, equal to the original force. If the two components are at right angles to each other, then,

$$F_x = F \cos \theta \qquad F_y = F \sin \theta$$

and

$$F^2 = F_x^2 + F_y^2$$

QUESTIONS 37, 38, 42

$\Sigma F = 0 \Rightarrow \Sigma F_x = 0$ and $\Sigma F_y = 0$

Each of the forces acting on the body can be resolved in this way. If the sum of all the F_x components is zero there is no force tending to accelerate the body in the x direction, and similarly for the sum of the F_y components. For equilibrium both sums must be zero.

Addition of forces: triangle and polygon of forces

EXPERIMENT A17
Forces on systems in equilibrium

The resultant effect of several forces acting through a point is found by adding them, taking account of both magnitude and direction. Each force can be represented by a line whose length and direction represent the magnitude and direction of the force. The lines are drawn head to tail. Suppose there are three forces and these lines form a closed triangle: then their resultant is zero and the forces are in equilibrium (figure A21). If the triangle does not close, the set of forces does have a resultant and the body will accelerate.

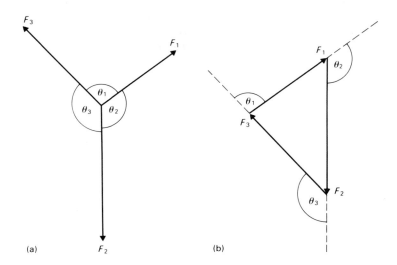

Figure A21
(a) Space diagram.
(b) Force diagram.

(a)

(b)

QUESTIONS 44, 45

With more than three forces the same technique applies: the resultant force can be found by drawing a polygon of forces, each side of the polygon representing one of the forces. The resultant force is represented by the line joining the first to the last point. If the polygon is closed, then the resultant force is zero.

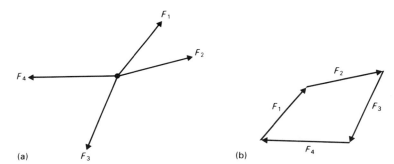

Figure A22
(a) Space diagram.
(b) Force diagram.

(a)

(b)

Moments and couples

The *moment* or turning effect of a force about a given point is the product of the force and the perpendicular distance from the point to the line of action of the force.

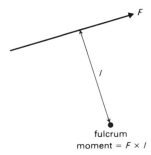

fulcrum
moment = $F \times l$

Figure A23

A *couple* consists of two forces, equal in magnitude but opposite in direction, which do not pass through a point. The effect of a couple on a body is to change its rate of rotation: a couple does not cause any linear acceleration.

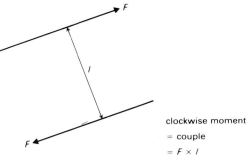

clockwise moment
= couple
= $F \times l$

Figure A24

EXPERIMENT A19
'Weighing' a retort stand
$\Sigma(\text{moments}) = 0$

To test for equilibrium we add all the moments. For equilibrium, the sum of the moments tending to produce clockwise rotation must equal the sum of those tending to produce anti-clockwise rotation.

Conditions for equilibrium of coplanar forces
Both conditions discussed above are necessary for equilibrium:
i the sum of the resolved components of all the forces in any two perpendicular directions is zero, and
ii the resultant moment about any point in the plane of the forces is zero.

If condition *i* is true the body will not accelerate; if condition *ii* is true its rate of rotation (if any) will not change. If all the forces pass through one point (concurrent forces) then condition *i* is sufficient. Condition *ii* is also required for non-concurrent forces.

These are the rules used to calculate the forces which exist within any rigid structure, such as bridges, towers, cranes, electricity pylons, and so on.

Static friction
When two surfaces in contact are at rest, the frictional force between them, parallel to the contact plane, opposes their relative motion and can have any value up to a limiting value. The ratio of the maximum frictional force, F, just as motion begins, to the normal force, N, is called the coefficient of static friction, μ.

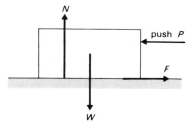

For no motion $F = P \leqslant \mu N$

Figure A25

EXPERIMENT A17c
Coefficient of friction

Friction plays an important role in all mechanical situations, some beneficial, some inconvenient. Dynamic (moving) friction usually causes translational or rotational energy to be transformed into thermal energy. Friction is reduced by ball-, roller-, or fluid-bearings, by lubrication, or even by special surface coatings such as PTFE ('Teflon'). Without static friction life would be intolerable, since the contact force at a 'smooth' (frictionless) surface is normal, that is, at right angles to the surface. There would be no grip between surfaces in contact. For example, normal walking would be impossible, no ordinary wheel would roll, and a ladder leaned against a wall would slide to the ground.

Composite materials

Few materials used by the engineer are pure substances. Many pure metals, for instance, are soft and relatively weak.

QUESTIONS 46 to 49

HOME EXPERIMENT AH3
The jelly column

HOME EXPERIMENT AH4
Cement

READING
Materials used in architecture (page 33)

See Section A2 page 12

Composites are combinations of materials where each retains its own properties but the combination leads to a wider range of applications. For example, ordinary glass fibres are strong, but brittle when scratched. Encasing the fibres in a resin produces a tough material, glass fibre, which resists cracks. Glass fibres are flexible, but the material glass fibre is stiff. Carbon fibres in a polymer resin produce a material which is lighter and stronger than steel. Concrete is strong in compression but weak in tension. Through the use of steel rods, a stronger more useful composite material, reinforced concrete, is made.

Alloys are a mixture of a pure metal with other elements. For example, steel is formed by adding a small quantity of carbon to iron. Different amounts of carbon give the alloy different properties. Special steels for specific applications may contain elements such as chromium, tungsten, and nickel. The engine blocks of some modern motor car engines are made of an aluminium alloy which is as strong as steel, but much lighter. Strong modern plastics have also replaced metals in some applications, reducing both cost and mass.

In any branch of technology, progress depends on the availability of the right material for the job. The development and use of new substances and the application of existing materials through new machining techniques (for example, superplastic moulding of metals such as titanium) together make possible continuing advances.

Section A4 **MOMENTUM AND THE SIMPLE KINETIC THEORY OF GASES**

Newton's Laws of Motion, momentum, and impulse

Newton 1

Newton's First Law of Motion states that if no external resultant force acts on a body it will continue in its present state of rest or uniform motion. To change a body's state of motion, that is, to cause it to accelerate, a resultant external force must act on it. Newton's First Law is difficult to demonstrate because, in practice, there is nearly always some frictional force acting.

Linear momentum
$p = mv$

The *linear momentum, p,* of a body of mass m moving with velocity v is mv. Velocity, v, is a vector quantity, and so therefore is momentum.

Momentum is a useful and important quantity because it is a conserved quantity – it does not change. If no net force acts on a single body, then since its velocity does not change, neither does its momentum. For an isolated system of interacting bodies (for example the molecules in a jar of gas) the total momentum remains constant, though there may be an exchange of momentum between individual bodies. Isolated means that no unbalanced external forces act on the system.

$\Sigma p = \text{constant}$
$\Sigma \Delta p = 0$

In general the *Law of Conservation of Linear Momentum* can be written:

EXPERIMENT A22
Collisions on an air track

the total linear momentum of a system of interacting bodies remains constant provided there is no net external force.

Momentum is not conserved unless the system being considered does include all of the interacting bodies. For example, a swinging pendulum does not appear to conserve momentum, nor does a car accelerating. In both these cases the Earth is one of the interacting bodies and its motion must also be considered when applying the Law.

A

QUESTIONS 51, 53

If no resultant external force acts on the centre of mass of the system, then the vector sum of the momenta of all the constituent parts of the system relative to the centre of mass must be zero.

Newton 2

Newton's Second Law describes the effect of an external resultant force on a body. A force, F, causes the body's momentum to change at a rate given by $F = \Delta p / \Delta t$. Or, in words:

EXPERIMENT A21
Momentum change and impulse

the resultant force on a body is equal to the rate of change of linear momentum.

QUESTION 50

The product $F \Delta t$ is called the *impulse* of a force, and an alternative way of writing Newton's Second Law as $F \Delta t = \Delta p$ expresses the fact that

$\text{Impulse} = F \Delta t$

$F \Delta t = \Delta(mv)$

impulse = change of momentum.

The change in momentum of any body acted upon by a variable force can be found from the area under the force–time graph.

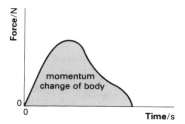

momentum change of body

Figure A26

QUESTION 52

As long as the mass of the body does not change

$\Delta p = \Delta(mv) = m \Delta v$

and so

$$F = m \frac{\Delta v}{\Delta t} = ma$$

$F = ma$ force = mass × acceleration

Newton's Third Law of Motion is concerned with the forces between two interacting bodies. Forces always occur in pairs, one acting on each of the bodies: the forces have equal magnitude but are in opposite directions. So, for example, when two bodies collide they exert equal and opposite forces on each other, for the same time.

QUESTION 54

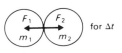

Figure A27

The Law of Conservation of Momentum follows from Newton's Third Law of Motion. Consider two bodies of masses m_1 and m_2 interacting for a time Δt. If the force on body 1 from body 2 is F_1, and the force on body 2 from body 1 is F_2, then $F_1 = -F_2$.

So the total impulse, $F_1 \Delta t + F_2 \Delta t = 0$.

And from the impulse–momentum change relationship $\Delta p_1 + \Delta p_2 = 0$.

The vector sum of the momenta of the two bodies interacting with each other is constant as the changes in momenta add to zero.

Elastic and inelastic collisions

Unit F, 'Radioactivity and the nuclear atom'

Evidence for the nuclear model of the atom, for the nature and mass of atomic particles, and explanations of the behaviour of matter described on an atomic scale come from the study of collisions.

EXPERIMENT A23
Newton's cradle

Collisions in which the total kinetic energy of the bodies before and after the collision is the same, are called *elastic collisions*. An electron can 'bounce off' a gas atom elastically exchanging little kinetic energy with the relatively massive atom. Alpha particles making elastic collisions with helium nuclei in a cloud chamber can transfer all of their energy to the helium nucleus if the collision is a head-on one. If it is oblique, then the particles will move apart at right angles, showing that they have equal mass. The mass of the neutron was determined by Chadwick (in 1932) by observing the elastic recoil of different nuclei making visible tracks in a cloud chamber.

EXPERIMENT A25
Collisions in two dimensions

QUESTIONS 55 to 59
Unit F, 'Radioactivity and the nuclear atom'

Situations where some of the kinetic energy is transformed to other forms of energy, such as internal energy (for example, random molecular motion), are called *inelastic collisions*. Two vehicles on an air track colliding and moving apart may interact inelastically. When an electron makes an ionizing collision with a gas atom some kinetic energy is lost in removing an electron from the atom. If the objects stick together, for example, an air rifle pellet fired into a lump of Plasticine, then the collision is said to be perfectly inelastic, but the total kinetic energy does not necessarily fall to zero.

EXPERIMENT A24
Speed of air rifle pellet

In all collisions where there are no resultant external forces acting the total momentum of the interacting bodies is conserved, regardless of whether the collision is elastic or inelastic.

The simple kinetic theory of gases

The ideal gas equation

On a *macroscopic* scale, an *ideal gas* exactly obeys the relationship

$$pV = nRT$$

$R = 8.31 \, \text{J mol}^{-1} \text{K}^{-1}$

n is the number of moles of gas molecules present, and R is a universal constant called the *molar gas constant*.

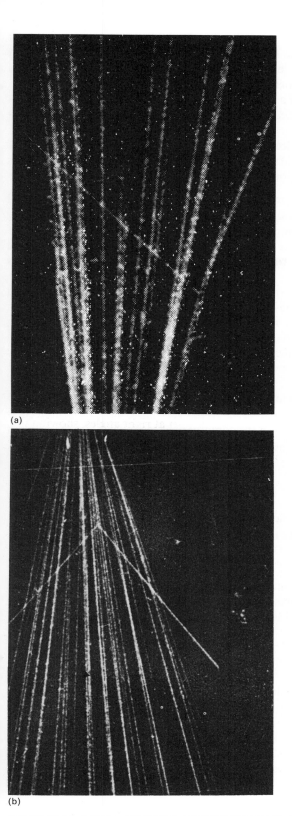

(a)

(b)

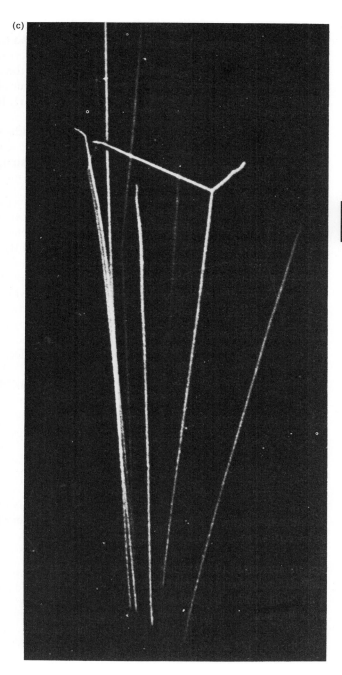

(c)

Figure A28
(a) Alpha particle tracks in wet hydrogen. One collided with a hydrogen nucleus, which recoiled forward and upward, making a thin track.
BLACKETT, P. M. S. Proc. Roy. Soc. *A. Vol. 107*, *Pl. 6(1)*, 1925.
(b) Collision of alpha particle with helium.
BLACKETT, P. M. S. Proc Roy. Soc. *A. Vol. 107*, *Pl. 6(3)*, 1925.
(c) Collision of an alpha particle with nitrogen.
BLACKETT, P. M. S.

QUESTIONS 60, 61

V_m is molar volume

Such a gas would obey Boyle's Law (pV = constant at constant temperature), and the Pressure Law ($p \propto T$, if volume is held constant) exactly. For one mole of ideal gas, $pV_m = RT$. This equation defines temperature on the ideal gas scale.

The behaviour of a real gas at low pressures and ordinary temperatures is very nearly ideal.

Assumptions of kinetic theory

On a *microscopic* scale we have a model for an ideal gas which embodies the following assumptions.

'Kinetic' means moving

i The ideal gas is made up of very large numbers of identical monatomic molecules. They are in constant random motion colliding by chance with one another and with the walls of the container.

ii These collisions are (at least on average) elastic, so that the total kinetic energy of the molecules remains constant; and the centre of mass of the gas is at rest.

iii The volume of the gas molecules themselves is negligible compared with the volume of the container.

iv The forces between molecules are negligible except during collisions. Collisions occupy a negligible time in comparison to the time between collisions; the molecules move with uniform velocity between collisions (gravitational effects are ignored).

QUESTION 64

The pressure in an ideal gas

QUESTIONS 62, 63

Applying Newtonian mechanics to the very large number of molecular collisions with the walls of the container per second enables the product pV for a quantity of gas to be related to the mean square speed, $\overline{c^2}$, of its molecules.

$$pV = \tfrac{1}{3} Nm\,\overline{c^2}$$

where there are N molecules each of mass m.

Also, since density $\rho = Nm/V$

$$p = \tfrac{1}{3}\rho\overline{c^2}$$

The root mean square (r.m.s.) speed of air molecules $\sqrt{(\overline{c^2})}$ at room temperature and pressure is about $500\,\mathrm{m\,s^{-1}}$. The speeds of the molecules are very varied and are distributed as suggested in figure A29.

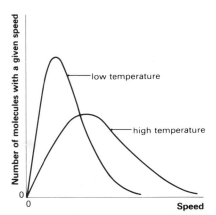

Figure A29

The kinetic interpretation of temperature

M is molar mass

The total kinetic energy of the molecules in one mole of gas is $\tfrac{1}{2}M\overline{c^2}$. Combining the equation $pV = \tfrac{1}{3}Nm\overline{c^2}$ and the ideal gas equation $pV_m = RT$, the total kinetic energy per mole $= \tfrac{3}{2}RT$.

A monatomic ideal gas can only store energy by its translational motion. (Diatomic gas molecules can store energy by rotating and vibrating too.) For an ideal gas the total kinetic energy is the internal energy of the gas, U.

The molar heat capacity of the ideal gas (the energy required to raise the temperature of one mole of the gas by one kelvin) is $\frac{3}{2}R = 12.5\,\text{J mol}^{-1}\,\text{K}^{-1}$. The monatomic gases helium and argon have molar heat capacities close to this value at room temperature.

QUESTIONS 65, 66

The mean kinetic energy of one molecule $= \frac{3}{2}\dfrac{R}{L}T = \frac{3}{2}kT$, where k is a universal constant called the *Boltzmann constant*.

$k = R/L$
$= 1.38 \times 10^{-23}\,\text{J K}^{-1}$

At the same temperature the mean translational kinetic energy per molecule is the same for all monatomic gases and is equal to $\frac{3}{2}kT$.

Behaviour of gases
The kinetic theory of gases offers an explanation of many of the known properties of gases. These include Avogadro's Law, which states that equal volumes of gases at the same temperature and pressure contain the same number of molecules, Dalton's Law of partial pressures, and Graham's Law of diffusion.

Effects of intermolecular collisions
The average distance travelled by a molecule between collisions is called the *mean free path, λ*.

By considering the passage of a molecule having a finite diameter, d, in a gas containing n molecules per unit volume down an imaginary cylinder of radius d (the greatest distance away another molecule may be for a collision to occur) it can be shown that

Figure A30
Mean free path of a molecule.

$$\text{mean free path } \lambda \approx \frac{1}{\pi d^2 n}$$

irrespective of the molecule's speed.

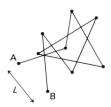

Figure A31
Molecular motion from A to B – a random walk.

For purely random motion between collisions, a molecule moves a distance L through the gas in time t given by $L = \sqrt{N}\,\lambda$, where N is the number of collisions made by the molecule in time t.

QUESTIONS 67, 68

At normal temperature and pressure, an air molecule of diameter about $3 \times 10^{-10}\,\text{m}$ has a velocity of about $500\,\text{m s}^{-1}$. It travels about $10^{-7}\,\text{m}$ between collisions, making about 5×10^9 collisions per second. It would take a molecule more than a week to diffuse across a large room of still air.

Work done in expansion

When a gas expands reversibly at constant pressure by an amount ΔV, the work done by the gas is $p\Delta V$. (Reversible refers to an ideal process in which there are no losses due to friction, etc., and where a very small change in conditions will reverse the direction of the process.)

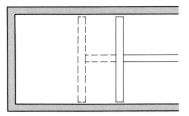

Figure A32

Randomness

The kinetic theory of gases is just one of the topics in physics which involve the ideas of randomness and chance. Others include radioactive decay, the movement of electrons which constitutes a current in a wire, and the exchange of energy between atoms in a solid. Although the behaviour of an individual particle in such a system can only be predicted as a probability, it is possible to say quite precisely what the properties of the whole collection will be – if a large enough number of particles is involved.

Liquids

The liquid phase is much less well understood than either the solid or gaseous phases, and is a subject of current research. As a rule a substance is less dense as a liquid than as a solid (water is an important exception). The molecules in the liquid are, on average, slightly further apart than they are in the solid, and they have fewer nearest neighbours. There is some short range order, but not the long range order of a crystalline solid. A gas of course is completely disordered. In the liquid phase molecules, or groups of molecules, are free to move about at random; the speeds are lower than in the gas.

READINGS

MECHANICAL TESTING OF MATERIALS

To study the mechanical properties of different materials various types of test have been devised. Some, such as the tensile and hardness tests, can be used on all materials; but for materials which are brittle or weak in tension the compression test may be used.

Figure A33
Tensile test machine.
Griffin & George.

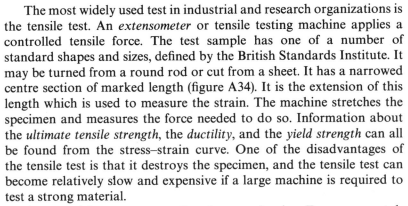

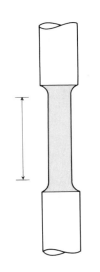

Figure A34

The most widely used test in industrial and research organizations is the tensile test. An *extensometer* or tensile testing machine applies a controlled tensile force. The test sample has one of a number of standard shapes and sizes, defined by the British Standards Institute. It may be turned from a round rod or cut from a sheet. It has a narrowed centre section of marked length (figure A34). It is the extension of this length which is used to measure the strain. The machine stretches the specimen and measures the force needed to do so. Information about the *ultimate tensile strength*, the *ductility*, and the *yield strength* can all be found from the stress–strain curve. One of the disadvantages of the tensile test is that it destroys the specimen, and the tensile test can become relatively slow and expensive if a large machine is required to test a strong material.

As materials get stronger they become harder. For some metals hardness is directly related to tensile strength. A hardness test can be quick and convenient. *Hardness* is defined as the resistance to penetration or abrasive wear. In most testers a small sphere or cone is forced into the surface of the specimen by a load for a fixed time. The hardness number is defined as the ratio of the applied load to the area of the indentation it causes. This number allows one material to be compared with others.

As well as struts and ties, rigid structures often contain beams or cantilevers which bear loads, tending to make them bend. Materials

Figure A35
Hardness tester.
Griffin & George.

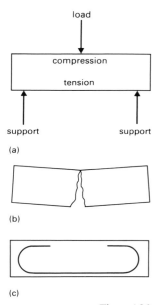

load

compression

tension

support   support

(a)

(b)

(c)

Figure A36
(a) The forces in a concrete beam.
(b) How a concrete beam would break.
(c) Reinforced concrete containing steel rods as shown.

must therefore be tested by bending. The specimen is subjected to both compressive and tensile forces at the same time and therefore bends. For example, concrete which has a high compressive strength but low tensile strength would fail the test shown. Steel rods strong in tension are embedded in the concrete for further reinforcement, hence the name *reinforced concrete*. The hooked ends of the steel rod transfer the load to the concrete. This is shown in figure 36.

Many substances suffer *creep*. A constant load is applied over a period of time. Creep is a very temperature-dependent property and becomes important at different temperatures for different materials. This can be critical if the material is bearing a load. On a hot day in a tropical country, the engineer could find that the structure he has built has collapsed because creep has become a significant effect. Creep tests are therefore carried out over a wide range of temperatures.

When a metal is put under stress, tiny cracks are likely to form within the material. If the stress repeatedly increases and decreases, the cracks will increase in number and may grow in size. Continuous stress cycles lead eventually to the cracks becoming visible. The strength of the material falls rapidly until fracture occurs. *Fatigue testing* normally uses a periodically varying stress for a long period of time. A simple tester might consist of a rod clamped in a ball race connected at the other end to a motor but slightly off axis. The rod flexes slightly as it rotates. After a large number of stress reversals, fracture can occur at forces well below the elastic limit of the material. At larger loads, fracture occurs more quickly. The time between the appearance of surface cracks and failure is usually very short.

Questions

a How many methods of mechanical testing are mentioned in the passage?

b Suggest how you might compression test a short ceramic pipe.

c In what units might hardness be measured?

d A diamond pyramid indentor measures a range of hardness values from 20 to more than 2000. Suggest materials that might be at either end of the scale. (*Hint:* Think of a girl in stiletto-heeled shoes walking over different floor surfaces in a house.)

e Why do you think a steel ball indentor cannot register accurately hardnesses of more than 350?

f How do the ends of the hooked rod in the concrete beam transfer the load to the concrete?

g Sketch a design for a simple fatigue tester.

LOOKING AT THE STRUCTURE OF MATERIALS

Of the hundred or more known elements, about 80 per cent are metals. Metals have a grain structure, formed when the liquid metal solidifies. Within each grain or crystallite the atoms are arranged regularly. The sizes and shapes of the grains of the material can vary greatly, depending on the purity and method of formation of the metal (figures A37 and A38).

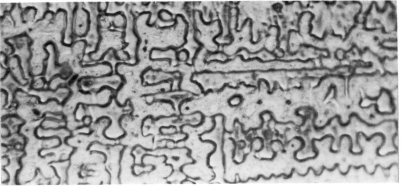

Figure A37
Cast bronze (5 % Sn; 95 % Cu), polished and etched (taken at × 400). Note dendrite-type structure (fern-like). If bronze is annealed, the structure looks like that shown for brass.
Dr A. A. Smith, Faculty of Engineering, King's College, University of London.

Figure A38
Brass (70 % Cu; 30 % Zn), polished and etched (taken at × 100).
Dr A. A. Smith, Faculty of Engineering, King's College, University of London.

Grains can be as small as 0.01 mm, but some are large enough to be seen with the naked eye (figure A39).

Optical microscopes, specially designed to observe reflected light from the chemically etched surface of metals, are used to examine the tinier grains, grain boundaries, and the structure of faulted and unfaulted materials. The incident light is often polarized to produce greater contrast and show the different orientations of the grains more clearly. Magnifications of up to × 2000 can be achieved, enabling details only 1 micron (10^{-6} m) apart to be distinguished. The parallel bands seen inside some of the grains in figure A38 are growth faults, called *twins*. The atomic arrangement within the band is the mirror image of the arrangement in the rest of the grain.

The different grain structures of steel, depending on the amount of carbon present and on the heat treatment during composition, can be seen in figures A40 to A43. Also contrast the grain structures of a metal after cold-working and after annealing (figure A44).

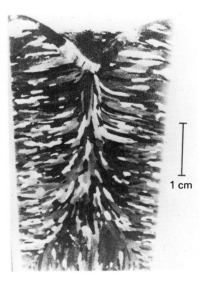

1 cm

Figure A39
The macrostructure of a cast metal. Large grains have formed at the centre of the ingot.
Dr R. T. Southin, School of Metallurgy, University of New South Wales.

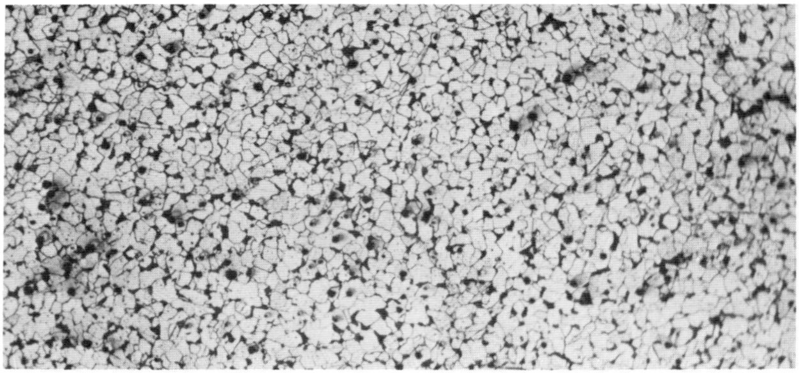

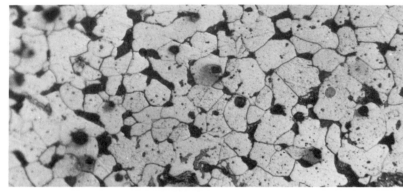

Figure A40
Steel (0.12 % C, taken at × 100). Note the
ferrite grains and small dark regions of
pearlite in the grain boundaries.
*Dr A. A. Smith, Faculty of Engineering,
King's College, University of London.*

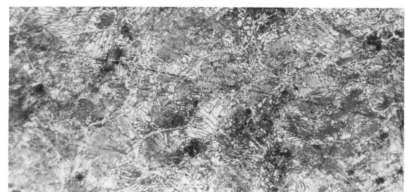

Figure A41
Steel (0.12 % C, taken at × 400).
*Dr A. A. Smith, Faculty of Engineering,
King's College, University of London.*

Figure A42
High carbon steel (1.2 % C), polished and
etched (taken at × 400).
*Dr A. A. Smith, Faculty of Engineering,
King's College, University of London.*

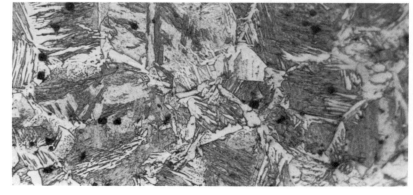

Figure A43
Steel (0.12 %C), quenched (taken at
× 400). Note the changes in structure. The
dark regions are oxide.
*Dr A. A. Smith, Faculty of Engineering,
King's College, University of London.*

When a metal is plastically deformed the layer of regularly-spaced atoms within each grain can slip over each other, like shearing a pack of cards. These slip planes show up on the surface of a metal as a series of parallel edges, and can be seen in figure A45.

To detect imperfections on the atomic scale, such as dislocations in metal crystals, we need to use higher magnifications, that is × 20 000 to × 50 000. An electron microscope is used for this purpose. An extremely thin foil of metal is placed in the path of an electron beam and an image of the imperfections is produced on a fluorescent screen. Figure A46 shows the dislocation structure of a sliver of metal cut before it had been deformed. Figure A47 shows how the dislocations are entangled with each other after plastic deformation of the original aluminium specimen, causing work hardening. The dislocations are so interlocked that they are no longer able to move around.

A

Figure A44
(a) Grain structure of a metal specimen after cold-working.
(b) Grain structure of a metal specimen after annealing.
Welsh Laboratory, British Steel Corporation, Port Talbot.

(a)

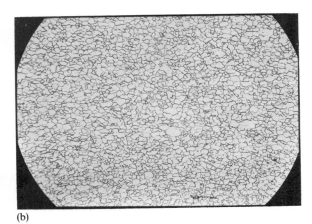

(b)

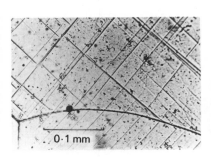

Figure A45
A specimen of aluminium showing slip.
Dr J. W. Martin and Dr J. M. Dowling, Department of Metallurgy and Science of Metals, University of Oxford.

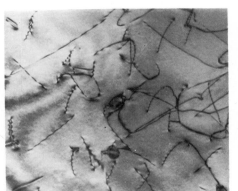

Figure A46
Dislocation structure of a metal specimen before plastic deformation.
Welsh Laboratory, British Steel Corporation, Port Talbot.

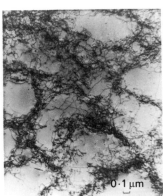

Figure A47
Electron micrograph of a thin foil taken from a crystal of plastically deformed aluminium showing dislocations tangled into 'cell' walls.
Dr J. W. Martin, Department of Metallurgy and Science of Metals, University of Oxford.

The electron micrograph of figure A48 is at a much greater magnification and shows up rows of atoms.

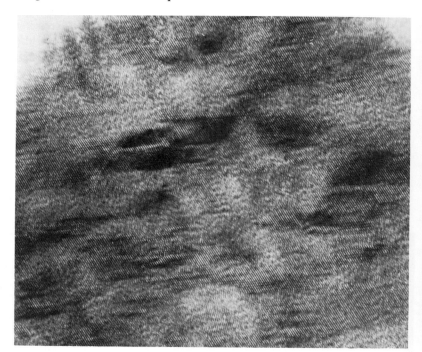

Figure A48
Electron micrograph of titanium dioxide. The planes of atoms are visible as patterns of lines in the photograph. Taken at × 3 400 000.
Dr Harvey Flower, Department of Metallurgy and Materials Science, Royal School of Mines, Imperial College of Science and Technology, University of London.

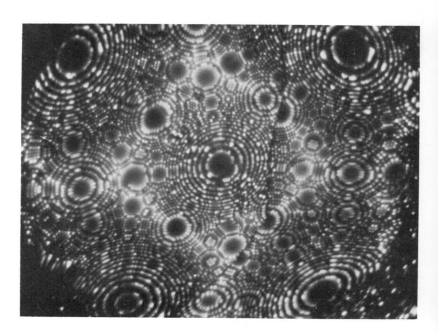

Figure A49
Field-ion micrograph of iridium showing a grain boundary.
Professor B. Ralph, Department of Metallurgy and Materials Science, University College, Cardiff.

The field-ion microscope has provided evidence of the regular arrangement of atoms in metal crystals. Gas ions are created close to the sharp point of the metal specimen, then accelerated away radially to hit a fluorescent screen and produce an image of the metal surface. More ions are produced near prominent atoms, so these points show up as bright spots on the screen (figure A49). A grain boundary between two crystallites is visible. Within each grain of the metal, the atoms are arranged in the same regular structure, but the relative orientation of the atoms between crystal grains is different.

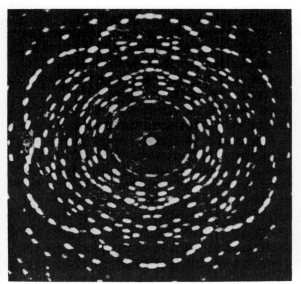

Figure A50
Von Laue pattern from an intermetallic compound of aluminium, iron, and silicon in the form of a single crystal.
Professor P. J. Black, Centre for Science and Mathematics Education, Chelsea College, University of London.

Figure A51
X-ray diffraction pattern for a single crystal (sucrose).
Department of Crystallography, Birkbeck College, University of London.

Figure A52
X-ray diffraction pattern for a polycrystalline metal (copper wire).
Department of Crystallography, Birkbeck College, University of London.

Other techniques, for example X-ray diffraction (see experiment A13 and Unit J, 'Electromagnetic waves'), show up the regular structure within each grain through a series of sharp rings or regular arrays of dots. The crystal structure and spacing between atoms can be determined from measurement of these patterns. Figure A50 shows the Laue pattern of an inter-metallic compound of aluminium, iron, and silicon in the form of a single crystal. Figures A51 and A52 are Bragg diffraction pictures of a single crystal (sucrose) and of a polycrystalline metal specimen (copper wire). X-ray patterns for glass and distilled water are shown in the next two photographs (figures A53 and A54).

The very fuzzy ring pattern or haloes indicate the lack of order within these substances, showing that glass is rather like a 'solid liquid'.

The final pictures (figures A55 and A56) are electron diffraction patterns (see experiment L7) of unstretched and stretched rubber. When rubber is stretched the jumbled-up long chain molecules are stretched out and line up partially, producing some measure of orderliness. This would explain the existence of the spots in figure A56 which do not exist in figure A55.

Figure A53
X-ray diffraction pattern for glass.
Pilkington Brothers P.L.C.

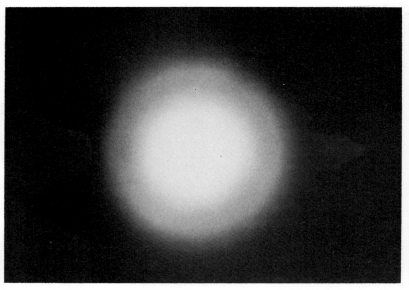

Figure A55
Electron diffraction pattern for unstretched natural rubber, showing only amorphous haloes.
The Malaysian Rubber Producers' Research Association.

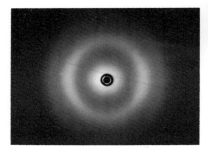

Figure A54
X-ray diffraction pattern for distilled water.
Pilkington Brothers P.L.C.

Figure A56
Electron diffraction pattern for highly stretched natural rubber, showing evidence of crystallization.
Professor E. H. Andrews, Department of Materials, Queen Mary College, University of London.

Further reading

MARTIN and HULL, *Elementary science of metals*. (Chapters 2 and 4.)
NUFFIELD REVISED ADVANCED CHEMISTRY Special Studies, *Metals as materials*. (Chapters 2, 3, and 5.)

Questions

a Twins are a common feature of pictures of the microstructure of metals. Why do twins have two parallel sides, unlike grain boundaries, which are irregular? How does this structure arise? (Look in one of the references.)

b Figures A40 to A44 include examples of the microstructure of metals which have been cold-worked, quenched, or annealed. How are the properties of the metal changed by these processes?

c In figure A45, why do the slip planes appear in different directions although the stress is only in one direction?

d What does an electron microscope look like? How large is it compared with an optical microscope? How large in area and how thick are the very thin metal specimens?

e Find out how a field-ion microscope works. (You may not fully understand the principle until you have studied Unit E, 'Field and potential'.) In figure A49 you are not looking at atoms directly. What are the bright spots on the picture?

f Explain why fuzzy rings and haloes in the X-ray diffraction pictures indicate lack of order, while sharp rings and spots indicate a degree of order in a solid structure.

MODELS

The following passage about models is an extract from Eric M. Rogers *Physics for the inquiring mind: the methods, nature, and philosophy of physical science* Copyright © 1960 by Princeton University Press and Oxford University Press. Excerpt reprinted by permission of Princeton University Press. (Chapter 24.)

We do not necessarily believe that the picture of nature we thus form *is* the real world. Many scientists say it is simply a *model* that works. It is easy to see that our picture of atomic structure is only a model – the invisible atom described in terms of large visible bullets and baseballs and large forces that we can feel like weights and the attraction of magnets. Yet it is uncomfortable to realise that we do not know what an atom is 'really like', and can only say that it 'behaves as if ...'. Moreover, with the progress of invention, microscope ... electron microscope ... ion microscope ... , you may decide that we can see real atoms and not just a model of them. Yet all such 'seeing' of the micro-world, however clear its results, is quite indirect: the images we obtain must be interpreted in terms of the models that guided our use of apparatus. In casual talk we gladly say, 'Now we know what the atoms are really like, how they are arranged and how they move about'; but in serious discussion, most scientists say, 'We have only shown that our model serves well, and we have obtained some measurements of parts of our model.' In a way, we use models in almost all our scientific thinking: atoms, molecules, gravity, magnetic fields, perfect springs, ...

Questions

a There are some models whose job is to remind one of what things are like: a model of a petrol engine, a circuit diagram, or a sketch map, for example. Do the models described above have anything in common with this type?

b Are there scientists who do not make use of models at all? Are there ways of using models other than in the way described above?

c In a sentence or two, indicate the differences between model and reality in some of the following examples:
i A model railway layout.
ii A railway map showing the lines neatly straightened out.
iii A model boat hull built for tests in a tank before building the real structure.
iv Molecular models used by chemists.
v The use of a computer to simulate the random flow of traffic, so as to improve the design of a road system.
vi Balls joined by springs to represent vibrating atoms in a solid.
vii An atom as a hard sphere.

d Is a mathematical equation of the vibration of two isolated atoms, for example an iodine molecule, a model in the same way that two air track vehicles, connected by a spring oscillating about their common centre of mass, are? Are all mathematical equations of physical situations only models?

THEORIES – TRUE OR NOT?

The following extract is by Karl Popper, who is a philosopher. He explains his view of scientific truth.

The way in which knowledge progresses, and especially our scientific knowledge, is by unjustified (and unjustifiable) anticipations, by guesses, by tentative solutions to our problems, by *conjectures*. These conjectures are controlled by criticism; that is, by attempted *refutations*, which include severely critical tests. They may survive these tests; but they can never be positively justified: they can neither be established as certainly true nor even as 'probable' (in the sense of the probability calculus). Criticism of our conjectures is of decisive importance: by bringing out our mistakes it makes us understand the difficulties of the problem which we are trying to solve. This is how we become better acquainted with our problem, and able to propose more mature solutions: the very refutation of a theory – that is, of any serious tentative solution to our problem – is always a step forward that takes us nearer to the truth. And this is how we can learn from our mistakes.

As we learn from our mistakes our knowledge grows, even though we may never know – that is, know for certain. Since our knowledge can grow, there can be no reason here for despair of reason. And since we can never know for certain, there can be no authority here for any claim to authority, for conceit over our knowledge, or for smugness.

From the author's preface to K. R. Popper *Conjectures and refutations*, Routledge & Kegan Paul Ltd, 1963.

Questions

a Rewrite the main points of the argument of this passage in your own words.

b 'Theories are nets: only he who casts will catch.' (*Novalis*.) Does Popper agree with this statement? If so, where in the passage?

c Discoveries come through learning to ask the right questions. How do we learn to ask the right questions?

MATERIALS USED IN ARCHITECTURE

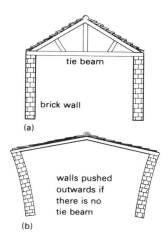

Figure A57(a) is a sketch of a conventional brick house, with a tiled roof carried on timber rafters. The brick walls are in compression, supporting the roof. The tie beam across the roof that prevents the roof from pushing the top of the walls outwards – see figure A57(b) – is in tension. An alternative way of preventing walls from bending uses compression, by providing buttresses outside the walls. The beautiful flying buttresses used by some cathedral builders are a fine example of the principle. (See figure A58.)

Figure A57

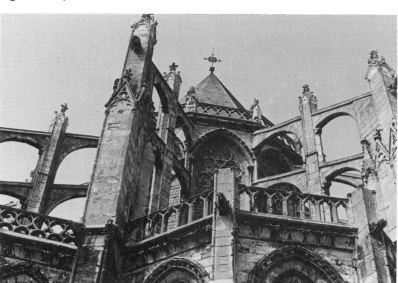

Figure A58
Tours Cathedral.
Clive Hicks.

The architecture of compression

Much early architecture relies mainly on compression for the stability of its buildings, because stone and brick are brittle materials good in compression but poor in tension. (Wood was for a long time the only tough material, good in tension, that was available in quantity. But wood rots, and was often avoided in buildings that were to be as permanent as possible.) The 'Romanesque' arch is a fine example of the architecture of compression, the compression often being expressed visually in the massive, thick pillars and short arches. (See figure A59.) Greek temples used stone columns in compression, carrying short thick stone beams bridging the tops of the columns. (See figure A60.)

Figure A59
Blyth Priory.
Clive Hicks.

(a)

Figure A60
The Theseion, Athens.
J. Allan Cash Ltd.

The architecture of tension

The introduction of steel as a structural material meant that parts of a structure could be in tension without risk of fracture, for steel is tough, not brittle. The most obvious example is the suspension bridge.

A less obvious example of the architecture of tension is the modern tall tower block of offices or flats. In a conventional house, the walls hold up the roof; in one design of tower block the roof holds up the walls. The block has a central spine from which cantilever arms of steel or steel-reinforced concrete stick out. The top cantilever arm carries the roof, and the walls, often of glass in alloy frames, are hung between the lower arms. The arms also take the load of the successive floors, by withstanding bending forces. (See figures A61(a) and (b) and A62.)

(b)

Figure A61
Construction photograph showing
(a) Top suspended structure of Commercial
Union building in the City of London.
(b) Completed building.
The Architectural Press.

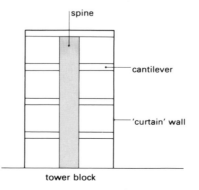

Figure A62

Questions

a Why is part of a beam that is being bent in tension? Sketch one beam and indicate which parts are in tension and which are in compression.

b Why had the stone beams joining the tops of columns in a Greek temple to be short and thick?

c Figure A63 shows a pre-stressed concrete beam being used to support both a floor and adjoining balcony in a building. Steel reinforcing bars are set in the concrete and tightened before the beam is put in position. Copy the figure and indicate where, in the beam, two rods should be put and justify the positions you have chosen. What advantage does this pre-stressed concrete have over unstressed concrete?

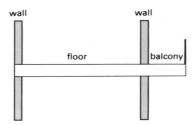

Figure A63

d The arch structure of figures A58 and A59 makes use of the compressive strength of stone. Indicate the forces acting on an arch and hence justify the use of flying buttresses.

e What properties would you consider desirable in choosing a material or materials for a road bridge?

f Give some other examples of the use of compression and tension in architectural structures.

g Fit the modern motorway flyover into this pattern.

h How are large domes built? Are the internal forces similar to those in an arch?

i 'The tree is nature's precursor of the modern office block.' Does this make sense?

LABORATORY NOTES

EXPERIMENT

A1a The stretching and breaking of metals to compare strength, ductility, and hardness

Selection of materials:
stainless steel wire, 1 m lengths, 0.08 mm diameter
copper wire, 1 m lengths, 0.28 mm diameter
other metal wires, *e.g.*, iron, nichrome, fuse wire, etc., as available
selection of metal strips about 100 mm × 10 mm × 1 mm, *e.g.*, copper, steel, aluminium

G-clamp, small
Hoffman clip
2 wooden blocks, about 20 mm × 20 mm × 10 mm, and 2 strips, about 60 mm × 10 mm × 5 mm
2 dowel rods, 1–2 cm diameter; 10–15 cm long
retort stand base, rod, boss, and clamp
centre punch
plastic, card, or metal tube (about 30 cm long, to take centre punch)
hand lens
safety spectacles

Safety note: Take care when doing experiments which involve stretching wires or breaking brittle materials. If the specimen suddenly breaks it may fly up into your face. Wear eye protection.

Samples of different metals are available in wire and thin sheet form. Apply forces to the specimens to observe their behaviour in different situations: stretching (tensile forces), flexing, scratching, and hitting (impact hardness).

Examine, with a hand lens, the broken ends of a wire or the impact or scratch on a sheet.

Various terms are used to describe the different behaviours of the metals. For each sample you have tested, classify it using some of the following terms: strong, weak, hard, soft, ductile, brittle, rigid, tough, malleable, flexible, elastic, plastic. (Some definitions are given on pages 3 and 4, or use a reference book.) For example, a steel wire is strong in tension; it is ductile, elastic, and suffers brittle fracture.

Suggest a use for each of the specimens and say why the chosen metal is particularly appropriate for this use. What metal might be suitable for toothpaste tubes (malleable) but not suitable for beer cans?

(a)

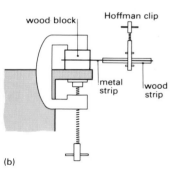

(b)

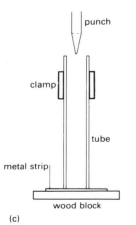

(c)

Figure A64
Simple testing of metals.
(a) Stretching.
(b) Bending.
(c) Hardness.

EXPERIMENT

A1b A preliminary study of the force–extension relationship for lengths of various materials

rubber bands
thin rod of Plasticine
paper
polythene (*e.g.*, from a food bag)
nylon line
strip of expanded polystyrene
steel wire ⎫
copper wire ⎬ as in A1a
2 dowel rods ⎭
safety spectacles
plastic ruler

Pull the various materials provided, using the two dowels where necessary, and sketch graphs of applied force against extension for each specimen. You should wear safety spectacles.

Compare the forces required to stretch the various materials and then break them. Classify the materials using the terms listed in experiment A1a.

EXPERIMENT

A2a A study of how the length of a rubber band varies with the applied force

size 32 rubber band, 75 mm × 3 mm × 1 mm unstretched
hanger and slotted masses, 0.1 kg, up to 2 kg
retort stand base, rod, boss, and clamp
metre rule

Stretch and release the rubber band a number of times very rapidly. Do you notice any physical change(s)?

Hang the rubber band from the clamp. Attach masses in units of 0.1 kg and measure the extension for forces up to 15 N. (If the band breaks below this applied force, repeat up to a force below the breaking force.) Remove the masses one at a time, measuring the extension as the load is reduced. Plot a graph of applied force against extension.

Study the shape of your graph. Did the band have a permanent extension at the end of the experiment? Were the extensions as the force decreased equal to the extensions for the same value of applied force when it was increasing? Can you suggest any significance for the size and shape of the resulting loop? If you have time, repeat the experiment loading the band to only 0.4 of the maximum load you used before. Compare the shape of the two graphs.

Some of the terms applied to the metals of experiment A1a can also be used to describe the behaviour of the rubber. Which? Do the properties of the rubber change as it is stretched?

EXPERIMENT

A2b The stretching of a nylon fishing line or a strip of polythene

nylon fishing line (breaking load about 10 N), 1.5 m long
retort stand base, rod, boss, and clamp
hanger with slotted masses, 0.1 kg
polythene strip, 200 mm × 10 mm, (e.g. 500 gauge, about 0.1 mm thick)
2 polarizing filters, 50 mm × 50 mm
adhesive tape
metre rule
safety spectacles

Nylon

Safety note: Take care when stretching filaments. If the specimen suddenly breaks it may fly up into your face; it is advisable to wear eye protection.

Tie together the two ends of the fishing line to make a loop about 0.75 m long. Make the knot carefully, and test it to see that it does not slip. Hang the loop from the clamp and carefully attach masses in steps of 0.1 kg up to 0.8 kg, measuring the extension each time. Remove the masses one by one, again measuring the extension.

Draw a graph as in experiment A2a, and answer the same questions.

How will the behaviour of an angler's fishing line differ from the behaviour of the short sample you have used? What factors will be the same? You might like to find the maximum extension of the nylon which will permit it to return to its original length. A more useful quantity to know is the strain, the extension divided by the original length.

Imagine what would happen if you tried the same experiment using a strand of wool (try it if you wish). You can now suggest why nylon garments last longer than woollen ones. Other properties, however, make wool a better material for some applications. Suggest some.

Polythene

Make sure the polythene is cut with clean and not ragged edges. Devise some simple method to hang up the strip and stretch it using the masses. Watch what happens very carefully. When you have stretched it, put the strip between crossed polarizers and look for stress patterns.

EXPERIMENT

A3 Measurement and prediction of the breaking force for aluminium samples

strip of aluminium (kitchen) foil, 100 mm × 10 mm
micrometer screw gauge or Vernier calipers
ruler
strip of aluminium sheet, 100 mm × 10 mm × 1 mm
adhesive tape
spring balance, 10 N, or hanger with slotted masses, 0.1 kg

Use paper and adhesive tape to make 'handles' to enable you to find a value for the force required to break the aluminium foil by pulling along its length.

Measure the thickness of the aluminium foil and of the sheet and estimate how many strips of foil would be required to make the sheet. Hence estimate the force which might be required to break the sheet. Estimate an uncertainty in this value as follows. What is the maximum percentage uncertainty in your estimates for:

i the foil breaking force

ii the foil thickness

iii the sheet thickness?

Add all three of these together to find the maximum percentage uncertainty in the calculated force. (See 'Calculation of uncertainty' on page xvi.)

To estimate the force you assumed the breaking stress would be the same for both the sheet and the foil. Suppose you had been given a sheet of twice the width, how would the breaking force have varied? Would the breaking stress still have been the same?

Can you suggest any reasons why your prediction is likely to be too high or too low? Later, when you have studied how defects in the structure of materials affect their strength, you may be able to give a better answer to this question. The uncertainty you were asked to calculate above is a *random* error. Some of your suggestions may involve *systematic* errors.

EXPERIMENT

A4   Measuring the breaking strength of a glass fibre

soda glass rod, 0.2 m long, about 3 mm diameter
retort stand base, rod, boss, and clamp
Bunsen burner
micrometer screw gauge
hanger with slotted masses, 0.1 kg
hand lens
string
safety spectacles

Safety note: Remember to wear safety spectacles.

Hold the top of the glass rod vertically in the clamp so that it is about 0.75 m above the bench or floor. Tie a 1 kg mass to the lower end with string. (A clove-hitch is suitable.) Support the mass with one hand and heat the middle of the glass over as short a length as possible until it is red hot. The lower part of the rod may need to be supported. Remove the flame and release the mass. The result should be a straight glass fibre with a length of glass rod at each end.

Hang masses from the fibre to find the breaking force. By how much does the fibre stretch?

Observe the broken ends of the glass fibre using a hand lens and contrast the point of fracture with that seen for a copper wire in experiment A1a. How do the two fractures differ? Explain the different behaviours using the terms for describing materials introduced in experiment A1a. Measure the diameter of the fibre carefully at the place where it broke.

Using the same method as for experiment A3, estimate the maximum load which the glass rod could support. The result of the calculation should show you that glass is very strong. Is glass used as a strong material? If so, where? What is the major difficulty with glass?

EXPERIMENT
A5 Measuring the strength of paper

mass, 1 kg

paper cylinder

platform

bench

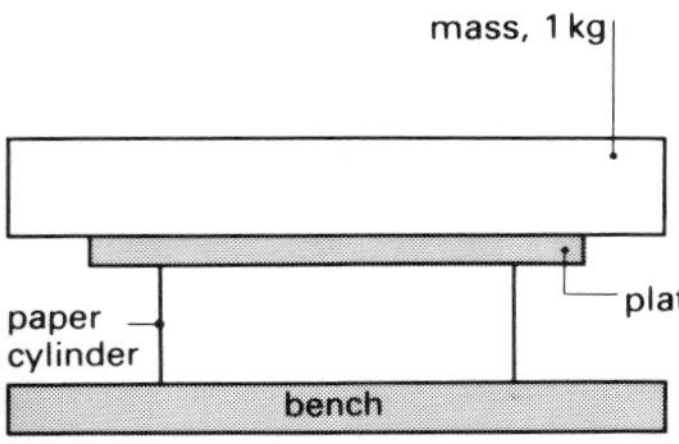

Figure A65
Measuring the strength of paper.

sheet of thin paper, A4
sheet of corrugated paper, at least 30 mm × 200 mm
hardboard square, about 80 mm × 80 mm
2 masses, 1 kg
10 slotted masses, 0.1 kg
adhesive tape
raw potato
paper drinking straw
metre rule

Cut several strips of paper about 30 mm wide, and make paper cylinders about 50, 60, and 70 mm in diameter.

Using the corrugated paper (or plain paper folded concertina-fashion), make a similar cylinder about 60 mm in diameter.

Using the hardboard square as a platform, load each of the paper cylinders centrally, carefully, and gently with masses until it collapses. Repeat with the corrugated paper cylinder.

What factors affect the strength of each cylinder? Why is the corrugated paper cylinder so much stronger?

What is the area of the paper which is supporting the load in each case? Is there a relationship between the load and the area of paper? Could you define a term *crushing strength* for the paper? Do you have enough data to decide? If not, how would you proceed to be able to give an answer? How would you re-design the experiment to avoid some of the present difficulties?

Can you predict the load that a paper cylinder of diameter 120 mm would support? Is your prediction likely to be too large or too small?

Take a raw potato. Hold it between your fingers and thumb so that the top and bottom are clear. Hold a paper straw firmly in your other clenched fist and thrust it very rapidly and firmly at the potato with a 'karate' chopping action. Do not check your swing but make a determined effort to drive the straw right through the potato.

Estimate the force exerted by the paper straw. Why did it not buckle as would be predicted by the investigation above on the strength of paper?

EXPERIMENT

A6   A short investigation of spring behaviour

4 expendable steel springs
hanger with slotted masses, 0.1 kg
retort stand base, rod, boss, and clamp
short length of stiff wire (*e.g.*, from coat hanger, or 0.9 mm diameter copper)
stop watch

i Each spring should obey Hooke's Law, $F = kx$, for small extensions, x, of the spring. By plotting a graph of applied force, F, against extension, x, find the spring constant, k (restoring force per unit displacement), for one of the springs. Do not exceed the elastic limit of the spring.
ii Connect two springs in series. Find how the spring constant, k_s, for this system (that is $F = k_s x$) is related to k for the single spring.
iii Connect two springs in parallel as shown in figure A66, using a 'zigzag' piece of wire to couple them to the mass, and measure the spring constant, k_p, for this system. How is this spring constant related to k for the single spring?

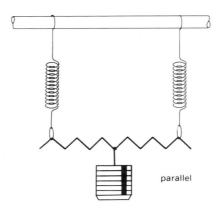

series

Figure A66
Springs in 'series' and in 'parallel'.

parallel

Find a general rule relating the spring constant of a system of springs connected in series or parallel to the spring constant of a single spring.
iv Hang 0.2 kg from the end of a single spring, pull down and release. Find the time period for the vertical oscillation. Connect four springs in series and repeat. How do you think the period of oscillation is related to the spring constant? Check your prediction by using two springs in series.
v Compare the periods for the vertical oscillation of 0.4 kg and 0.1 kg masses suspended from the end of two springs connected in series. How is the period related to the mass? Check your prediction using other masses.

You have now determined how the period of oscillation, T, varies with m, the mass, and k, the spring constant for a spring. What are the units of m and k? Does your expression for T in terms of m and k have the units of time? If it does not, your prediction is either wrong or incomplete, that is, the period of oscillation depends on other measurable quantities, such as the length of the spring.

DEMONSTRATION
A7 The effect of cracks

Bunsen burner
3 lengths of soda glass, 0.1 m, about 3 mm diameter
file, *e.g.*, triangular
matchstick
2 strips of polythene, 100 mm × 10 mm (*e.g.*, 500 gauge, about 0.1 mm)
scissors
safety screen
Perspex strip, 250 mm × 10 mm × 6 mm (clear faces 6 mm wide)
2 polarizing filters, 50 mm × 50 mm, or liquid crystal display filters, 105 mm × 35 mm
slide or overhead projector
safety spectacles

A7a A glass rod can be cut by making a file mark and bending the rod across the crack. Why does this work? Why does glass snap cleanly?

A7b Stress concentration may be shown by projecting the image of a nick cut in a strip of polythene or adhesive tape and held between crossed polarizers. A strip without a nick is stretched first, to show that colours develop as the strip is stretched.

A7c Why are equally-spaced lines seen across the Perspex strip when it is flexed between crossed polarizers? How can this help to explain why a bent glass fibre may break if touched on the outside but not on the inside?

EXPERIMENT
A8 Measurement of the Young modulus and breaking stress of a wire

G-clamp
2 wooden blocks
single pulley on clamp
metre rule
adhesive tape or gummed paper tape
hanger with slotted masses, 0.1 kg, up to 2 kg
micrometer screw gauge
safety spectacles
lengths of iron wire, 0.20 mm diameter
lengths of copper wire, 0.28 mm diameter
lengths of steel wire, 0.08 mm diameter

Choose one of the metal wires. Devise a careful, simple (and safe) experiment to obtain a stress–strain graph, and values within stated limits for the Young modulus and the breaking stress of the metal.

Arrange your apparatus horizontally across the bench. What length of wire should you stretch?

Steel, especially, is springy. Make sure you have a safety device to stop an end whipping back when the wire breaks; and wear safety spectacles.

How will you measure the extension of the chosen length of wire accurately? Why should you measure the diameter of the wire in several places and take readings at right angles in each place?

From your graph find the Young modulus for the wire and estimate the uncertainty in your value. Which of the measurements you have made gives the largest contribution to the uncertainty? What is your estimate for the breaking stress of the wire?

Would you expect the Young modulus for a bar of the metal you have used to have the same value as you have found for the wire?

On your graph, identify the elastic limit and estimate the percentage strain the wire can withstand before its elastic limit is reached.

Decide whether it would be better to design two different sets of apparatus to investigate:

i the properties of the wire up to its elastic limit, and
ii its properties from its elastic limit to its breaking stress.

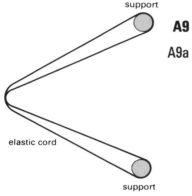

support

elastic cord

support

Figure A67

EXPERIMENT
A9 **Testing the formula for translational kinetic energy**

A9a Catapulting an air track vehicle

air track and accessories
air blower
timer, resolution 1 ms
photodiode assembly with light source
metre rule
3 elastic cords for accelerating vehicle
thin card
balance, 1 kg

An elastic catapult is made from a rubber band or cord by stretching it tightly between two firm supports as shown in figure A67, at such a height that it will project a vehicle along the air track. Catapult the air track vehicle with one, two, or three elastic bands or threads all at the same stretch, so that the vehicle is given one, two, or three units of kinetic energy.

Measure the speed of the vehicle. Also alter its mass to test the expression for kinetic energy: $\frac{1}{2}mv^2$.

A9b Measurements with potential energy changing to kinetic energy

either
i
Apparatus as experiment A9a (air track, etc.) with:
hanger and slotted masses, 0.01 kg
string
single pulley
or
ii
runway
2 dynamics trolleys
single pulley on clamp
string
hanger and slotted masses, 0.1 kg
ticker-tape vibrator
carbon paper disc
ticker-tape
transformer
balance, 2 kg

i Air track: level the air track carefully, and then use a falling mass to accelerate the air track vehicle. Use the photodiode assembly and timer to measure the vehicle's speed, v, at different distances, h, fallen by the mass.

ii Trolley and runway: compensate the trolley runway carefully for friction without the string connected. Attach 0.4 kg to the end of the string and accelerate the trolley using the falling mass. Make sure you catch the trolley before it comes off the edge of the bench. Use the ticker-tape to find the speed of the trolley, v, at different distances, h, fallen by the mass. You only need one ticker-tape.

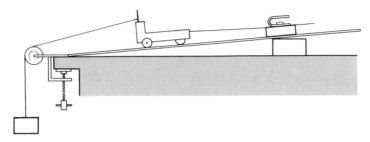

Figure A68

i and *ii* Plot graphs of v against h, and v^2 against h. How does your straight line graph show that the kinetic energy is proportional to v^2?

What is the total mass being accelerated in your experiment?

Change the value of the falling mass and/or use more trolleys stacked together (or a train of air track vehicles) to show how the kinetic energy depends on mass.

Does the small change in gravitational potential energy as the trolley moves down the sloping runway lead to an error in the kinetic energy? If not, where has this energy gone?

EXPERIMENT

A10 Measuring the elastic strain energy stored in a spring

Method 1

runway
dynamics trolley
expendable steel spring
spring balance, 10 N
retort stand base, rod, boss, and clamp
G-clamp

either
ticker-tape vibrator
carbon paper disc
ticker-tape
transformer

or
timer, resolution 1 ms
photodiode assembly with light source
metre rule

Compensate the trolley runway carefully for friction. G-clamp the retort stand to the bench and use the rod and clamp to hold a trolley peg vertically over the middle of the runway. Its lower end should be a few millimetres higher than the peg in the trolley. Hook the spring over the two pegs, pull back the trolley, and release it. The spring should fall from the fixed peg as the trolley passes.

Measure the extension of the spring, x, and the final speed of the trolley, v, for extensions of the spring up to 120 mm. Calculate the kinetic energy given to the trolley in each case.

Use the spring balance to obtain a force–extension graph for the spring over the range of x used.

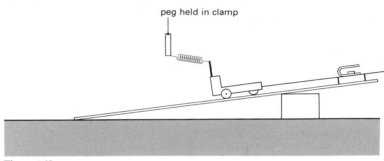

Figure A69

Method 2

Apparatus as for method 1 except for:
dynamics trolley with spring plunger
balance, 2 kg, or masses, 1 kg
wooden block at least 8 cm high
adhesive tape
graph paper

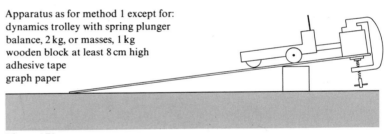

Figure A70

Calibrate the trolley's spring plunger as follows. Make a scale on the plunger (using adhesive tape and a graph paper strip for example). Note the zero compression position, and measure the force required to compress the plunger through various distances. Plot a graph of force against compression distance.

Set up a friction-compensated slope and firmly G-clamp a block to it. The trolley spring can be compressed to known distances, x, by pushing against the block. Measure the maximum velocity, v, attained by the trolley for several compression distances and then calculate its kinetic energy.

For either method compare the values of $\frac{1}{2}mv^2$ with the area under the appropriate part of the graph of force against distance. Are they equal within the limits of the experiment? If the graph is a straight line, show that the maximum velocity, v, should be proportional to the spring compression or extension, x. Plot a graph to check whether this is true in your experiment.

Suggest why a part of the elastic energy originally stored in the spring is not transformed into kinetic energy of the trolley. Estimate the fraction of the elastic strain energy not transformed into kinetic energy.

EXPERIMENT

A11 Changing elastic strain energy into gravitational potential energy or translational kinetic energy

A11a Apparatus as for experiment A9a, plus:

either
spring balance, 10 N
or
hanger and slotted masses, 0.1 kg

Repeat experiment A9a, using only one elastic band. Repeat the experiment for various extensions of the catapult. Find the force required to pull the band back the various distances x, using the spring balance or hanger and masses. From the area under the force–distance graph, find the stored strain energy in the band. Compare this with the final kinetic energy of the vehicle.

Are the two energies equal for each value of x?

A11b steel nail, 100 mm, or equivalent mass of wire, 10–15 g, bent into shape of horseshoe staple
retort stand base, rod, boss, and clamp
card or board
metre rule
hanger and slotted masses, 0.1 kg
rubber band (size 32, *i.e.*, 75 mm × 3 mm × 1 mm unstretched)
safety spectacles
balance, 0.1 kg

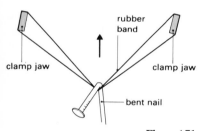

Figure A71

Safety note: Wear safety spectacles: the nail may fly up in to your face.
i Stretch the rubber band between the jaws of the clamp so that it makes a narrow horizontal loop. Place the bent nail across both strands and fire it vertically upwards about 0.5 m to 1.0 m.

Use the hanger and masses to obtain a force–distance graph for the rubber band and hence find the energy stored in the band from the area under the curve.

Compare the maximum gravitational potential energy of the nail with the elastic strain energy of the stretched rubber band.
ii Instead of firing the nail into the air, stretch the rubber band over the end of the rule to make a narrow vertical loop, so that when you release the band it rises about 1.0 m. Compare the maximum gravitational potential energy of the band with its elastic strain energy when stretched, found by a similar method to part *i*.

Can you suggest why the force–extension graphs for the rubber band in parts *i* and *ii* are different?

Contrast the two versions of experiment A11b. Which, do you think, will be more

a accurate,

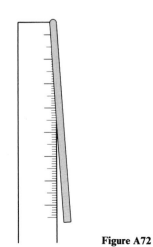

Figure A72 *b* reliable, that is, give consistent results?

One student measured the force F_{max} required for maximum extension x_{max} in experiment A11bii and estimated the elastic strain energy from $\frac{1}{2}F_{max}x_{max}$. How close would you expect the calculated and measured heights reached by the rubber band to be? Justify your answer.

In any of experiments A11a, bi, or bii the energy stored as elastic strain energy is likely to be larger than the maximum kinetic or gravitational potential energy. Why is this?

EXPERIMENT
A12 Energy absorbed in deformation

iron wire, 1.0 m long, 0.71 mm diameter
plastic tube, wooden rod, or similar, about 35 mm diameter, and at least 10 cm long
hanger with slotted masses, 0.01 kg
retort stand base, rod, boss, and clamp
metre rule

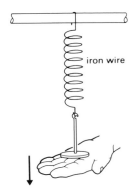

iron wire

Figure A73

Wind the iron wire around the former to produce a closely wound 'spring' with a suitable hook at either end. Suspend it from a clamp and attach 0.1 kg to the lower end, supporting the mass so that the 'spring' is unstretched. Release the mass and measure the extension. Calculate the potential energy lost by the mass.

Did the mass oscillate before coming to rest?

Rewind the coil to its original shape and plot a force–extension graph by loading it with small masses up to about 0.2 kg. Compare the area under the graph up to the extension caused by the original 0.1 kg, with the potential energy lost by that mass. How close you would expect the two values to be? Justify your answer.

How is this experiment relevant to real devices such as climbing ropes and car safety belts, or to situations where energy is absorbed by crash barriers and car bodies?

EXPERIMENT/DEMONSTRATION
A13 An optical analogue of X-ray diffraction

either
m.e.s. bulb, 2.5 V, 0.3 A, in holder
cell holder with 2 cells

or
lamp, 12 V, 24 W, in holder
transformer
card
pin

colour filters
leads
hand lens

set of Nuffield diffraction grids
and/or
piece of material with regular structure, *e.g.*, cotton handkerchief, or piece of fine mesh nylon, or Terylene net, or umbrella fabric

microscope slide dusted with lycopodium powder

Experiment

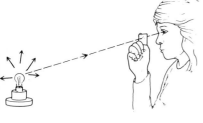

Figure A74

Set up a point light source. Place the material close to your eye and look at the light source.

What happens to the pattern you see if you stretch the material? Rotate it? Tilt it? How is the pattern you see related to the structure and the spacing of the threads? Given a photograph of the pattern, could you have guessed the structure of the object diffracting the light?

Look at the lamp through the glass slide covered in lycopodium powder. How does the pattern you see differ from the pattern made by rotating the handkerchief about the axis of the light beam? (Imagine a time exposure made as the handkerchief is rotated.)

How is the lycopodium powder arranged on the slide?

Repeat the experiment with the Nuffield diffraction grids, if available. Use the hand lens to observe the 'particle' arrangement in each grid and compare with the pattern you see when looking through the grid.

For at least one of the objects you have been looking through, compare the effect seen with red and with blue light. What happens to the pattern you see as the wavelength of the light is decreased?

Demonstration

Compare the width of the diffraction pattern on the screen made by wires of known diameter, with the width of the pattern made by the lycopodium powder. Estimate the size of the powder particles. How would the patterns have been different if a laser emitting blue light had been used?

This experiment is an optical analogue to the X-ray diffraction method used by von Laue to look at the structure of crystals. X-rays used have wavelengths of about 0.1 nm. What, approximately, is the wavelength of the visible light emitted by the laser? Would the pattern have been wider or narrower if a 'laser' emitting X-rays had been used? Would smaller particles give wider or narrower rings?

To produce a pattern with X-rays the same size as was produced with light, what size particles would have to be used? How does this compare with the size of an atom?

Suggest two reasons why X-ray diffraction photographs of real materials will be more complex than the diffraction patterns from the simple regular (or completely irregular) single sheets you have used.

DEMONSTRATION/EXPERIMENT
A14 How atoms are arranged in metals

either
40 expanded polystyrene spheres, 50 mm diameter (demonstration)
or
40 glass marbles or other small spheres (experiment)

4 books (or wooden battens 0.25 m long)

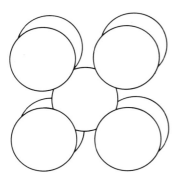

Figure A75
Part of a body-centred cubic crystal
structure.

Put 20 balls within the battens to form a 5 × 4 ball rectangle. How many nearest neighbours does each ball have?

Add three more layers of balls to form a pyramid. How many nearest neighbours does each ball have in one of the faces of the pyramid?

In which planes are the balls most closely packed — the base of the pyramid or its faces?

Use a triangle of battens to make a base layer of 15 balls in the close packed hexagonal arrangement (that is, as in the faces of the pyramid). Add further layers.

The model you have built represents a common crystal structure for metals, in which the atoms are packed together as closely as possible. How many nearest neighbours does an atom in the middle of one of these structures have?

In another possible structure for metals the layers of atoms are arranged in more open square arrays. Those in the first and third layers are at the corners of a cube, which has an atom from the second layer at its centre (figure A75). How many nearest neighbours does each have in this arrangement?

DEMONSTRATION
A15 An intermolecular force model using a linear air track

air track and accessories
air blower
elastic cord
2 small magnets to be attached to vehicles

Two identical air track vehicles represent two molecules interacting with each other. They are connected by an elastic cord and each vehicle has a magnet attached repelling its neighbour. What force does the elastic cord represent? What force do the magnets represent?

How does the force of attraction vary for two molecules as they move apart? How does the force in the cord vary as the vehicles move apart?

Another model could be constructed where the magnets attract and springy buffers replace the elastic cord. Would this be a better model for the intermolecular force? Explain.

What does the oscillating situation represent?

At what position does the system have minimum potential energy? The potential energy is nearly zero when the vehicles (molecules) are at large distances as the attracting force is very small. When the separation is less than the equilibrium distance you have to apply more and more force to push the vehicles (molecules) closer and closer together. Relate the main features of the potential energy–separation curve to those of the force–separation curve.

EXPERIMENT/DEMONSTRATION

A16 Chain molecules, dislocations, and alloys

A16a Splitting of stretched rubber/polythene

2 squares of balloon rubber, 50 mm × 50 mm
pin
polythene strip, 100 mm × 10 mm, 250 gauge (about 0.05 mm, *e.g.*, from a food bag)

Hold a piece of rubber and 'stir' the middle with a pin to pierce it. Now stretch the second piece of rubber and 'stir' near one corner with the pin. Now stretch the rubber at right angles to the first direction, and 'stir' in another corner.

How does the piercing and the hole change when the rubber is stretched? Can you explain this in terms of the molecular structure in rubber?

Split a strip of stretched polythene. Is the effect the same as with the rubber?

A16b Bubble raft model: dislocations, grain boundaries, and foreign atoms

Petri dish
length of rubber tubing to fit gas tap
hypodermic needle, 25 gauge
2 L-shaped pieces of wire, about 2 mm diameter and 10 cm long
bubble solution
Hoffman clip
Bunsen burner

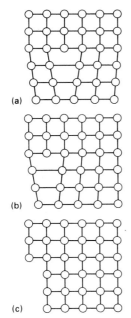

(a)

(b)

(c)

Figure A76
Movement of a dislocation.

Practise making bubbles one to two millimetres in diameter in the dish. Use the Hoffman clip and alter the depth of the needle in the solution for fine adjustment. Passing the Bunsen burner quickly across the surface will clear it for another attempt.

Make a raft of identical bubbles. How many bubbles surround each single bubble? Of which crystal structure(s) is the raft a two-dimensional model?

Make a bubble raft large enough to squeeze between the pieces of wire. Observe the movement of grain boundaries.

Look for dislocations running along a row of bubbles as the raft is deformed. Does this form of slipping happen more easily than the slipping of whole layers together? Explain the analogy: to adjust the position of a carpet, it helps to form a ruck and kick it across the carpet, thus moving the carpet a few centimetres. To pull the carpet the same distance bodily is very hard.

Make more dislocations by bursting some bubbles with a hot wire. Does it become easier or harder to compress or stretch the raft?

Make a raft which includes a few larger bubbles (foreign atoms). How do these affect the movement of dislocations?

Why do small soap bubbles act as good models of a two-dimensional crystal? Think of the attractive and repulsive forces between atoms. What forces act between soap bubbles?

A16c Heat treatment of steel

either
4 steel sewing needles
or
5 cm wire from expendable steel spring

2 pairs of pliers
Bunsen burner
safety spectacles

Safety note: Wear eye protection.

A steel sewing needle will bend before it snaps. Its properties are
changed by heating depending on how the alloy atoms of carbon are
distributed in the iron. If the needle is cooled slowly after heating it can
be bent easily. If quenched (cooled rapidly) it is very hard and brittle. If
tempered (reheated and cooled slowly) it becomes ductile again, is fairly
hard and much tougher. Tempered steel is a form in common use.

EXPERIMENT
A17 Measuring the forces on systems in equilibrium

2 retort stand bases, rods, bosses, and clamps
2 G-clamps
2 spring balances, 10 N
mass, 0.5 kg
0.5 m rule or lath
protractor
string

A17a Triangle of forces and resolution

Tie the mass to the centre of a length of string and make a loop at either
end. Attach two spring balances as shown. Measure T, F, and θ for
various values of θ.

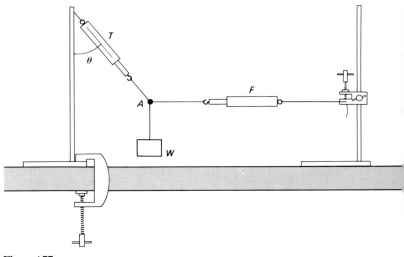

Figure A77

For each set of measurements draw a space diagram of the situation with arrows representing the magnitude and direction of the forces acting at point A. Draw another diagram in which the force vectors are joined head to tail (force diagram) to verify that they form a closed triangle when in equilibrium. Show that the 'closed triangle' condition is equivalent to

$$W = T \cos \theta$$

and

$$F = T \sin \theta$$

Show that for small angles θ, F is proportional to the horizontal displacement of the weight W.

A

A17b Strut or tie?

Clamp the retort stand close to the edge of the bench so that the rule can be pivoted freely at the edge as shown. Tie the mass to the end of the rule, making sure that it cannot slip. Connect a string and spring balance between the end of the rule and the retort stand. Measure the tension in the string when the rule is horizontal.

What is the magnitude and direction of the force in the rule? Is this situation identical to experiment A17a, replacing F by the rule? If not, how is it different?

For three non-parallel forces to be in equilibrium their lines of action must pass through a point. Use this idea to decide which of the arrangements in figure A79 is stable. Then check your predictions by trying to set them up.

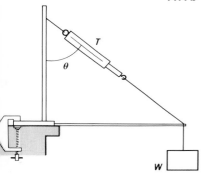

Figure A78

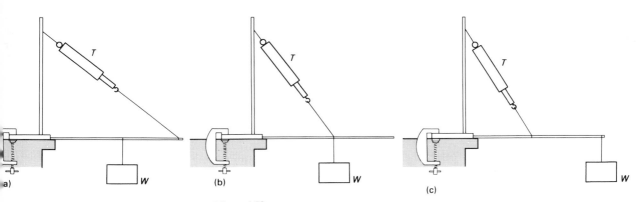

Figure A79

Why is arrangement (a) stable? What force(s) besides W and T act on the strut? What equilibrium condition is obeyed by this arrangement?

Do any of the sketches show unstable situations? What would have to be done to make them stable? What equilibrium condition could then be satisfied?

A17c Determination of the coefficient of static friction for two wooden surfaces

trolley runway
wooden block with screw eye
several 1 kg masses
hanger and slotted masses, 0.1 kg
single pulley on clamp
string

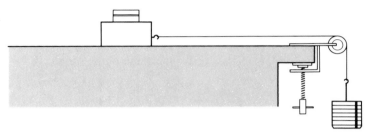

Figure A80

A small horizontal force applied to the block will not make it move. A frictional force sufficient to prevent motion exists. When the block just begins to move the frictional force has increased to a limit known as the limiting friction. The limiting friction between two surfaces depends on the normal force between them.

Ensure that the two surfaces in contact are horizontal. (Why is this necessary?) Increase the load on the wooden block to investigate the relationship between normal force and the limiting friction (that is, the minimum force required to move the block).

You should find that the limiting friction is proportional to the normal force. Obtain a value for the constant of proportionality, μ, which is known as the coefficient of friction.

μ can also be found using only the runway and wooden block. If the runway is tilted to an angle, θ, but the block does not move, what three forces in equilibrium act on the block?

By resolving these forces parallel and perpendicular to the runway show that, when the block does just begin to move, $\mu = \tan \theta$.

Measure this value of θ and compare the values of μ obtained by the two methods.

What forces are acting on the block when it does move?

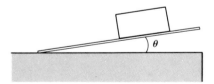

Figure A81

EXPERIMENT
A18 Investigating the forces at the supports of a loaded bridge

2 spring balances, 10 N
2 retort stand bases, rods, bosses, and clamps
metre rule
2 hangers and slotted masses, 0.1 kg
string

Use the spring balances (as supports) to hang the rule (bridge) horizontally from the two clamps. The rule should have its faces vertical. The supports need not be at the very ends of the bridge.

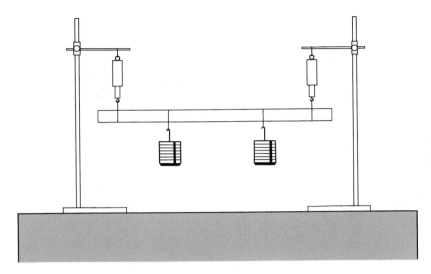

Figure A82

Hang masses from the bridge placed a suitable distance apart to act as the loads of the front and rear axles of a lorry crossing the bridge.

Investigate the way in which the forces on the bridge supports change as the lorry moves across.

How is the model over-simplified? For example, consider the weight of the lorry compared with that of the bridge, the areas over which the load is distributed, the rigidity of the bridge structure, etc.

EXPERIMENT
A19 'Weighing' a retort stand

2 retort stand bases and rods
boss and clamp
mass, 0.5 kg (or 1 kg)
string

Using only the apparatus given, find the mass of the retort stand.

EXPERIMENT
A20 Investigating the forces in a roof truss

2 dynamics trolleys
Plasticine
2 laths about 0.5 m long (or 0.5 m rules) with holes at 5 cm intervals
mass, 1 kg
spring balance, 10 N
bolt with nuts and washers (about 40 mm long, 5 mm diameter)
string

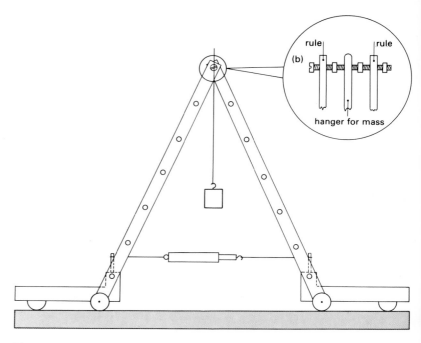

Figure A83

Bolt the two rules together as shown in figure A83, so that they can pivot freely and there is space between to hang the 1 kg mass. The screw which fixes a trolley wheel to the trolley's body passes through a hole in the lower end of each rule. The rules should pivot freely at each end. Attach the spring balance to the two dynamics trolleys. Measure the force in the 'tie bar' and at the apex angle. You may need to steady the apex to prevent the structure toppling sideways.

Are the struts in tension or compression?

Draw space and force diagrams for each corner of the structure to find the force in each strut and the normal force on each trolley.

Suppose the tie bar were made longer, so that the 'roof' was less steeply pitched. Would the forces in the struts change? Lengthen the tie bar with a piece of string and check your prediction.

If the tie bar was placed across the mid points of the two rules, what would you expect the force in it to be (assuming the apex angle to remain unchanged)?

The walls of a house are not mounted on trolleys. Why in this experiment were the struts on trolleys? What extra force(s) would be acting if the struts had been placed directly on the bench top? Would the tension in the tie be greater, smaller, or the same? The walls of a house are rather firmly attached to the ground: what effect does this have on the tension in the roof joists?

EXPERIMENT
A21　Measurement of momentum change and impulse

Apparatus as for experiment A9b(*ii*)

Repeat experiment A9b to make a single ticker-tape if you have not already done so.

Sketch the graph of the force acting on the trolley against the distance moved by this force. What does the area under the curve represent? You have already answered this question when doing experiment A9b.

Sketch the graph of the force acting on the trolley against the time for which the force acts. What does the area under this curve represent?

Use your ticker-tape to plot a graph of v against t, where v is the speed of the trolley after a time t, starting from rest. What is the slope of this graph? How can this graph be related to the force–time one?

Suppose the force acting on the trolley had not been constant but had varied. Why would you still expect the area under the force–time curve to represent the same quantity as before?

DEMONSTRATION
A22　Collisions on an air track

air track and accessories
air blower
2 photodiode assemblies and light sources
card
2 timers, resolution 1 ms

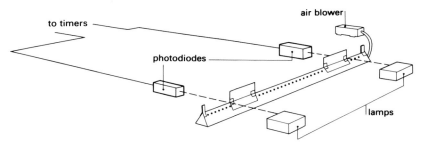

Figure A84
Collisions on an air track.

The track and vehicles need to be kept clean, and vehicles should be stored and handled in such a way that they do not become deformed.

Use one vehicle first on the air track to check that it is level.

Vehicles can be fitted with nose-pieces, for example elastic bands, magnets, needles, corks, Plasticine, etc., so that elastic or inelastic collisions can be achieved. Any load on a vehicle must be balanced, or it will lift at the less loaded end and be driven along by air from the track.

Vehicles of the same or different masses can be used with one vehicle stationary, or both moving, for both elastic and inelastic collisions. A maximum of four separate times must be measured to calculate the speeds of the vehicles before and after collision. An explosive separation can also be investigated by compressing a spring between the two vehicles and then releasing them.

Calculate the momentum, mv, and the kinetic energy, $\frac{1}{2}mv^2$, of each vehicle before and after the collision.

Find the change in the total mv and in the total $\frac{1}{2}mv^2$ of the two vehicles caused by the interaction.

Is the total momentum of the vehicles conserved to within the uncertainty of the measurements?

When is the total kinetic energy conserved? What becomes of it when it is not conserved?

DEMONSTRATION
A23 Newton's cradle: investigation of a line of colliding balls

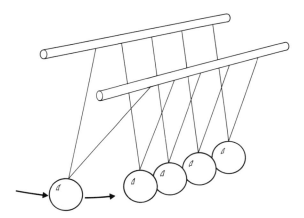

Figure A85
Newton's cradle.

Pull one, two, or three balls to one side and release them. Observe the subsequent motion of all the balls.

Explain this motion by considering a series of separate elastic impacts between a moving ball and a stationary one of the same mass. Show that there is only one result which conserves both momentum and kinetic energy.

Why is it important that the centres of the balls should lie in a horizontal line? Why should the balls just touch? What will happen if either of these conditions is not fulfilled?

What would happen if three or four of the balls were stuck together?

What would happen if the middle ball was much denser but of the same size as the others?

Why is it necessary to suspend the balls on a bifilar suspension? Why not roll marbles along a narrow track?

DEMONSTRATION

A24 Measurement of the speed of an air rifle pellet using a ballistic pendulum

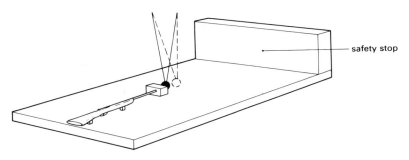

Figure A86
Measuring the speed of an air rifle pellet.

Any pellet which misses, or passes through the Plasticine, must enter some soft, absorbing material.

The apparatus is set up so that the Plasticine bob hangs in front of the muzzle of the air rifle and about 2 cm from it. The pellet is fired into the Plasticine and the maximum height through which the bob rises is measured.

Use the Law of Conservation of Energy (that is, initial kinetic energy transformed into gravitational potential energy) to calculate the initial speed of the Plasticine bob.

Use the Law of Conservation of Linear Momentum to find the speed of the pellet in terms of the masses of pellet and bob and the speed of the bob. Hence calculate the muzzle velocity of the pellet.

Estimate the uncertainty of your result.

EXPERIMENT

A25 Collisions in two dimensions using pucks

CO_2 pucks kit
CO_2 cylinder
dry ice attachment
camera
motor-driven stroboscope
safety spectacles

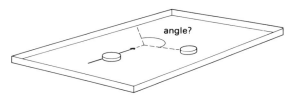

Figure A87
'Collision' between two magnetic pucks.

Safety note: Lumps of solid carbon dioxide should always be handled with tongs and/or thick gloves, 'snow' with a spatula or spoon. Eye protection should be worn. Before crushing, lumps should be covered with a cloth.

Clean the glass plate very carefully with methylated spirit and/or window-cleaning liquid.

Level the plate carefully with the wedges. Use a puck to check that the plate is level.

Try a head-on 'collision' between two magnetic pucks of equal mass with one initially stationary.

Now try a collision where both pucks move off in different directions, one puck having been stationary before the collision (figure A87).

Observe the angle between the directions of the two pucks after collision. Does it depend on the direction of the incident puck?

Repeat the experiment with pucks of unequal mass (that is, by adding masses or a second puck on top of the first) and draw some general conclusions about the angles after collision for each case. How are these results applicable to the analysis of cloud chamber photographs where elastic collisions between fast and very slow moving particles occur?

You may be able to take multiflash photographs of these events, and of collisions between two moving pucks. Take measurements from the photographs to test whether momentum is conserved in two dimensions.

HOME EXPERIMENTS

AH1   Saved by a hair!

Find the force required to break a human hair. Estimate the number of hairs on a human head. Could the Prince have climbed up Rapunzel's hair as suggested in the fairy tale? (Assume that a suitable anchorage for the hair was available, such as a window tie, and that magic enables hair to grow rapidly until it is long enough to reach the ground.)

AH2   An accurate newtonmeter

In this practical task you should attempt to construct a newtonmeter from a home-made spring that can resolve forces to 0.001 N with a range of 0 to 1 N. Check the accuracy of your mechanism by 'weighing' an object of known mass. Your spring must obviously extend by a measurable amount when the load is applied but you might consider ways of 'amplifying' the extension when measured against the scale of your device. Compare your device and its accuracy with others in the class.

AH3   The jelly column

Use ordinary dessert jelly from the grocers to make two identical columns with a base diameter of 4 to 6 cm. Use concentrations recommended by the manufacturer and arrange matters so that it is easy to remove each column from the mould you have used.

Before one of the moulds has set, mix in a small quantity of some fibrous material such as wool or hair to form a rudimentary 'composite material'. After removing the moulds, compare the columns as you think a scientist might. You might make a qualitative comparison between the elastic properties of the two materials. You might test each column to destruction and closely observe how failure occurs in each. Watch particularly for the way in which cracks are propagated.

Compare your effort with others in the class.

AH4   Cement

Cement, used in the making of mortar and concrete, is a most important structural material and it is consumed in vast amounts in the industrial areas of the world. In this task you should attempt to make 1 kg of cement using, as far as possible, raw materials obtained in or around your home and school. You might begin this task by reading about the process behind the manufacture of cement. The pottery department might provide you with help in processing your cement mixture.

After processing, see if your 'cement' will set after mixing with water and, maybe, sand. Compare your product with others in the class and with ordinary builders' cement.

QUESTIONS

The variety and behaviour of materials

1(I) Imagine that you are blindfolded and a piece of string is put into each of your hands. You are told that the strings are tied to the ends of a specimen which you can stretch. As you start to pull, it hardly extends at all, but when you pull a little harder it starts to stretch very easily and quite rapidly. Then you relax and it contracts again, but not back to its original length.

Sketch a graph to represent these observations.

2(E) The graphs in figure A88 show how the lengths of various materials vary as the force extending them is increased. For each graph, describe the behaviour of the substance. Hazard a guess as to what each might be. Each specimen was originally 100 mm long and the area of cross-section was 1 mm².

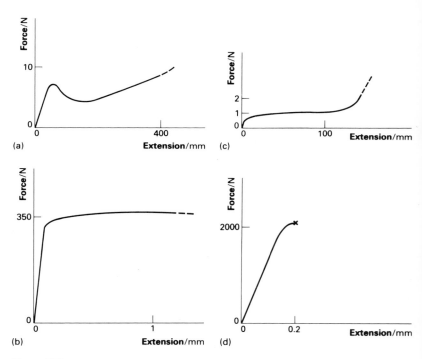

Figure A88

3(L) Read the definitions of the words 'ductile', 'elastic', 'plastic', 'strong', 'tough', and 'brittle' on pages 3 and 4. Using these words, classify the following substances: polythene, rubber, copper wire, brick, lead, Plasticine, cast iron, steel, biscuit, human skin, glass, bone.

4(P) Suppose that the graph (figure A89b) is the load–extension curve for rubber webbing used as the base of a chair.

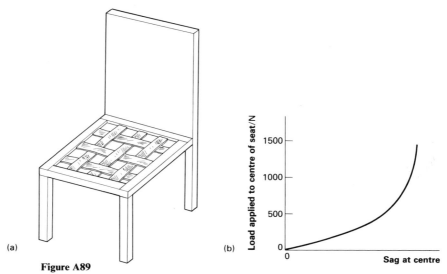

(a)   (b)

Figure A89

a The following are predictions about what will happen when a person sits on the chair.
i He will sink only a little, unless he is very heavy, when he will suddenly sink a lot.
ii He will sink a moderate distance and then be supported at about the same level whether he is heavy or light.
 Explain which is the correct prediction.

(Short answer paper, 1971)

b Describe the motion of a person who 'flops' into the chair.

Stress and strain

5(L) For a steel spring, obeying Hooke's Law, the extension, x, is proportional to the applied force, F, such that

$F = kx$, where k is called the *force constant* of the spring.

a A light vertical spring is extended 10 mm by a load of 1.0 N. What is its force constant?

b A second, identical spring is connected to the end of the first and the same load is hung from the pair. What is the extension? What is the force constant of the pair together?

c The two springs are disconnected and hung side by side from the same support. The ends are connected together to the same load. What is the extension? What is the force constant of the pair together?

d k is a measure of the stiffness of the system. How is the stiffness of the single spring related to the stiffness of the pair in 'series' and in 'parallel'?

e To answer part **b** you have used the fact that the extension per spring is the same for a given stretching force. In part **c** double the force would be required to produce the same extension as for a single spring. How do these experiments illustrate the importance of the quantities *stress* and *strain*?

6(P)a *Tensile stress* is tension per unit area of cross-section. A long strip of rubber whose cross-section measures 12 mm by 0.25 mm is pulled with a force of 3 N. What is the tensile stress in the rubber? Give the units.

b *Tensile strain* is the increase in length as a fraction of the original length. A strip of rubber originally 90 mm long is stretched until it is 120 mm long. What is the tensile strain? Why has the answer no unit?

c The greatest tensile stress which steel of a particular sort can withstand without breaking is about $10^9 \, \mathrm{N\,m^{-2}}$. What is the greatest load that can be supported on a wire of cross-sectional area 0.01 mm^2 made of this steel? (Use the original cross-section in calculation, as if it were not reduced when the load is applied.)

d Estimate the diameter of a single steel wire which would suspend a car without breaking.

e A wire 2 m long is given a strain of 0.01. By how much has its length increased?

f Rubber needs a stress of roughly $10^6 \, \mathrm{N\,m^{-2}}$ for each unit increase in strain. What is the tension in a rubber band 1 mm thick and 3 mm wide stretched to three times its original length? (The question is ambiguous. Explain why.)

g A student finds that a particular steel wire needs a force of $10^2 \, \mathrm{N}$ to increase its length by one per cent, and records this as $10^4 \, \mathrm{N}$ per unit strain. Another student thinks this result means that $10^4 \, \mathrm{N}$ are needed to double the length of the wire. What is wrong with this interpretation? Is the record sensible in the form given?

7(P)a A specimen of rubber of cross-sectional area 2 mm^2 is extended in length from 0.1 m to 0.15 m by a force of 0.4 N. Use these results to predict the force needed to extend a piece of the same material with 4 mm^2 cross-section from a length of 0.50 m to 0.75 m.

b On the basis of the above data, either calculate exactly or estimate roughly the force needed to stretch the second piece of rubber
i from 0.50 m to 0.55 m
ii from 0.50 m to 3.50 m.
Comment on your answers.

c If a 20 kilogram mass hanging on a steel wire of 1 mm^2 cross-section produces a 0.1 per cent strain in the wire, what mass hanging on the wire would give it a strain of 10 per cent?
Comment on your answer.

8(P)

Material	Young modulus/ N m^{-2}
cast iron	15×10^{10}
mild steel	21×10^{10}
wood, spruce:	
along grain	1.0–1.6×10^{10}
across grain	0.04–0.09×10^{10}
glass	8×10^{10}
Perspex	0.6×10^{10}

Table A1

a The Young modulus for glass is higher than that for Perspex. What does this tell you about the ease of stretching glass compared with Perspex? Which material is stiffer?

b The Young modulus for wood is different along the grain compared with across the grain. How does this show up when you flex a thin sheet of wood along the grain direction and then across the grain?

c Wood has a much lower value of Young modulus than steel. Which would be the easier to stretch?

9(E) In Arthur C. Clarke's novel *The fountains of paradise*, a material is developed which is so strong that, when in the form of a thread so fine as to be invisible unless looked for very carefully, it will support the weight of a human.

a Estimate the minimum breaking stress of this material.

b Compare this with the breaking stress of steel, about $1 \times 10^9 \, \mathrm{N\,m^{-2}}$.

c For an engineer *yield stress* is as important a quantity as breaking stress. It is the stress at which permanent or plastic deformation starts. For safety reasons any wire, beam, or strut in a system must withstand four times the maximum stress for which it is designed. The yield stress for steel is about $3 \times 10^8 \, \mathrm{N\,m^{-2}}$. What diameter steel wire is needed to hoist a person safely?

Dimensions

10(L) The dimensions of a physical quantity are found in terms of [mass] $= M$, [length] $= L$, and [time] $= T$. Thus, the dimensions of velocity are $L\,T^{-1}$, written [velocity] $= L\,T^{-1}$. Dimensional analysis is a powerful method of checking whether an algebraic expression is likely to be correct.

a Write down the dimensions of force (remember $F = ma$).

b What are the dimensions of the force constant or stiffness of a spring, k (*i.e.*, from Hooke's Law $F = kx$)?

c What are the dimensions of frequency?

d A student remembers that either the frequency or the period of oscillation of a spring is proportional to $\sqrt{\dfrac{k}{m}}$. Use a dimensional check to decide which it is.

e The constant of proportionality in part **d** is $1/2\pi$. This is known as a dimensionless constant. Can you suggest why it is so called? In Hooke's Law, k is a constant of proportionality too, but as you have shown it has dimensions.

f What are the dimensions of stress, of strain, and hence of the Young modulus?

g In an experiment to measure the Young modulus for glass, a glass slide acted as a simple cantilever. The theory of the bending of beams relates the Young modulus, E, to the force, F, acting on the cantilever causing a deflection, y, by the expression:

$$E = \frac{4L^3 F}{bd^3 y}$$

where L is the length, b the width, and d the thickness of the glass slide. Check that the righthand side of the equation has the same dimensions as the Young modulus.

Elastic energy

11(L) Some babies play in a 'bouncer'. This is a harness attached to a rubber cord hung from a door frame, and the baby is suspended so that its feet just brush the floor. Babies do actually seem to enjoy it, although it sounds, and looks, like an elaborate form of torture!

a Suppose a baby has a mass of 10 kg. What is its weight in newtons?

b This weight may stretch the rubber cord by 20 cm (0.2 m). If the force exerted by the cord were proportional to the amount it is stretched, what force would it exert when the baby has been attached to the unstretched rubber cord and then lowered by 2 cm (0.02 m)?

c What is the average force exerted by the rubber as the baby is gently lowered the full 20 cm?

d What is the energy stored in the rubber when the baby has been lowered 20 cm? (Use the average force, and the total distance.)

e If the greatest force exerted is F, and the greatest extension is x, what is the energy stored if the force is proportional to the extension?

f In fact, however, for rubber the proportionality assumption made above is wrong. The graph of force and extension will be as in figure A90. If the force at 20 cm is still 100 newtons, is the energy stored more, or less than that given by the answer to **e**? Explain your reasoning.

$(g = 10\,\mathrm{N\,kg^{-1}}$ approximately.)

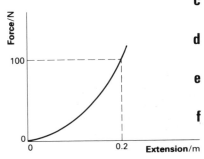

Figure A90

12(P) Figure A91 shows the apparatus with which the stroboscopic photograph (figure A92) was made. The photograph shows the distance scale marked in 0.1 m intervals and below it a series of 'glimpses', at 0.1 s intervals, of the peg attached to the trolley. An elastic cord was held at one end on a fixed peg (whose shadow obscures the righthand mark 2 on the scale) and at the other end by the peg on the trolley. The cord was just taut when the trolley peg was at the zero mark on the scale. Then the trolley was pulled back so that its peg was below the lefthand mark 5, that is, half a metre extension of the elastic. At this point the cord pulled the trolley with a force of 0.44 N. The mass of the trolley was 1.0 kg. The trolley was then released from rest, starting from the mark 5 of the scale. As the trolley peg passed the zero mark, the cord became slack and dropped off the fixed peg.

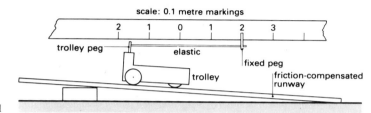

Figure A91

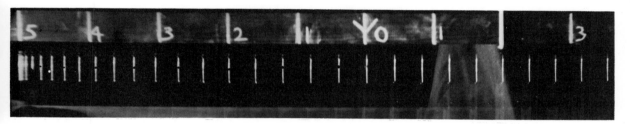

Figure A92

a From mark 5 to mark 0 the trolley accelerates. Would you expect uniform acceleration?

b Calculate the speed of the trolley after the cord has stopped pulling and hence its kinetic energy.

c How much energy would have been stored in the elastic cord had the force been proportional to the extension?

d Compare the answers to **b** and **c**. Is the supposition made in **c** likely for a rubber cord? How might the comparison of **b** and **c** be explained?

e Find the kinetic energy of the trolley at a number of different distances to the left of the zero mark; that is, for several different extensions of the rubber. (Find the kinetic energy from the speed close to that distance.)

  Now find the elastic potential energy at each point. For this, you can use the fact that the total energy is constant, and is equal to the kinetic energy after the cord goes slack, since there is then no elastic potential energy. At all points, the total energy is the sum of potential and kinetic energy. Draw a graph of elastic potential energy against extension of the cord.

f Has the curve of elastic potential energy against extension roughly the shape you might expect?

13(R) A trolley of mass 0.80 kg is held in equilibrium between two fixed supports by identical springs (S_1 and S_2) as shown in figure A93(a); each spring has an extension of 0.10 m. In figure A93(b) the trolley is shown moved to the right a distance 0.05 m.

   The relation between force F in newtons and the extension x in metres for *each* spring is given by

$$F = 20x.$$

a What is the *change* in force exerted by spring S_1 caused by moving the trolley to the right as in figure A93(b)?

b If the trolley is now released what will be the magnitude of
i the resultant force acting on the trolley at the moment of release, and
ii the initial acceleration of the trolley?

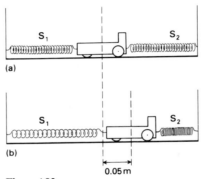

(a)

(b)

0.05 m

Figure A93

c Showing the steps in your calculations, determine the total energy stored in S_1 and S_2 when the springs are stretched
i as in figure A93(a), and
ii as in figure A93(b).

d What is the kinetic energy of the trolley as it passes through the equilibrium position?

(Short answer paper, 1981)

14(L) The energy stored in a rod of cross-sectional area A, stretched by a length x, less than its elastic limit, is given by $\frac{1}{2}Fx$ where F is the force to produce the extension x.

a What is the tensile stress in the rod under these conditions?

b The rod's original length is l. What is the tensile strain?

c Substitute for F and x in the formula for energy to obtain an expression for the energy stored per unit volume in terms of stress and strain.

d Typical working stresses and strains are given for four different substances in table A2.

	Strain/ %	Stress/ $\mathrm{N\,m^{-2}}$
steel	0.3	700×10^6
wood (yew)	0.9	120×10^6
tendon (in ankle)	8.0	70×10^6
rubber	300	7×10^6

Table A2

Which material stores the greatest energy per unit volume, in these typical working conditions?

e The energy stored per unit volume for each of these substances is large compared with that for many other materials. Suggest a situation for each one where this is put to use.

15(E) This question shows why it is a wise precaution to wear safety spectacles when doing experiments to stretch elastic materials until they break.

When a steel wire breaks, about 25 % of the stored elastic energy is transformed to kinetic energy. Show that the speed, v, of the wire when it breaks is given by $(E_p/2\rho)^{\frac{1}{2}}$, where ρ is the density of steel and E_p is the energy stored per unit volume (see question **16**). Use the data from question **14** and $\rho = 8 \times 10^3\,\mathrm{kg\,m^{-3}}$ to estimate the speed of the end of the wire.

16(P) The energy stored per unit volume of a solid which obeys Hooke's Law is $\frac{1}{2}$(stress × strain). Check that this formula is dimensionally correct. Note that the method of dimensions cannot prove that any relationship *is* correct, only that it is possible. Note also that the method gives no information about the numerical constants in a relationship.

17(R) Figure A94 shows a load–extension graph for a length of steel wire.

a Use the graph to estimate:
i how much energy could be stored in the wire and also recovered from it, and
ii the extra energy which would have to be supplied to fracture the wire.

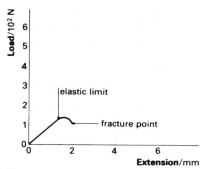

Figure A94

b Copy the diagram and draw carefully on it, the graph as it would have been
i if the wire had been twice as long (label it A), and
ii if instead the wire had had three times the cross-sectional area (label it B).

c Why are very long cables used for towing oil platforms out at sea?

(Short answer paper, 1979)

Atoms and molecules

18(L) A spherical oil drop of diameter d spreads out to form a circular patch of radius R on a clean water surface.

   a Show that the thickness, h, of the oil is $d^3/6R^2$.

   b In an experiment, the following measurements were made:
      i $d = 0.5 \pm 0.05$ mm by holding the drop on a cotton thread against a scale and viewing with a hand lens.
      ii $R = 9.3 \pm 0.1$ cm using a ruler.
        Find the value of h.

   c Calculate the percentage uncertainties in d and R and hence find the uncertainty in h.

   d Why is the value of h found in part **b** only the upper limit to the length of a molecule?

   e Other experiments indicate that the diameter of a single atom is about 2×10^{-10} m. About how many atoms long is an oil molecule?

   f Estimate how many atoms there are in the drop on the cotton thread in part **b**.
     (The area of a circle, radius R, is πR^2; the volume of a sphere, radius r, is $\frac{4}{3}\pi r^3$.)

19(P) Light of wavelength 600 nm diffracted by lycopodium particles of diameter 30 μm causes a pattern of fuzzy rings on a screen placed at some distance beyond the powder sample along the axis of the beam.

   a If X-rays of wavelength 0.1 nm are diffracted at the same angle from a glass sample, what is the ratio of the size of the powder particles to the glass molecules?

   b Estimate the size of the glass molecules.

   c In another X-ray experiment with a different material the diffraction angle was larger. Are the molecules of the second sample larger or smaller than glass molecules?

20(L) The *mole* is a convenient unit for measuring the quantity of matter when considering atomic or molecular properties. The Avogadro constant, L, is approximately 6×10^{23} items per mole. (More precisely, 6.022×10^{23} mol^{-1}.)

   a Use a book of data to find the *molar mass* (the mass of one mole) of sodium atoms and of chlorine atoms.

   b What is the molar mass of NaCl?

   c How many pairs of Na$^+$ and Cl$^-$ ions are there in 0.0585 kg of sodium chloride?

   d If one mole of sodium atoms and one mole of chlorine atoms combine to form NaCl, one mole of electrons is transferred from sodium atoms to chlorine atoms. How much electric charge is carried by a mole of electrons? (Charge on one electron $= 1.6 \times 10^{-19}$ C.)

e Estimate the volume of a mole of salt.

21(L) Estimate the number of atoms in 1 cm³ of copper by the following method:

a The molar mass of copper is $0.064\,kg\,mol^{-1}$. How many atoms are there in 1 kg of copper?

b The density of copper is $8900\,kg\,m^{-3}$. How many atoms are there in $1\,m^3$ of copper? How many atoms in $1\,cm^3$?

c If all the atoms stacked neatly as tiny cubes, what would be the 'volume' of a copper atom? Estimate the 'diameter' of a copper atom.

d Gold is one of the densest metals and aluminium one of the least dense. Find the molar volume (the volume of one mole of substance) of gold and of aluminium atoms. (Use tables to find the molar mass and density of each metal.) What conclusion can you draw about the sizes of the metal atoms?

22(E)a Estimate the mass of a single grain of salt. What would be the mass of a mole of grains of salt?

b Write the letter 'O' with a carbon pencil. Estimate how many million atoms have been rubbed from the pencil onto the paper.

c Maximum wear of a car tyre occurs when the car corners or brakes. Assuming that, on average, a layer one molecule thick is worn away per revolution of each wheel, estimate how far the car will travel before each tyre loses a mole of rubber molecules. A rubber molecule is about $10^{-9}\,m$ across.

23(I)a Which of the diagrams in figure A95 represents a layer of soap bubbles?

b Explain why soap bubbles take up this pattern.

c Since layers like the soap bubble layer occur in many metals, for example copper, what might be concluded about the forces between the atoms in such metals? (Think about their directions.)

d The ions in an NaCl crystal take up an arrangement rather like **1**. Suggest why this might be.

24(L) The Avogadro constant can be found from knowledge of the size of atoms and how they are arranged in crystals. The molar mass of copper is $0.0636\,kg\,mol^{-1}$. The density of copper is $8930\,kg\,m^{-3}$.

a Calculate the molar volume, that is, the volume occupied by one mole of copper atoms.

Experimental work with X-rays shows that copper atoms have a *diameter* of $2.55 \times 10^{-10}\,m$. To estimate the Avogadro constant we need to work out the volume effectively occupied by a single copper atom.

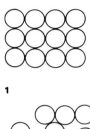

1

2

3

4

Figure A95

b *i* Calculate the volume occupied by a spherical copper atom.
ii Use this value for the volume and your answer to **a** to estimate the Avogadro constant, L, the number of atoms in one mole.
iii Comment on this estimate.

In a real solid there is some empty space between atoms.

c Assume that atoms are arranged in a 'square' array and that each one occupies the volume of a cube, side 2.55×10^{-10} m.
i Calculate the volume occupied by each copper atom on this assumption.
ii Use this value to estimate L.
iii Comment on your answer.

In fact we know from X-ray work that the atoms in a copper crystal are arranged in hexagonal layers, and that these layers are stacked together as closely as possible. In this arrangement about $\frac{1}{4}$ of the total volume is empty space between the spherical atoms.

d Use this information and your answer to **b** to make a more accurate estimate of L.

Intermolecular forces

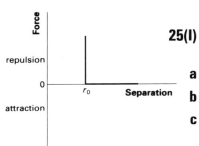

Figure A96

25(I) Figure A96 shows the force–separation curve for two identical perfectly hard spheres.

a Explain the shape of the curve.

b What is the significance of the distance r_0?

c If the theoretical spheres were replaced by two billiard balls making a line of centres collision, how would the shape of the curve change?

d Sketch the shape of the *potential* energy–separation curve for the two billiard balls as they collide.

26(L) Two identical air track vehicles have springy buffers and strong attracting magnets mounted on facing ends as shown diagramatically in figure A97.

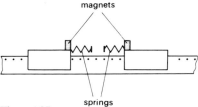

Figure A97

a Imagine that the buffers are removed, and sketch a curve to represent the force–separation graph between the vehicles.

b Now imagine that the springy buffers are replaced, and the magnets are removed. On the same axes sketch a curve to represent the force–separation graph between the vehicles.

c Add the two curves together to indicate the force–separation curve when both attractive and repulsive forces are present.

The force–separation graph for two molecules has the shape shown in figure A98.

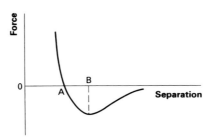

Figure A98

d Does the curve you have sketched for the air track model have a similar shape?

Explain the shape of the curve and the importance of the two separations 0A and 0B.

e Suggest a suitable value for 0A in figure A98.

27(P) The potential energy–separation graph for two atoms is shown in figure A99.

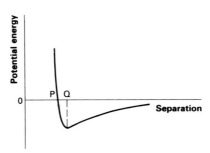

Figure A99

a Explain the shape of the curve and the importance of the two separations 0P and 0Q.

b Using the same separation axis draw figure A99 and figure A98 on the same graph to indicate how the important features of each curve are related to each other.

c The two atoms are not at rest but will have kinetic (thermal) energy and so will oscillate about their equilibrium position. How will the potential energy vary? (Use the figure in your answer.)

d If enough energy is given to the two atoms, they can break free of each other's attractive force (the binding energy). How can the binding energy be found from the graph?

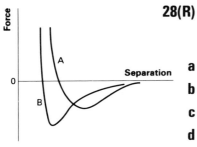

Figure A100

28(R) Figure A100 shows the force–separation curves for a pair of atoms of two crystalline solids A and B. Both solids have the same structure. Which solid

a has larger atoms?

b is stiffer?

c has the larger Young modulus?

d has the greater breaking stress?

29(P) This question is useful preparation for the ideas of the next one.

A school laboratory has a demonstration model, shown in figure A101, which is meant to illustrate the stretching of bonds between atoms as a piece of material is stretched. The model has four horizontal planes of small balls linked horizontally by rods and vertically by springs. There are six balls in each horizontal layer. All the springs are identical and may be taken to be 50 mm long. The dimensions are shown, all in millimetres. When a vertical stretching force of 12 N is applied, the length of 150 mm increases to 165 mm.

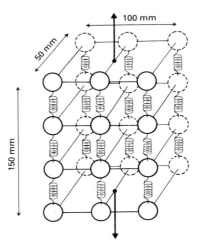

Figure A101

a What is the restoring force for 10 per cent strain for a single spring?

b What is the force constant (restoring force for unit displacement) for a single spring?

30(L) This question makes a link between the Young modulus of a metal (stress/strain), and the springiness of bonds between atoms.

Imagine layers of atoms in square array, each atom distance r_0 from its nearest neighbours both in its own layer and in layers above or below, as shown in figure A102.

Figure A102

Suppose a long wire, with many layers, is stretched a little so that each layer is now $r_0 + \Delta r$ from those above or below it.

a If there are n layers in the length of the wire, how long is the wire before it is stretched?

b By how much has the wire extended?

c What is the *strain* in terms of r_0 and Δr?

Now think of the bonds holding each atom as being like a spring, so that there is a force of $k\Delta r$ pulling a pair of atoms together.

d k is the stiffness of the spring. What are its units?

e If there are m atoms in each layer, what is the force pulling adjacent layers together?

f The stress is the force per unit area. In terms of the spacing r_0, how many atoms are there per square metre of a layer?

g What is the stress in terms of k, Δr, and r_0?

h What is the elastic modulus in terms of k, r_0?

i For steel, the elastic modulus is 20×10^{10} newtons per square metre. The spacing r_0 is about 3×10^{-10} metre. What is the stiffness, k, of the springy bond?

Non-metallic materials

31(L) This question is about rubber, but it uses an idea from the kinetic theory of gases.

A gas molecule bouncing about from collision to collision in steps of length x will not usually be a distance nx from its starting point after making n steps, for that would need all the steps to be in a straight line. It turns out that the most probable distance is about $x\sqrt{n}$.

A rubber molecule is a long chain of n links, with the links free to swivel so that the chain points hither and thither along its length.

a If each link is of length x, how far will one end of the chain be from the other, on average?

b How long can the chain possibly be, fully stretched out?

c If $n = 100$, at what strain will rubber become very hard to pull out any further?

32(E) This question is about using words well enough to be understood in discussion.

A 'Now let's get this clear. You say that this grain of NaCl here is a little block of one crystal.'
B 'Yes.'
A 'And this piece of copper wire is also crystalline.'
B 'Yes, but it's polycrystalline.'

A 'Ah, so there is crystalline and polycrystalline. Is rubber polycrystalline?'

B 'No. Rubber is amorphous until you stretch it.'

A 'Amorphous–you mean like a liquid?'

B 'Yes.'

A 'A liquid doesn't become crystalline when you stretch it?'

B 'Perhaps we'd better go back to the beginning.'

Can you help B?

33(E)a A metal is a crystalline substance, so is sugar. Metal bends, does sugar?

b When sugar is made into toffee it behaves like glass. Glass shatters, does toffee?

c How do you distinguish a glass-like substance from a crystalline substance?

d Why are some crystalline substances ductile and others not?

Imperfections in structure

34(E) 'So here is the dislocation, and now imagine that we try to slide the top this way and the bottom this way – the atoms around the dislocation can rearrange themselves so in the next picture ...'.

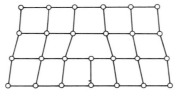

Figure A103

Draw the next picture. And the next. And the next. Write out the rest of the talk which might accompany the pictures.

35(E) A student writes the following short passage. Rewrite it to explain each of the points made, giving greater detail.

'Cracks occur in all substances. Cracks only propagate in materials with no long-range molecular order so materials with no long-range order undergo brittle fracture. Cracks are a macroscopic phenomenon.

'A few dislocations, foreign atoms, etc., can stop cracks from propagating by making a material ductile. Dislocations are a microscopic phenomenon. Many dislocations or foreign atoms can make a material harder. Harder materials are more likely to suffer brittle fracture.

'Temperature changes can also produce dramatic changes in properties. Most materials become more brittle at lower temperatures. For example, a rubber ball cooled in liquid nitrogen will shatter like a piece of china if dropped on the floor.'

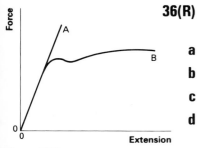

Figure A104

36(R) The force–extension curves of two steel wires of identical dimensions are shown in figure A104.

a Which wire is stronger?

b Which wire is tougher?

c Which one is made of high carbon steel and which one of mild steel?

d Briefly explain how the change in the concentration of carbon causes the difference in properties of the two wires.

Resolutions of forces and moments

37(P) A picture of weight W hangs from a hook by a single cord connected to two rings, as shown in figure A105, on the back of the frame. The picture does not touch the wall.

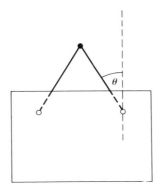

Figure A105

a Draw diagrams indicating all the forces acting on the picture and on the hook, showing their directions and points of application. The tension in the cord is T.

b How does the tension vary with the length of the cord? Should you use a short or long length to hang the picture?

38(P) Figure A106 shows a dynamics trolley of weight W placed on a runway at an angle θ to the horizontal. It is held in place by a horizontal force, F. N is the force on the trolley normal to the runway.

a Resolve horizontally to find an expression for F in terms of N.

b Resolve vertically to find an expression for W in terms of N.

c Combine **a** and **b** to find an expression for F in terms of W.

d Resolve parallel to the runway to check your expression for F part **c**.

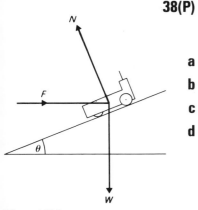

Figure A106

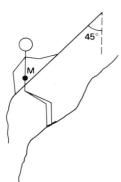

Figure A107

39(R) A climber rests for a moment as he climbs a fixed rope on a 45° rock face.

a If the rock is wet and there is virtually no friction between his soles and the rock, explain how he can stay in the position shown in figure A107.

b The climber weighs 707 N, which may be taken to act through his centre of mass, M. What is the tension in the rope in this situation? What is the force between the man and the rock?

c If there is enough friction between his boots and the rock so that he will not slip, which way should he lean to reduce the tension in the rope? The rope remains at the same angle to the vertical.
    Is the force on the rock increased or reduced? Explain.

40(R) A five-bar gate is 2 m long and is supported by hinges 1 m apart. The gate weighs 1000 N. Assume the centre of gravity is 1 m from the pivot line.

a Sketches of the hinges are shown in figure A108. Which one is the top and which the bottom one? Explain.

b Sketch the gate and indicate the three forces acting on it.

c Find the magnitude and directions of the forces at the two hinges.

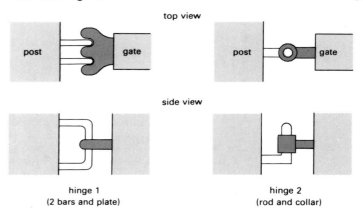

Figure A108

hinge 1
(2 bars and plate)

hinge 2
(rod and collar)

41(R) A tractor and trailer move across a single span bridge supported at points 21 m apart. The axle loads and position of the vehicles are as shown in figure A109. What are the vertical forces at each of the supports caused by this load on the bridge?

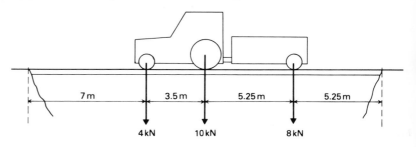

Figure A109

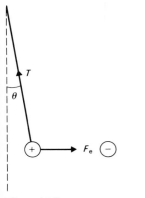

Figure A110

42(P) A light plastic ball (mass m) is suspended on a string.

a What is the tension in the string when the ball is hanging vertically?

The ball is now given an electric charge and figure A110 shows what happens when a second, oppositely charged ball, is brought near.

b Is the tension in the string now greater, less, or the same as before?

By resolving the forces on the ball vertically and horizontally respectively obtain expressions for:
i The tension in the string, T.
ii The electric force on the ball, F_e.

c For a small horizontal displacement of the ball, d, show that $F_e \propto d$.

43(R) A rigid beam of negligible mass is hinged to a wall, and held horizontal by a string as shown. Calculate:
i the tension, T, in the string
ii the compression force in the beam
iii the force exerted by the hinge on the beam in each of the following situations:

a A weight of 200 N is hung from the far end of the beam as shown in figure A111(a).

b The weight is moved to the mid-point of the beam – figure A111(b).

c A second 200 N weight is now added at the far end of the beam–figure A111(c).

d With the weights still in position, the string is shortened and tied to the mid-point of the beam, which remains horizontal. The string remains attached to the same point on the wall – figure A111(d).
Hints:
i The force at the hinge will not necessarily be exerted normal to the wall. You may find it easiest to resolve the force at the hinge into two components: one along the wall and another at right angles to the wall.
ii Remember to apply both the conditions for equilibrium.
iii Sketches are essential; you may even wish to confirm the results of your calculations by making scale drawings showing the force parallelograms or triangles.

44(P) A wall-mounted crane lifts a mass of 1 tonne (1000 kg). The angle between strut and tie (figure A112) is 45°.

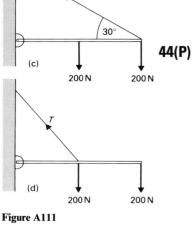

Figure A111

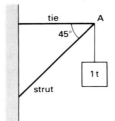

Figure A112

a Use the fact that the forces at A are in equilibrium to answer these questions.
   i Is the strut in tension or compression?
   ii Is the tie in tension or compression?
   iii Calculate the values of the force in the strut and in the tie.

b Draw a triangle of forces for point A.

c Which of the two, strut or tie, provides a vertical force to support the load?

45(P) Four hinged rods are mounted in a vertical plane as shown in figure A113. A weight *W* is suspended from point B.

a By considering the forces acting at points B and C, find out which of the two rods are in compression (struts) and which are in tension (ties).

b Which two provide a vertical force to support the load?

c What is the function of the other two rods?

d (*Harder*) Figure A114 shows a simple (Warren) girder bridge.

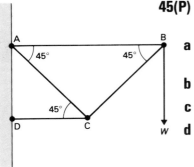

Figure A113

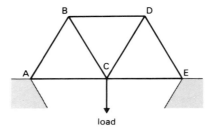

Figure A114

Which members are ties (in tension) and which are struts (in compression)?

Composite materials

46(F) The typical tensile strengths of some materials are given below in round numbers:

Material	Tensile strength/$N\,m^{-2}$
mild steel	10×10^8
wrought iron	5×10^8
wood, spruce:	
along grain	1×10^8
across grain	0.3×10^8
glass	$0.3\text{--}1.7 \times 10^8$
concrete	0.04×10^8

Table A3

a The upper value for glass compares quite favourably with iron and wood. Why isn't glass used in the same way that iron and wood are used (for example, why are railway lines not made of glass instead of wrought iron; or boats made of glass instead of wood)?

b Why is plywood made the way it is (see figure A115)? Where would you expect to see plywood being used? Why? Why isn't plywood used for all purposes where wood is employed?

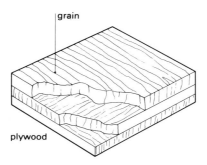

grain

plywood

Figure A115

c Concrete has the lowest tensile strength of all the materials listed. How do you explain the widespread use of concrete in building?

47(R) Civil engineers have found that concrete pipes laid deep in the earth have sometimes cracked as shown (in cross-section) in figure A116(a).

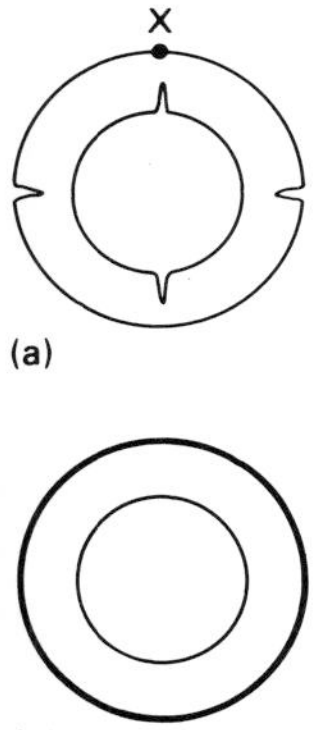

a Explain, in terms of the various forces acting and the behaviour of concrete under stress, why the pipes crack as shown in figure A116(a).

b Why would you *not* expect to find a crack forming at the point X on the outside of the pipe?

The engineers found a solution. They cast round the pipes a tough plastic layer which shrank as it solidified, squeezing the pipe all round, as shown in figure A116(b).

c Why did this prevent the pipes from cracking?

d Suggest one other incidental advantage of encasing the pipes in plastic.

(Short answer paper, 1973)

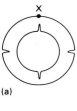

(a)

(b)

Figure A116

48(R) A vertical building is built from four very heavy identical concrete floor slabs, S, supported by light, slender steel pillars, P, as shown in figure A117. The architect needs to know the stress in the pillars on different floors.

Here are five possible values of the ratio of the stress in the pillars resting on the ground to the stress in the pillars supporting the uppermost slab:

A 4
B 2
C 1
D $\frac{1}{2}$
E $\frac{1}{4}$

Neglecting the weight of the pillars:

a What is the ratio if the rods at the bottom and top have the same diameter?

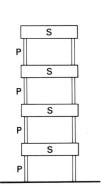

Figure A117

b What is the ratio if the pillars on the ground have twice the diameter of the pillars at the top?

(Coded answer paper, 1980)

49(R) In this question you are asked to use the information in the figure to compare the merits of steel, glass fibre (glass fibre-reinforced plastic), concrete, and wood as load-bearing materials in tension.

Figure A118(i) illustrates the strength of the materials in tension, by showing the cross-section that will just support a load of one tonne (10^3 kg, weight approximately 10^4 N).

Figure A118(ii) shows the relative cost of the materials per unit of strength, by giving the cost of one metre of material of the cross-section shown in figure A118(i).

Figure A118(iii) gives the cost of one cubic metre of each material, while figure A118(iv) illustrates their respective densities.

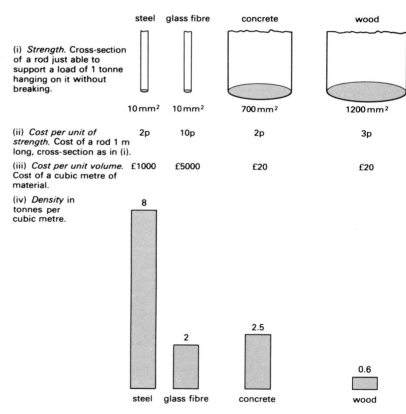

Figure A118
Strength, cost, and density of four materials.

a Say briefly what Figure A118(i) reveals about the variation in *tensile strength* from material to material.

b Comment, in the light of differences in the nature and structure of the materials, on the differences in their behaviour under stress.

c Discuss reasons for selecting or rejecting each material for each of **two** purposes. Choose your two purposes so as to *contrast* the properties of the materials, and to illustrate *different* aspects of the problem of choosing between the materials. Use the data given, and any further ideas of your own.

(Long answer paper, 1974)

Momentum and impulse

50(I) A ball of mass 0.1 kg is dropped from a height of 3.2 m onto a flat surface. It rises to a height of 1.8 m on the rebound (see figure A119). Calculate:

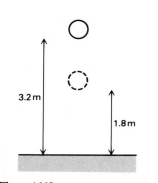

3.2 m

1.8 m

Figure A119

a the speed of the ball at the instant of impact,

b the speed of the ball immediately after leaving the surface,

c the change in energy of the ball caused by the impact,

d the change in momentum of the ball between the time of just touching the surface and leaving it (remember that momentum is a vector quantity),

e the average force exerted on the ball by the surface for an impact time of 0.04 s.

51(P) A long flat-topped railway truck of mass M which can move freely on a horizontal track, is at rest. A man of mass m is standing on the truck. The man suddenly runs at velocity v parallel to the direction of the track for time t and then stops.

a Describe the motion of the truck.

b What is the velocity of the truck whilst the man is running?

c What is its velocity after he stops?

d How far does the truck travel whilst the man is running?

52(R) A car of mass 600 kg travelling at a steady speed crashes into a rigid wall. From the time at which the car hits the wall until it stops, the engine compartment is crushed so the main body of the car continues to move forwards. The graph (figure A120) shows how the retarding force on the main body of the car varies with time from the initial impact ($t = 0$).

a What impulse is required to stop the car?

b By how much does the car's momentum change during the impact?

c Estimate the velocity of the car at the instant of impact.

d Sketch a graph of the velocity of the main body of the car against time until it came to rest. Explain how your graph is related to the one shown in figure A120.

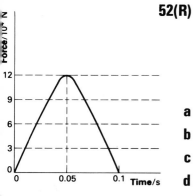

Figure A120

53(P)   Two air track vehicles with masses and velocities as shown in figure A121 move freely together and adhere on impact.

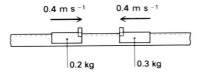

Figure A121

a   What is the momentum of each vehicle before impact?

b   What is the total momentum after impact?

c   What is the change in momentum of each vehicle caused by the impact?

d   What is the impulse on each vehicle at impact?

e   The collision took 0.1 s. What is the average force on each vehicle during the collision?

f   What is the final velocity of the two vehicles?

g   What is the kinetic energy of each vehicle before impact?

h   What is the final kinetic energy of the two vehicles?

i   What has happened to the remainder of the kinetic energy?

54(E)   In using Newton's Third Law of Motion difficulties often arise in deciding which forces are associated pairs.

a   A freely falling object is acted on by a force W, the gravitational pull of the Earth. According to Newton's Law there should be another force, also equal to W, in the opposite direction. On what body does this act?

b   A book rests on a table. There are two forces acting on this book. What are they? These forces are equal and opposite but are not the associated pair of Newton's Third Law. Explain this statement.

c   A body falling towards the Earth is continually gaining momentum. Explain how the Law of Conservation of Momentum is not violated.

d   A train starts from rest – momentum zero – and accelerates, gaining momentum, so again momentum does not appear to be conserved. How can the momentum conservation law be justified in this situation?

Elastic and inelastic collisions

55(L)   When an object A of mass M and velocity u collides head-on elastically with a stationary object B of mass m, B moves off with a velocity v_B and A has a velocity v_A (figure A122), where

$$v_A = \frac{M - m}{M + m} u$$

$$v_B = \frac{2M}{M + m} u$$

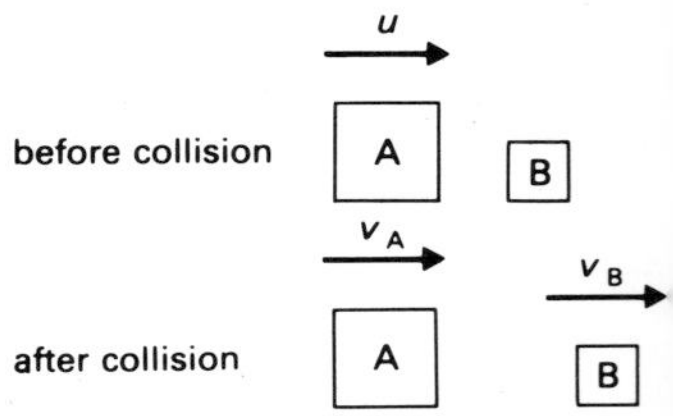

Figure A122

a Use these formulae to show that a golf ball will be driven from the tee at approximately twice the speed of the head of the golf club that hits it. Assume that the mass of the golf club is much greater than the mass of the ball.

b An alpha particle makes a head-on collision with a stationary helium nucleus. Show that the alpha particle will stop and the helium nucleus move off at the original speed of the alpha particle. (An alpha particle is a helium nucleus.)

c A fast-moving electron makes a head-on elastic collision with an atom. Show that it bounces back with almost its original speed. (The mass of the electron is tiny compared to the atomic mass.)

56(L)a Use the formula for v_B in question **55** to write down an expression for the kinetic energy of B in terms of m, M, and u.

b Divide your expression by $\frac{1}{2}Mu^2$ to show that kinetic energy of B is
$$\frac{4Mm}{(M+m)^2} \times \text{initial kinetic energy of A.}$$

c Show that very little kinetic energy is transferred between A and B if A and B are of very different mass, but that all the kinetic energy is transferred from A to B if A and B are of equal mass.

d An electron will lose a negligible amount of kinetic energy when it collides elastically with a gas atom, some 10 000 times more massive than itself. Estimate how accurately the electron energies would have to be measured to detect any kinetic energy loss.

57(R)a When considering collisions between alpha particles and atoms, or other scattering experiments, we ignore the effects of any electrons. An alpha particle makes a head-on elastic collision with a stationary electron. What is the ratio of its velocity after the collision to its velocity before? What is the percentage change in velocity? Is the approximation justified?
    (The mass of the alpha particle is 7500 times greater than the mass of an electron.)

b A gas molecule moving at $400\,\mathrm{m\,s^{-1}}$ collides elastically head-on with the stationary piston of a gas syringe. What is its speed of recoil? You may assume that the recoil of the piston is negligible. The piston is now moved inwards at $1\,\mathrm{m\,s^{-1}}$. What is the new speed of recoil of the gas molecule? What will happen to the kinetic energy of all the gas molecules as the piston is moved inwards?

58(R) A 0.1 kg bullet travelling at $400\,\mathrm{m\,s^{-1}}$ and a 2000 kg car travelling at $0.02\,\mathrm{m\,s^{-1}}$ each collide with identical blocks of wood. The momentum of each before the collision is the same, but the effects produced are very different. Explain why this is so, even although in both collisions momentum is conserved.

(Special paper, 1974)

59(P) The simulated multiflash photograph in figure A123 shows a moving puck colliding with a stationary one of equal mass.

    Take suitable measurements from the diagram to show either by scale drawing or by calculation that the total momentum in any direction is conserved.

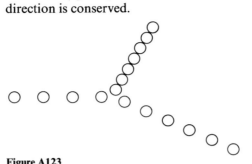

Figure A123

Kinetic theory of gases

60(P) The *molar volume* of a substance, V_m, is the volume occupied by one mole of its molecules. The molar volume of any gas is $2.24 \times 10^{-2}\,\mathrm{m^3\,mol^{-1}}$ at standard temperature and pressure (s.t.p.), which are 273 K and 1.01×10^5 Pa respectively.

a The density of air at s.t.p. is $1.29\,\mathrm{kg\,m^{-3}}$. What is the molar mass of air?

b The ideal gas equation for 1 mole of gas is $pV_m = RT$. Calculate the value of R, the molar gas constant. What are the units of R?

c One mole of substance always contains the Avogadro number, L, of molecules, $6.02 \times 10^{23}\,\mathrm{mol^{-1}}$. How many molecules are there in 1 kg of air?

d On average, how far apart are air molecules at s.t.p.? How does this compare with the diameter of an atom $3 \times 10^{-10}\,\mathrm{m}$?

e The Boltzmann constant, k, is the gas constant per molecule. What is its value?

61(P)a A cylinder of volume $1.0\,\mathrm{m^3}$ contains nitrogen gas at a pressure of 1 atmosphere (1.0×10^5 Pa) at room temperature (290 K). How many moles of gas does the cylinder hold? ($R = 8.3\,\mathrm{J\,mol^{-1}\,K^{-1}}$.)

b The molar mass of nitrogen is $0.028\,\mathrm{kg\,mol^{-1}}$. Gas is pumped into the cylinder until the pressure is increased to 6 atmospheres. What mass of gas does the cylinder now contain?

c The pressurized gas is released into the atmosphere. What volume of space will it now fill?

62(L) A molecule of mass m bounces backwards and forwards between the ends of a box of side l at a constant speed c_x, shown in figure A124.

    Neglecting any recoil of the box and the finite time of each collision, find expressions for

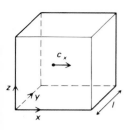

Figure A124

a the number of normal collisions the molecule makes on one end in one second;

b the change in momentum of the molecule at each collision;

c the total change in momentum at one end in one second;

d the mean force exerted by the molecule on the end.

In a real box there will be N molecules moving at random at various velocities c. Each velocity can be resolved into components parallel to three perpendicular edges of the box, such that $c^2 = c_x^2 + c_y^2 + c_z^2$. The average value of c^2 is $\overline{c^2}$ which equals $3\overline{c_x^2}$ as $\overline{c_x^2} = \overline{c_y^2} = \overline{c_z^2}$.

e What is the mean force exerted by all the molecules on one wall of the box in terms of $\overline{c_x^2}$?

f What is the mean pressure exerted on one wall of the box in terms of $\overline{c^2}$?

g The density of the gas is total mass/volume. Write down an expression for density in terms of l, N, and m. Rewrite your answer to **f** in terms of ρ and $\overline{c^2}$.

h The density of air at room temperature is about $1.2\,\mathrm{kg\,m^{-3}}$. Atmospheric pressure is 1.0×10^5 Pa. What is the mean square speed of molecules of air? What is their root mean square (r.m.s.) speed?

63(L) Consider six chosen molecules, each of mass 5×10^{-26} kg, of a gas contained in a box. Three are moving to the right with speeds 200, 300, and $500\,\mathrm{m\,s^{-1}}$, and three are moving to the left with speeds 100, 400, and $600\,\mathrm{m\,s^{-1}}$ (see figure A125).

a What is their mean or average speed?

b What is their mean velocity?

c What is their mean kinetic energy?

d What is the speed of a molecule which has this mean kinetic energy?

e What is their root mean square speed?

f What is their mean momentum?

g If we had considered all of the molecules in the box, not just the chosen six, which of the quantities calculated in parts **a** to **f** above, would you expect to be zero?

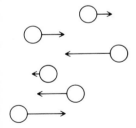

Figure A125

64(E) The following passage appears in a textbook:

'This indeed is how a physicist works; by trying to reason out the easiest problems first. In fact, he has to invent them. Think of it this way. The world is a very complicated place, and a physicist tries to see patterns of order and reason in all this complexity. His technique is never to try to deal with all aspects of a physical situation as it really

A

is. Instead, he deals with simplified models of selected aspects of reality.'

Discuss whether the ideas in this passage do, or do not, in your opinion, give a good description of the theoretical argument in the kinetic theory of gases used to derive the expression $pV=\frac{1}{3}Nm\overline{c^2}$.

You should include in your discussion

a   the extent to which the ideas in the passage apply to the derivation of $pV=\frac{1}{3}Nm\overline{c^2}$, giving reasons for your view;

b   an account of aspects of the situation which are being simplified or selected;

c   a description of the ways in which the argument might give wrong answers in a real concrete situation because of the simplifying assumptions it makes.

(Long answer paper, 1976)

65(P)   Use the following data to answer the questions below: the Boltzmann constant, k, is $1.38 \times 10^{-23}\,\mathrm{J\,K^{-1}}$; the r.m.s. speed of nitrogen molecules at 273 K is $490\,\mathrm{m\,s^{-1}}$.

a   What is the mean translational kinetic energy of a nitrogen molecule at 273 K?

b   What is the mass of a nitrogen molecule?

c   What is the mean kinetic energy of an oxygen molecule at 273 K? Of a helium molecule at 273 K?

d   A nitrogen molecule is 7 times more massive than a helium atom. What is the r.m.s. speed of helium atoms at 273 K?

66(P)   The speed of smoke particles in thermal equilibrium in air can be measured from Brownian motion observations. It is found to be about $10\,\mathrm{mm\,s^{-1}}$. Of course, the direction of motion changes many times per second.

a   Estimate the speeds of air molecules at room temperature, given that the mass of an air molecule is $5 \times 10^{-26}\,\mathrm{kg}$ and that of an average smoke particle is $1 \times 10^{-16}\,\mathrm{kg}$.

b   Compare this result with the value obtained from kinetic theory using the following data (again given only to one significant figure): atmospheric pressure is $1 \times 10^5\,\mathrm{Pa}$; molar mass of air is $3 \times 10^{-2}\,\mathrm{kg\,mol^{-1}}$; molar volume at s.t.p. is $2 \times 10^{-2}\,\mathrm{m^3\,mol^{-1}}$.

67(L)   When bromine liquid is released into a tube which has been evacuated, the bromine vapour fills the tube almost instantaneously.

a   Find the r.m.s. speed of bromine molecules at room temperature. The molar mass of bromine gas is $0.16\,\mathrm{kg\,mol^{-1}}$.

b   How long will it take to fill an evacuated tube 0.5 m long?

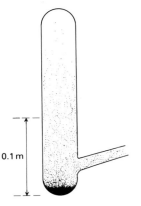

Figure A126

When bromine liquid is released into the tube containing air at atmospheric pressure, it takes about 500 seconds for an average bromine molecule to have travelled 0.1 m from the bottom of the tube. The top of the tube is still colourless, the bottom dark brown. And this point, 0.1 m up, is about 'half brown' in colour. (See figure A126.)

c The direct distance the bromine molecule travels through the air is calculated theoretically to be $\sqrt{N}$ times the mean free path, where N is the number of collisions the molecule makes along its path. What total distance did the molecule actually travel in 500 s? Hence find the number of collisions, N, the molecule made in 500 s.

d What is the mean free path, λ, of the bromine molecule, that is, the average distance between collisions?

e Theory shows that the mean free path $\lambda \approx \dfrac{1}{\pi n d^2}$, where n is the number of molecules per unit volume of the gas and d is the molecular diameter. The Avogadro constant is $6.0 \times 10^{23}\,\mathrm{mol^{-1}}$ and the molar volume is $0.022\,\mathrm{m^{-3}\,mol^{-1}}$. Use this information to estimate the radius of a bromine atom.

f Has it been assumed that bromine and air molecules are about the same size? Explain.

68(P) The bromine diffusion through air experiment enabled us to find the mean free path of bromine in still air in question **67**.

a Suppose some liquid bromine was dropped in the corner of a large room, say, 6 metres wide. How long would it take for an average bromine molecule to diffuse across the room?

b Your answer to part **a** is very large. If somebody dropped a stink bomb in the corner the smell would take only a minute or two to cross the room. Explain.

69(L) Gas at pressure p occupies a volume V in the cylinder shown in figure A127. The area of the piston is A.

a What is the force on the piston?

b The gas expands, pushing the piston back by a distance Δl. What is the work done by the gas?

c Write your answer to **b** in terms of the change in volume, ΔV.

d Suppose the gas is compressed and the volume decreases by ΔV. How much work is done on the gas?

e Mention two assumptions that you had to make to answer these questions.

f When you pump up the tyres of a bicycle the air in the pump gets hot. When you let the air out suddenly by removing the valve it cools. Relate these observations to the work done on or by the gas.

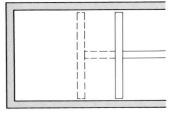

Figure A127

A

Unit B
CURRENTS, CIRCUITS, AND CHARGE

Nigel Wallis
Archbishop Holgate's School, York

Mark Tweedle
The Grammar School, Batley

B

SUMMARY OF THE UNIT

SUMMARY OF THE UNIT

INTRODUCTION

This Unit is about the nature of electric charge, how it can move in conductors and in empty space to form currents, and how it can be stored at rest, for example on capacitor plates.

These ideas are central to the rest of the course, not least because approximately half the mass of the Universe (including you!) is composed of electrically charged particles. More specifically, the circuit ideas will be used again in electronics, in Units C, 'Digital electronic systems' and I, 'Linear electronics, feedback and control'; charge and the concept of potential difference will be used in Unit E, 'Field and potential', Unit F, 'Radioactivity and the nuclear atom', and Unit J, 'Electromagnetic waves'.

Section B1   **THINGS WHICH CONDUCT**

Flow

The effects of an electric current are quite familiar, but the processes actually going on in the conductor are far from obvious. There is usually nothing to see, and even when there is it doesn't help us much to understand more about what is happening (for instance, the glowing of a light bulb filament or the light emitted by an L.E.D.). We are forced to make inferences from circumstantial evidence, to build up a picture from a number of clues rather than look to the result of one single conclusive experiment. (You might, for example, ask yourself how you know that molecules exist and that they move – are you convinced by the evidence, granted that you will never see one?)

Electric current is charge on the move – the charge may be of either sign and can be a 'fundamental' particle like an electron (which we shall be looking at again at the end of the Unit) or a charged atom or group of atoms called an *ion*. Some further experiments which lead to this view, based on capacitors, are considered in Section B3. Here we concentrate on the aspect of flow.

Experiments with simple series circuits show that the *order* of the components does not matter: a rheostat to control the brightness of a lamp can be placed on either side of the lamp; an ammeter reads the same wherever it is put, showing that current is not 'used up' in the circuit.

Similarly, in a parallel circuit where there is a branch (figure B1) the reading of the meter in the main part, I, is the sum of the readings in the branch circuits.

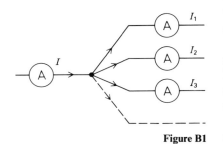

Figure B1
Multi-branching in a circuit.

Kirchhoff's First Law

$$I = I_1 + I_2 + I_3 + \ldots$$

This is usually referred to as Kirchhoff's First Law.

In its full form the Law states:
At a junction in a circuit,
Σ(current arriving) $= \Sigma$(current leaving)
which you will often find written more formally in textbooks as
Σ(current arriving) $= 0$, taking account of the sign (direction) of current.

These statements about series and parallel circuits are, of course, very useful in helping us to do calculations and to make quantitative predictions about circuit behaviour. They are also important, though, in helping to establish the idea of a current as a flow of something indestructible which just keeps circulating round. It would be difficult to think of any other hypothesis which would explain Kirchhoff's First Law.

DEMONSTRATION B1
Conduction by coloured salts

The movement of charge in one particular case can be demonstrated – the migration of coloured ions in a liquid as current flows. This brings out two major points:

B

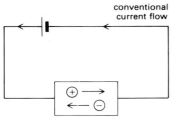

i *The direction of motion.* For a given current direction (which is conventionally taken from the positive battery terminal to the negative round the circuit) the charge flow can be either way depending upon its sign (figure B2). We believe that in a *metal* the charge carriers are negative (electrons) and therefore move in the opposite direction to the conventional current.

Figure B2
Movement of ions.

ii *The speed of the ions.* One might intuitively think that the charges move extremely quickly as they carry the current. Although electrical effects seem to happen instantaneously – a lamp comes on as soon as a switch which may be quite far away is closed – observation shows that the charges actually move quite slowly. An important equation which relates these facts is:

QUESTION 1

$$I = AvnQ$$

(A is the cross-sectional area of the conductor, n the density of charge carriers, v their speed, and Q the charge carried by each one.)

This equation can be used to estimate the speed of charge carriers in a liquid. Of course, if there is no resistance to motion (for example an electron beam in a vacuum) the charges *can* accelerate to very high velocities. This equation can still be used, so that for a given current large values of v imply small values of n, the charge carrier density.

Energy transformation

When a current flows through any conductor energy is transformed. The amount of energy transformed between two points per unit charge passed is called the *potential difference* (p.d.) between the points.

$$V = \frac{W}{Q}$$

DEMONSTRATION B2
Measuring p.d. without a voltmeter

(V is the p.d., W the energy, and Q the charge; in units, 1 volt $= 1$ joule per coulomb.)

This can also be expressed as

QUESTIONS 3 to 6 $W = VQ$

Dividing both sides by the time t,

$$\frac{W}{t} = \frac{VQ}{t}$$

or

$$P = VI$$

(P is the power, I is the current; in units, 1 watt $= 1$ volt $\times 1$ ampere)

Resistance

EXPERIMENT B3
Two-terminal boxes

The current through a component depends on the p.d. across it. The graph of such a relationship is called the 'characteristic' of the component, and it is a useful way of summarizing how the component behaves (figure B3).

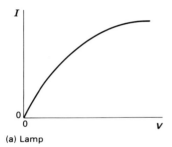

(a) Lamp

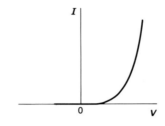

(b) Diode

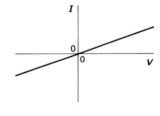

(c) Metal wire

Figure B3
Current–p.d. graphs for some conductors.

For any component the ratio V/I is called the *resistance*. In general it will not be constant (see lamp and diode graphs), but for some conductors, particularly metals, I is proportional to V (linear graph through origin). This is a statement of Ohm's Law. (The temperature should be constant for the Law to apply.)
For ohmic conductors:

$$R = \frac{V}{I} = \text{constant}$$

QUESTIONS 7 to 11

Using this definition of resistance we can now write the power converted in a resistor (VI) in two alternative and equivalent forms:

$$P = I^2 R$$

and

$$P = V^2/R$$

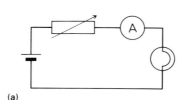
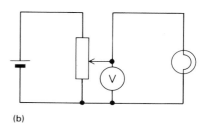

Figure B4
(a) Rheostat controlling current.
(b) Potentiometer controlling applied p.d.

(a)　　　　　　　　　　　　　　　　　(b)

Control of current and p.d.

EXPERIMENT B4
Comparison of rheostat and potentiometer

A rheostat (variable resistor) can be used to control the current in a circuit *between maximum and minimum limits* set by the resistance of the rheostat.

A potentiometer (potential divider) can be used to 'tap off' a certain proportion of a battery's voltage. The full range, from zero to maximum p.d., is possible.

Note that a potentiometer does not have to deliver current, but there has to be current flowing through the rheostat.

B

QUESTION 10

Combination of resistors

It is useful to consider two simple but very useful combinations of resistors, *series* and *parallel* (figure B5). If R is the value of the single resistance which can replace the combination:

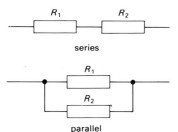

Figure B5
Resistor combinations.

series　　$R = R_1 + R_2$

parallel　　$\dfrac{1}{R} = \dfrac{1}{R_1} + \dfrac{1}{R_2}$

The formulae generalize to any number of resistors.

What does resistance depend on?

DEMONSTRATION B5
Effect of size and material on resistance

Three factors determine the resistance of a conductor: what it is made of, its dimensions, and (usually) its temperature. So for a given material at a fixed temperature only the shape is important.

Simple experiments or a theoretical argument (based on the series and parallel resistance formulae) show that for a conductor of length l and cross-sectional area A (figure B6)

QUESTION 15　　$R \propto \dfrac{l}{A}$

Figure B6

HOME EXPERIMENT BH1
The resistor

Notice that since R is inversely proportional to A, fat wires will conduct better than thin ones (other things being equal). Conductors which have to carry high currents (for example, to the starter motor of a car) are thick; those which carry only small currents – say on a printed circuit board in a radio – can be much thinner.

The resistance of a wire will change when it is stretched. Its length increases and cross-sectional area decreases: both these changes cause an increase in resistance. An important application of this effect is the *strain gauge*, much used in engineering to monitor and measure stresses in machinery and structures.

Resistivity

We can introduce a constant of proportionality into the relationship $R \propto l/A$:

QUESTION 17

$$R = \rho \frac{l}{A}$$

What are the units of ρ?

where ρ is a constant for a given material at a given temperature. It is called the *resistivity*.

QUESTIONS 13, 14

The range of values of ρ varies enormously by a factor of about 10^{26}. For a good conductor, copper, ρ is about $10^{-8}\,\Omega\,\mathrm{m}$; for an insulator such as PTFE it is about $10^{18}\,\Omega\,\mathrm{m}$. An important class of materials lies in the middle of the range – the *semiconductors*.

DEMONSTRATION B6

Effect of temperature on resistance

The resistivity also varies with temperature according to this classification. For plastic materials and other good insulators ρ decreases as they become hotter; metals behave in the opposite way and they become poorer conductors. The resistivity of semiconductors decreases markedly with rising temperature, and this effect is used, for example in the thermistor, for detecting or measuring small temperature changes.

Section B2 **CURRENTS IN CIRCUITS**

Electromotive force (e.m.f.)

EXPERIMENT B7

Four-terminal boxes

Electromotive force is a property of a power supply and is measured in volts (joules/coulomb). (Note that it is *not* a force!) However, unlike the p.d. across, say, a heating coil, e.m.f. describes the energy transformation *into* electrical energy. An e.m.f. $\mathscr{E}$ delivering current I is supplying power *to* the circuit at a rate $\mathscr{E}I$.

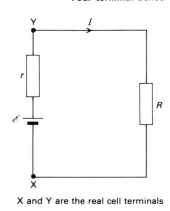

X and Y are the real cell terminals

Figure B7

Internal and external resistance.

A real power supply will have a resistance of its own (its *internal resistance*), and can be represented by an equivalent circuit having the internal resistance r in series with an e.m.f. $\mathscr{E}$ (figure B7). The power delivered by the supply is $\mathscr{E}I$, and the power transformed in the circuit is $I^2 r$ (in the supply itself) and $I^2 R$ (in the load resistance, R). By the Law of Conservation of Energy,

$$\mathscr{E}I = I^2 R + I^2 r \qquad \text{or}$$

$$\mathscr{E} = I(R + r)$$

Re-arranging,

$$IR = \mathscr{E} - Ir$$

$$V_{\mathrm{load}} = \mathscr{E} - Ir$$

QUESTIONS 21, 22

where V_{load} is the p.d. across the external resistance R.

Effect of a load on the p.d. of the supply

DEMONSTRATION B8
Drop in terminal p.d. of source on load

HOME EXPERIMENT BH2
The voltaic pile

EXPERIMENT B9
Comparison of voltmeters

The above equation shows that when a load draws current the actual p.d. across it is *less* than the e.m.f. by an amount Ir (the potential drop across the internal resistance).

This effect is important in two situations:

i If r is appreciable (or if a high current is drawn), then the actual p.d. across the circuit will be appreciably less than the e.m.f. of the source. Always use a voltmeter to check the actual p.d.!

ii A voltmeter will only give an approximate indication of e.m.f. If the voltmeter's resistance is not *much* higher than the internal resistance, it will draw an appreciable current, and so the reading will be less than the e.m.f. of the source. (The voltmeter reading will be correct – but it will not be equal to the e.m.f.) The equation $V_{load} = \mathscr{E} - Ir$ shows that only if $I = 0$ ('infinitely high' meter resistance) will the measured value equal the e.m.f.

The maximum current which can be drawn from a source is $\mathscr{E}/r$ (when $R = 0$), but this size of current (the *short-circuit* current) would rapidly damage many types of source.

DEMONSTRATION B10
High resistance voltmeter

It is important to choose the appropriate measuring instrument (voltmeter, oscilloscope, electrometer, etc.) for a particular job.

Digital meters have higher resistances than moving coil meters.

Power delivered to a load

QUESTION 22

'Systems' in the Reader *Physics in engineering and technology*

If the load resistance is either very high or very low the power transferred to it from the supply will be very small. The power is a maximum when $R = r$, and under this condition the source is working at it most efficient (as far as delivering useful power is concerned).

The potentiometer

EXPERIMENT B11
Potentiometer balancing an e.m.f.

QUESTIONS 20, 23 to 25

If a battery supplies a p.d. V across a uniform wire, then the p.d. across every centimetre length of the wire is the same. There is a uniform potential gradient along the wire. The potential drops in the direction of the current (from positive to negative): the potential gradient is therefore positive (rising) in the *opposite* direction to the current flow. Note that it is very often convenient to take point X, say, as being at zero potential (though any other point could be used). We can then give a value to the potential at any other point such as Y. In figure B8, Y is at a higher potential than point X.

If a cell is added in series with the voltmeter (figure B9), then by adjusting the length l a position can be found at which the voltmeter reading is zero. In this state of balance there is no p.d. across the meter; the potentials on either side of it are equal. No current flows through the meter or the cell, and the length l is a measure of the cell's e.m.f. Because it does not draw any current, the potentiometer is an excellent method for accurate measurements of e.m.f.

QUESTION 24

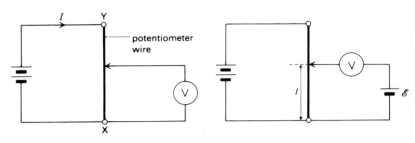

Figure B8
A rising potential gradient from X to Y.

Figure B9
A balanced potentiometer.

Bridge circuits

EXPERIMENT B12
Using two potentiometers to make a
bridge circuit

If the cell in figure B9 is replaced by a second potentiometer, then the arrangement can still be balanced (no current through the voltmeter), as in figure B10. The two negative terminals are joined to provide a common zero potential point for both sides. Two power supplies are not necessary, since both potentiometers can be driven from the one (see figure B11). This arrangement is called the Wheatstone bridge.

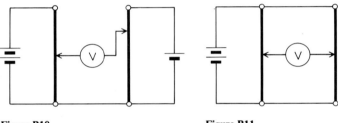

Figure B10
Two potentiometers in balance.

Figure B11
Wheatstone bridge.

EXPERIMENT B13
Detection of small resistance changes

If one of the two sliding contacts is adjusted slightly, then some current will flow through the meter. If the voltmeter is replaced by a sensitive galvanometer, then small out-of-balance movements can easily be detected.

READING
Applications of Wheatstone bridge
circuits (page 108)

Since the wire is uniform, equal movements of the contact correspond to equal changes in resistance. For small changes in resistance, the out-of-balance current is proportional to the change in resistance. This gives an extremely sensitive method of detecting small resistance changes in quite large resistances, and is the usual circuit to use with, for example, resistance thermometers or strain gauges.

Kirchhoff's Second Law

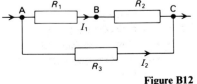

Figure B12
A circuit loop.

If we imagine a clockwise journey round the loop circuit of figure B12, starting and finishing at A, then the total change in potential must be zero. From A through B to C, there is a potential drop of $(I_1 R_1 + I_1 R_2)$. From C back to A through R_3 there will be a rise (against the current) of $I_2 R_3$. Hence $I_1 R_1 + I_1 R_2 - I_2 R_3 = 0$. This can be put in a very concise

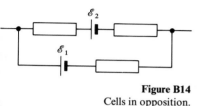

Figure B13
A loop containing a source of e.m.f.

QUESTION 27

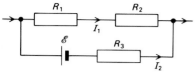

Figure B14
Cells in opposition.

form, to say that if we sum round any complete loop in a circuit in which there is *no* source of e.m.f.,

$$\Sigma IR = 0$$

remembering that individual 'IR' terms will be negative if the sense is against the current flow in that branch.

If the loop includes a source of e.m.f. $\mathscr{E}$, of negligible internal resistance, as in figure B13, then a similar argument shows

$$I_1 R_1 + I_1 R_2 - I_2 R_3 = \mathscr{E}$$

or again, round any complete loop

$$\Sigma IR = \Sigma \mathscr{E}$$

where the righthand side is the total net e.m.f. in the loop. The sign of $\mathscr{E}$ is taken as positive if the imaginary journey passes *from* the positive cell terminal around the rest of the circuit (*i.e.*, from negative *to* positive inside the cell). In figure B14 $\mathscr{E}_1$ is positive and $\mathscr{E}_2$ negative, for a clockwise circulation.

B

Section B3 ELECTRIC CHARGE

QUESTION 28

Electric charge, like mass, energy, and momentum, is a quantity which is conserved. When an object is charged, nothing is created, the available charge is simply re-distributed.

An electric current is a movement of charge, and the current I is the flow of charge per second. For steady currents the charge flowing around a circuit can be calculated using the equation

Charge flowing (in coulombs)
= Rate of flow of charge (in amperes) × time (in seconds)

$$Q = I \times t$$

EXPERIMENT B14
Capacitors and charge

QUESTIONS 29, 30

In experiments on charging and discharging capacitors, surges of current carry charges around the circuit. These currents rapidly fall to zero showing that the capacitor will only let a certain quantity of charge flow. Using an oscilloscope shows that the current carrying the charge around the circuit changes in the way shown in figure B15. The charge flowing can be calculated from the area under the current–time graph. This can be shown by dividing the area under the graph into vertical strips. Each strip has the same narrow width, representing a short time interval, Δt. During this short period of time the height of the strip representing the current I changes very little. The area of the rectangular strip represents the small charge ΔQ flowing in time Δt (figure B16).

$$\Delta Q = I \Delta t$$

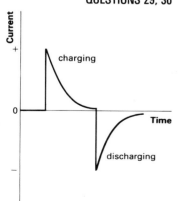

Figure B15

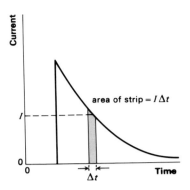

Figure B16

Summing the area of all the strips gives the total charge flowing during charge or discharge. In figure B17 the ammeters show equal pulses of current during charging or discharging, indicating that as much charge flows into one plate as off the other. The capacitor stores no net charge (figure B18).

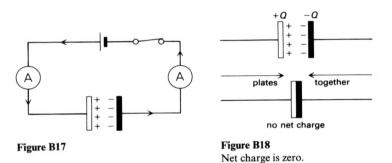

Figure B17

Figure B18
Net charge is zero.

QUESTIONS 31 to 33

DEMONSTRATIONS B15 and B16
Charging a capacitor at a constant rate
Spooning charge

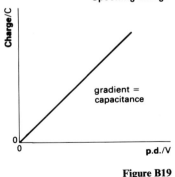

gradient =
capacitance

Figure B19
$Q \propto V$.

Charging a capacitor using a constant current, or 'spooning' charge from a high voltage electrode, shows that the potential difference, V, across the capacitor plates is proportional to the quantity of charge, Q, which has moved around the circuit (figure B19). The constant of proportionality is called the capacitance, and is given the symbol C.

$$Q = CV$$

The unit of capacitance is the coulomb per volt, or farad (F). The farad is a very large unit, and most practical capacitances are measured in microfarads (μF, 10^{-6} F) or picofarads (pF, 10^{-12} F). The value of capacitance indicates the amount of charge on each plate per volt of potential difference across the plates. Both plates store the same quantity of charge, but since there is positive charge on one plate and negative charge on the other, the net charge stored is zero. To say that a capacitor stores a charge of Q means that one plate has gained charge Q and the other has lost Q.

Capacitors in parallel and series

When capacitors are connected in parallel, the potential difference across both capacitors is the same, and the charges on the plates can be added. So twice the charge moves around the circuit in figure B20(b) as in figure B20(a), and the capacitance of the circuit is doubled.

QUESTIONS 34 and 35

Capacitances in parallel add:

$$C = C_1 + C_2$$

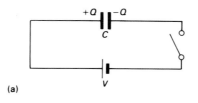

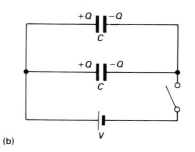

(a)

Figure B20
Capacitors in parallel.

(b)

When capacitors are connected in series across a supply voltage, the same current must charge both capacitors (Kirchhoff's First Law), and so the charge on each plate is the same – figure B21(a). Applying Kirchhoff's Second Law shows that the potential differences across the capacitors must add up to the supply voltage – figure B21(b).

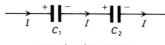

same charging current
for identical times

(a)

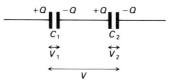

Figure B21
Capacitors in series.

(b)

$$V = V_1 + V_2$$

Dividing by Q (which is the same for both capacitors) gives

$$\frac{V}{Q} = \frac{V_1}{Q} + \frac{V_2}{Q}$$

and so

$$\frac{1}{C} = \frac{1}{C_1} + \frac{1}{C_2}$$

QUESTION 36 where C is the total effective capacitance of the capacitors in series.

Energy stored in a capacitor

QUESTION 37

DEMONSTRATION B17
Energy stored in a charged capacitor

DEMONSTRATIONS B18 and B19
Energy $\propto V^2$

When a capacitor discharges, charge moves through a potential difference as it flows off one plate and onto the other. If charge Q (in coulombs) moves through a potential difference of V (in volts), the energy transformed is QV (in joules). As the discharge continues V falls, and so the last charges to move around the circuit move through a smaller p.d. than the first. This can be compared with releasing a stretched spring, where the restoring force decreases as the spring contracts. The spring does less work in the last centimetre of contraction than in the first. The total energy stored by the capacitor is the area under the p.d.–charge graph (figure B22). The rectangular strip represents a tiny charge, ΔQ, flowing off the capacitor. During this small change the p.d. across the plates, V, remains approximately constant. The energy transformed is $V\Delta Q$, which is the area of the strip. Dividing the whole area into similar strips we find the total energy transformed during discharge is the area under the graph $=\frac{1}{2}Q_0 V_0$. Combining this equation with $Q = CV$ gives three expressions for the energy stored in a capacitor.

$$\text{Energy stored } = \tfrac{1}{2}Q_0 V_0$$
$$= \tfrac{1}{2}C V_0{}^2$$
$$= \tfrac{1}{2}\frac{Q_0{}^2}{C}$$

where Q_0 and V_0 are the initial charge and p.d.

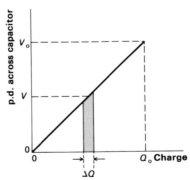

Figure B22
Energy as area of strip.

QUESTION 38 to 40

Capacitors are not viable as large-scale energy storage devices because of the large voltages required to store worthwhile amounts of energy. Furthermore, when supplying a current the voltage cannot be held at a steady value as can that of a battery or generator. The ideas linking charge, potential difference, and energy used here will be important in Unit E, 'Field and potential', when electric fields are studied.

Analogy between spring and capacitor

Spring   **Capacitor**

$$F \propto x \qquad\qquad V \propto Q$$

$$F = kx \qquad\qquad V = \frac{Q}{C}$$

$$\text{Energy} = \tfrac{1}{2}Fx \qquad \text{Energy} = \tfrac{1}{2}VQ$$

$$= \tfrac{1}{2}kx^2 \qquad\qquad = \tfrac{1}{2}\left(\frac{1}{C}\right)Q^2$$

Exponential decay of charge

When a capacitor is discharged through a resistor the current falls, producing the decay curve shown in figure B23. This curve has a constant ratio property: if the current is recorded at equal intervals of time, then $I_1/I_2 = I_2/I_3$. This sort of change is called exponential decay, and it crops up in a very wide variety of situations in virtually every branch of science.

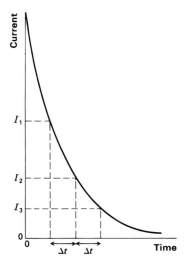

Figure B23
Exponential decay.

DEMONSTRATION B20
Decay of charge

    Why does the curve have this shape? The p.d. across the capacitor which drives the current through the resistor depends on the charge left on the capacitor plates. As the current flows, charge is removed from the plates causing the p.d. to drop, which in turn reduces the current. The current falls because of the current itself.

QUESTIONS 41, 42   p.d. across capacitor $V = Q/C$

$$\text{but } I = V/R$$

$$\text{so discharge current } I = \frac{Q}{RC}$$

Because RC is constant for a particular circuit, $I \propto Q$, and the graph of charge against time has the same shape as the graph of current against time (figure B23).

The current, I, is the rate of flow of charge

$$I = -\Delta Q/\Delta t$$

and so

$$\Delta Q/\Delta t = -Q/RC$$

the negative sign indicating that the flow of charge decreases Q.

The rate of flow of charge at any instant is proportional to the quantity of charge left on the capacitor. For any exponential growth or decay the rate of change of a particular quantity is proportional to the amount of that quantity. Exponential changes are important in many fields including population growth, use of resources, and radioactive decay. In the last case, for example, the number of nuclei decaying per second is proportional to the number of those nuclei remaining in the sample.

Unit F, 'Radioactivity and the nuclear atom'

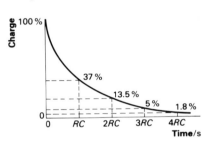

Figure B24
Exponential decay of charge.

QUESTION 44

The quantity RC is called the time constant of the circuit, and the units of the time constant are seconds. The time constant is a rule of thumb measure of the time taken for a capacitor to charge or discharge. After RC seconds only 37 per cent of the original charge is left on the plates (figure B24).

Logarithmic graphs

All exponential changes have the constant ratio property described above. The series:

$$10^1 \quad 10^2 \quad 10^3 \quad 10^4 \quad 10^5 \quad 10^6 \quad 10^7 \ldots$$

also has this property (it increases by a factor of ten each time). However, the logarithms of the numbers in that series:

$$1 \quad 2 \quad 3 \quad 4 \quad 5 \quad 6 \quad 7 \ldots$$

increase in equal amounts (a linear change).

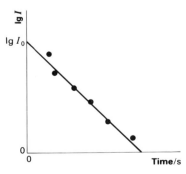

Figure B25
Log–linear plot of decay.

Plotting a graph of the logarithm of the discharge current against time produces a straight line (figure B25 – note that $\lg I = \log_{10} I$). If any quantity is thought to change exponentially, then looking for the constant ratio property or plotting a log–linear graph are useful tests.

QUESTION 43

Streams of charge

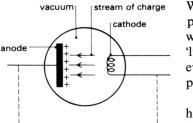

Figure B26
Thermionic emission.

DEMONSTRATION B21
Electron streams

We think that an electric current is due to the motion of small charged particles. We now consider in more detail some of the experiments which lead to the idea that charge can be transferred only in small 'lumps' on these particles. This is the beginning of atomic physics, and eventually brings us to the picture of the atom itself as composed of particles, some of which are electrically charged.

A stream of negative charge can be produced from a metal by heating it (*thermionic emission*) in a vacuum. The charge is accelerated by applying a p.d. between the metal (the *cathode*) and a nearby collecting plate (the *anode*) which is given a positive charge (figure B26). The stream of charge can form a shadow of an obstacle on a fluorescent screen, showing that it travels in a straight line between cathode and anode – figure B27(a).

It can also be deflected by an electric field – set up between two oppositely charged plates positioned either side of the stream, as in figure B27(b), and by a magnetic field – produced either by permanent magnets or a coil carrying a current – figure B27(c). The direction of the deflection in an electric field confirms the view that the stream carries negative charge and that it behaves like a conventional electric current (although no longer trapped inside a wire) whose direction is opposite to that of the velocity of the charge.

The stream can also be shown to heat a target in its path, evidence that it carries energy.

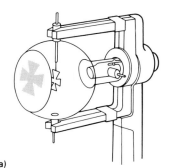

(a)

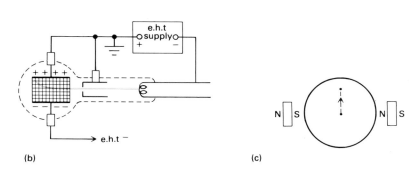

(b)   (c)

Figure B27
Streams of charge:
(a) forming a shadow;
(b) deflection by an electric field;
(c) upward deflection by a magnetic field (looking into beam).

The elementary unit of charge

EXPERIMENT B22
The Millikan experiment
(charge on an electron)

Experiments on electrolysis of different liquids (not part of this course) show that 1 mole (6.02×10^{23}) of atoms of an element, when ionized, always carries a charge of either $1 \times$, $2 \times$, $3 \times$, or $4 \times 96\,500$ C (the faraday). This suggests that each ion always carries a charge of exactly e, $2e$, $3e$ or $4e$, where

$$e = 96\,500/(6.02 \times 10^{23}) = 1.60 \times 10^{-19}\,\text{C}.$$

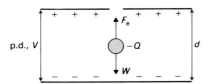

The fact that electric charge always occurs as a multiple of this value is shown independently by the *Millikan experiment*. Here a drop of oil of weight W, carrying a small charge of $-Q$, is balanced between a pair of horizontal plates with a p.d. of V between them (figure B28). If F_e is the electrical force upwards on the drop, then for equilibrium

Figure B28
Balancing an oil drop.

$$F_e = W$$

The weight W is found by measuring the steady downwards velocity of the drop under the action of gravity and air resistance when the p.d. is removed. (A calibration graph, figure B55 on page 131, relates W to this velocity.) F_e is shown from an energy argument to equal QV/d. Hence

QUESTION 48

$$QV/d = W$$

enabling Q, the charge on the drop, to be calculated.

Careful measurements show that the drop always carries a charge equal to a small multiple of e above, 1.60×10^{-19} C.

QUESTION 45

Experiments such as B21 and B22 support the view that negative charge is carried on a very small particle of definite mass (the *electron*) and that all electrons carry the same charge of -1.60×10^{-19} C.

Energy and speed of the electrons in a stream

If electrons of charge e and mass m are emitted from a cathode with negligible velocity, then we can apply the energy ideas of Sections B2 and B3 to calculate the velocity, v, with which they arrive at the anode (figure B29). The work done by the electric force F_e in pulling one electron across the space is equal to eV ($1\,\text{J} = 1\,\text{C} \times 1\,\text{V}$). This can also be considered as a *loss in energy* of the power supply. There will be a corresponding *gain* in kinetic energy of the electron of $\frac{1}{2}mv^2$. Thus

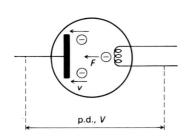

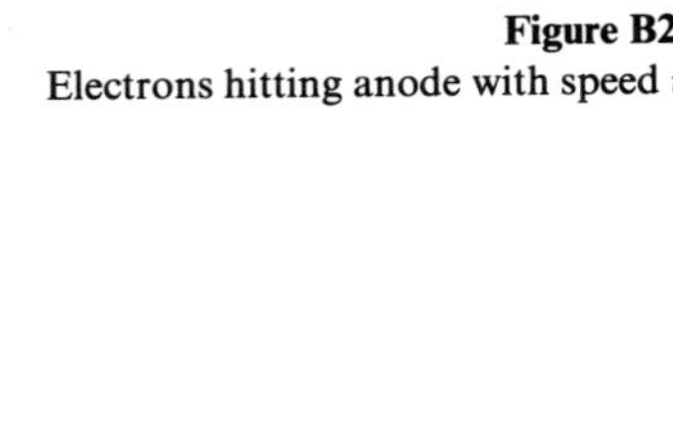

$$eV = \tfrac{1}{2}mv^2$$

Figure B29
Electrons hitting anode with speed v.

This is a fundamental equation governing the motion of any charged particle (in general e becomes ne, where n is a small integer). It will be used particularly in Unit H, 'Magnetic fields and a.c.'

Two points are worth noting about this equation:

i the velocity v ($= \sqrt{2eV/m}$) is independent of the distance between cathode and anode,

ii the velocity only depends upon the **ratio** e/m, the 'specific charge'. (The individual values of e and m are not needed.)

QUESTIONS 46, 47   An important unit of energy used when dealing with atomic-sized particles is the *electronvolt*. If a charge equal to *ne* is accelerated through a p.d. of *V*, it gains *nV* electronvolts of energy.

1 electronvolt (eV) $= 1.60 \times 10^{-19}$ J

It can also be shown that a stream of electrons carries *momentum*. This can be measured by observing the impulse given to a target when electrons are fired in bursts on to it. The way in which the impulse (and hence the momentum transferred) from a given number of electrons depends upon the accelerating p.d. (impulse $\propto \sqrt{V}$) provides confirmation of the view that an electron stream does consist of discrete particles which possess mass as well as charge.

B

READING

APPLICATIONS OF WHEATSTONE BRIDGE CIRCUITS

(Adapted from two articles by C. R. Sawyer and by D. Bridgewater in *Physics at work*. Reproduced by kind permission from BP Educational Service. BP International Ltd.)

Introduction

This part of the *Guide* describes three industrial applications of the Wheatstone bridge circuit:

  i the explosimeter
  ii the corrosometer
  iii a strain gauge pressure transducer.

All rely for their working on the detection of small changes in resistance. In *i* the change is produced by a change in temperature, in *ii* by a change in cross-sectional area, and in *iii* by a change in length and area. (Question 16, on page 142, goes through the relevant theory.)

In two of the examples (*i* and *iii*) the bridge is driven out of balance by the change in resistance: the p.d. produced across the bridge (or the current this p.d. can drive) is taken as an indication of the effect being detected. (Experiment B13 is a laboratory example of this principle.) In the other example (*ii*) the bridge is *restored* to balance, and the change in resistance in one of the arms needed to do this is used as a measure of the effect.

The circuit

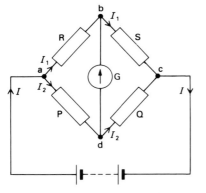

Figure B30
The Wheatstone bridge.

The basic circuit was developed by Professor Sir Charles Wheatstone in 1843, as a method for the accurate measurement of resistance. Four resistors, P, Q, R, and S, are connected as shown in figure B30, to form a bridge or network. A potential difference is applied across ac and a sensitive galvanometer is connected between b and d.

Suppose R is the unknown resistance, and P, Q, and S are variable resistances. The current, I, from the battery reaches junction a, where it will divide into currents I_1 through R, and I_2 through P. When I_1 reaches junction b it could divide, and similarly I_2 could divide at junction d. This would provide two opposing currents through G which, if resistors P, Q, and S are adjusted, can be made equal and effectively cancel each other out. No reading would result on the galvanometer and the bridge is said to be balanced. Under these balanced conditions all of I_1 goes through S and all of I_2 goes through Q, and a relationship exists between the four resistors as follows.

Since no current flows through the galvanometer, the potential at b must equal the potential at d. Therefore, the p.d. across R must equal the p.d. across P. Similarly, the p.d. across S must equal the p.d. across Q. That is,

$$V_{ab} = V_{ad}$$

and

$$V_{bc} = V_{dc}$$

Since $V = I \times R$, then

$$I_1 \times R = I_2 \times P \qquad \text{[equation 1]}$$

and

$$I_1 \times S = I_2 \times Q \qquad \text{[equation 2]}$$

Dividing equation 1 by equation 2 we have:

$$\frac{I_1 R}{I_1 S} = \frac{I_2 P}{I_2 Q}$$

Therefore:

$$\frac{R}{S} = \frac{P}{Q}$$

If P, Q, and S are known, the unknown resistance R can be calculated.

Applications

i The explosimeter

The explosimeter is a device that measures the concentration of flammable gases and vapours in the atmosphere. It uses the principle that the resistance of a conductor depends upon its temperature. This resistance is not measured directly or even calculated, but the Wheatstone bridge is used as a balanced circuit and it is the imbalance produced by a change in one of the resistances that is actually measured.

The explosimeter depends upon the heat developed by the combustion of an isolated sample of the atmosphere. The sample is drawn into the instrument and passes over a heated filament. Combustibles in the sample burn on the heated filament, so raising its temperature and increasing its resistance. The more combustibles that are present, the higher the temperature that results.

The filament forms one arm of a Wheatstone bridge arrangement which is balanced before any combustible gases are introduced (see figure B31). As the resistance of the filament changes, the balance of the bridge is upset, and a current flows through the meter. The higher the temperature the larger is the current. The reading on the meter scale indicates the concentration of combustible gases or vapours in the sample.

For obvious reasons the combustion of the test sample must be isolated from the atmosphere. To achieve this a metal gauze is used to dissipate the heat. By operating the aspirator bulb, the sample is drawn in through a filter then through the first gauze (the inlet flashback arrester). It passes over the filament before passing out through the outlet flashback arrester and two valves in the aspirator bulb.

The explosimeter is typically used in oil refineries and chemical plants as one means of preventing explosions.

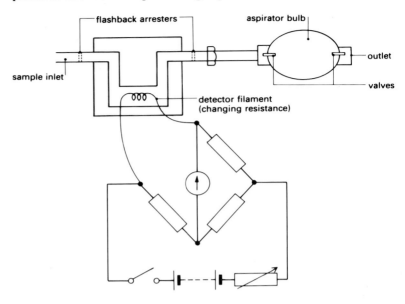

Figure B31
The explosimeter.

ii The corrosometer

The corrosometer is a device that measures the amount and rate of corrosion of metals in any chosen environment. It uses the principle that the resistance of a conductor depends upon its cross-sectional area. Again, the resistance is not measured directly or even calculated, but the imbalance in a Wheatstone bridge circuit produced by a change in one of the resistances determines the amount of corrosion that has occurred.

The corrosometer depends upon the fact that the conductivity of metals is high, while that of the majority of non-metals is negligible. The corrosion of metals produces non-metallic oxides, and so the resistance of the metal increases as the cross-sectional area decreases, because of the corrosion.

A U-shaped piece of wire made from the metal to be tested (the measuring element) projects from the end of a tube – see figure B32(b). One arm of the U is connected to a second piece of the test wire which is sealed inside the tube. This second piece of wire is protected from any corrosive materials throughout the test so that it remains totally unchanged and is called the reference element.

The measuring element, the reference element, and their tube holder are referred to as the probe. The probe is connected to two fixed resistors to form a network as shown in figure B32(a). This is a Wheatstone bridge arrangement, and before any corrosion of the measuring element occurs the variable resistance is adjusted so that no current flows through the meter.

The measuring element of the probe is placed in the corrosive environment. As corrosion proceeds the resistance of the measuring element gradually increases, while that of the reference element remains constant. This upsets the balance of the Wheatstone bridge and allows an ever increasing current to pass through the meter. By adjusting the variable resistor the meter is made to read zero again and the extent of the corrosion is read directly from a scale attached to the variable resistor.

In use the probe is left *in situ* over a long period of time and is only connected to the remainder of the circuit when a reading is to be taken.

Allowance is made in the design for the resistance due to the connecting wires of the probe.

Temperature has no effect on the readings taken as two resistors are in the probe, and the other two in the case, so maintaining a balance.

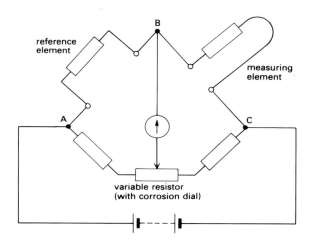

(a)

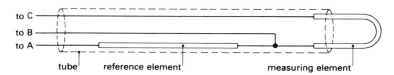

(b)

Figure B32
The corrosometer:
(a) the circuit;
(b) the probe.

iii The strain gauge pressure transducer

A strain produced in a conductor, usually metallic, causes a change in resistance. It would appear that a strain gauge is most likely to be used for considering structures under stress and indeed it is extensively used for this. However, stress depends on force applied and it is no great step from this to the use of the strain gauge as a pressure transducer. Pressure differences in a fluid flowing through an orifice are related to the velocity of the fluid and so indirectly the strain gauge can be used to measure the rate of flow of a fluid. Figures B33(a) and (b) show the construction of such a pressure gauge and the circuit. Initially the bridge is balanced. Note particularly the significance of the flexible cross-members and the rigid supports, these latter being independent of any changes in pressure.

Suppose the pressure on the diaphragm increased. If windings 1 and 2 were initially under tension they will shorten slightly and likewise windings 3 and 4 will lengthen slightly. The changes in the resistances of these windings will cause the bridge to become unbalanced, resulting in a p.d. occurring between X and Y. This p.d. is amplified and transmitted to a control room. The transmitted signal is a measure of the pressure applied to the diaphragm.

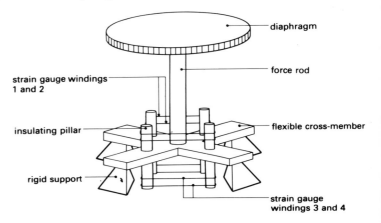

(a)

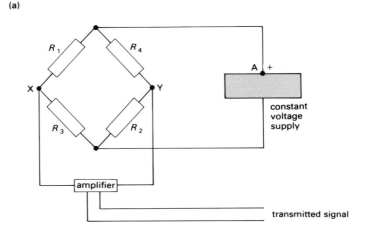

(b)

Figure B33
Strain gauge pressure transducer.

Questions

The explosimeter

a   Discuss how you could in principle set about calibrating the explosimeter, outlining any difficulties you might experience.

b   Is the calibration likely to change with usage?

c The flashback arresters are metal gauzes. How do they isolate the test sample from the atmosphere?

d The explosimeter is stated to be one means of preventing explosions. How could it be used in this way?

The corrosometer

e The resistance of a corroded wire is almost entirely that of the pure metal 'core'. Explain how this is consistent with the fact that the oxide coating has a very high resistance.

f A given length of uncorroded wire has a radius r and resistance R. The radius is reduced due to corrosion by an amount Δr and the resistance increases by ΔR. Show that provided Δr is small compared with r, the proportional change in resistance, $\Delta R/R$ is equal to $2\Delta r/r$. What significance does this result have for the operation of the corrosometer?

g Explain the sentence in the final paragraph that 'temperature has no effect on the readings'.

The strain gauge pressure transducer

h What does the word 'transducer' mean?

i Why are windings 1 and 2 and windings 3 and 4 connected diagonally opposite to each other in the bridge circuit?

j An increase in pressure on the diaphragm causes R_1 and R_2 to decrease, and R_3 and R_4 to increase. Consider carefully the effect of these changes on the p.d. between X and Y. (You may find it helpful to regard point A on the constant voltage supply as the positive terminal.)

k Suppose the strain gauge transducer is to be used to monitor a gas pressure in an environment where small changes in pressure are significant. This would mean that the transducer would need to be very sensitive. List and justify those factors in the design of the transducer as shown which you think determine its sensitivity.

l The explosimeter and the corrosometer detected respectively temperature changes and corrosion. What effect would these environmental changes have on the operation of the pressure transducer?

m The p.d. across XY, V_{XY}, can be shown to be proportional to ΔR, the change in resistance of the windings. Ideally V_{XY} should be proportional to the change in pressure Δp applied to the diaphragm (a so-called *linear* transducer), so that $\Delta R \propto \Delta p$. List any features of the design, and any restrictions under which the device would have to operate, which might make this proportionality a reasonable assumption.

LABORATORY NOTES

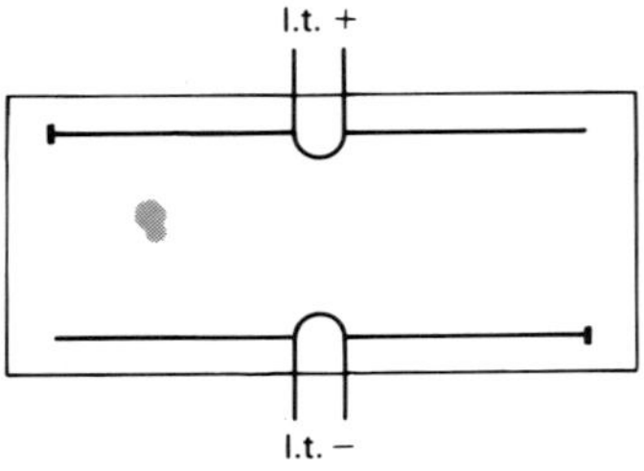

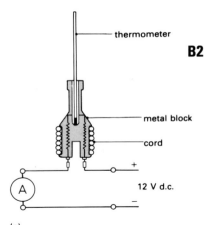

Figure B34
Movement of coloured ions.

DEMONSTRATION

B1 Conduction by coloured salts

Two different crystals are dissolved on to the filter paper. The purple manganate(VII) ion is negatively charged and the deep blue copper ammonium ion is positive. The motion of each ion is tracked by the colour trail left.

Questions

a Which is the conventional current direction?

b How does this relate to the motion of each ion?

c Roughly what value do you estimate for the speed of the ions?

d What happens when the supply terminals are reversed?

e What will happen to each ion when it arrives at the terminal? How will this contribute to the electron flow in the rest of the circuit?

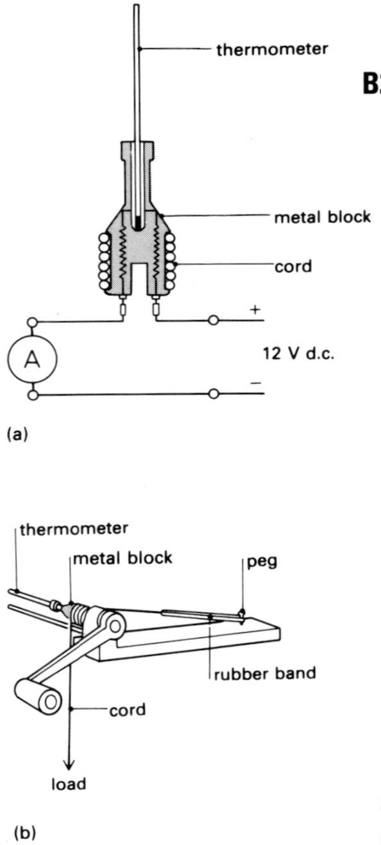

DEMONSTRATION

B2 Measuring p.d. without a voltmeter

apparatus for measuring joules per coulomb
ammeter, 1 A
clock
thermometer, 0–50 °C in 0.2 °C
slotted masses, 1 kg
metre rule
power supply, 12 V
leads

In the first part of the experiment, energy is supplied electrically to the block to raise its temperature by about 10 K. The cord should be wrapped round it 5 or 6 times so that the conditions are as nearly as possible the same as in the second part of the experiment. The current and time are recorded.

In the second part the block is heated by doing work against friction, using the cord as a friction band with a load of about 8 kg – figure B35(b). The temperature rise should be as in the first part.

The relationship *work done = force × distance* can be used to calculate the mechanical energy supplied against friction, using the tension in the cord, the circumference of the block, and the number of turns of the handle.

Questions

a Why should the temperature rise be kept fairly small?

b The assumption is made that the electrical energy supplied in the first part is equal to the mechanical work done in the second. What features of the experimental method might make this incorrect?

Figure B35
Apparatus for measuring a p.d. without a voltmeter.

c The second part of the experiment gives the energy supplied; the first part gives the charge passed which will supply this energy. Calculate these two quantities.

d Use the definition of the volt to calculate the applied p.d. in the first part of the experiment. Is this likely to be too high or too low? Check with a voltmeter.

EXPERIMENT
B3 Two-terminal boxes

milliammeter, 100 mA, occasional access to other meter ranges
2 cell holders with four cells
two-terminal boxes
leads
graph paper

You may put up to 12 V across the terminals of any box and the current will not exceed 0.1 A (it might be much less). You should investigate as many boxes as possible to see how the current passed depends upon the p.d. applied. Try to make a sensible choice of meter range, always starting with the highest value if you have no idea of the size of the quantity being measured.

Plot your *I–V* values for each box on a graph.

Questions

a Does the box conduct equally in both directions?

b Does the graph pass through the origin? If it does *not*, what is the significance of the intercept on the *V*-axis?

c Why might the ratio *V/I* sensibly be called the *resistance* of the box? Where *V/I* is constant, calculate it; where it is not constant, decide from the shape of the graph whether the resistance is increasing or decreasing with current.

d From the results of **c**, try to identify the contents of the box.

EXPERIMENT
B4 Comparison of rheostat and potentiometer as controllers

ammeter, 1 A
voltmeter, 10 V
rheostat, 10–15 Ω

diode and clip component holder
and/or
m.e.s. lamp, 2.5 V, 0.3 A, and holder

cell holder with two cells
leads

Set up the circuit as in figure B36(a). You will use either a diode or a lamp. Observe the range of current and p.d. values possible as the rheostat is adjusted from maximum to minimum resistance. Within this range you can make a much finer gradation of changes of I and V than was possible in experiment B3, and you might be asked to take a full set of pairs of readings of I and V.

Now change the circuit to that of figure B36(b). The adjustable component here is called a potentiometer (or potential divider). Note the range of variation of V and I obtainable with this circuit.

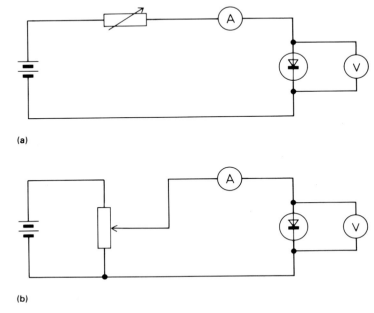

(a)

(b)

Figure B36
Rheostat and potentiometer as controllers.

Questions

a What determines the range of currents available in the first circuit?

b How does the p.d. obtainable in the second circuit depend upon the position of the sliding contact?

c Both arrangements enable the p.d. across a component to be varied. Compare the possible advantages and disadvantages of one method against the other.

DEMONSTRATION
B5   Effect of size and material on resistance

l.t. variable voltage supply
ammeter, 1 A
voltmeter, 10 V
constantan wire, two different thicknesses
nickel–chromium wire
leads
micrometer

(Covered wire can be used if the insulation is removed at appropriate points.)

A p.d. of 2 V is applied across 0.2 m of the thinner constantan wire and the current noted. The p.d. is then increased in steps of 2 V, the length of wire being increased so as to keep the current the same.

The readings are repeated with the thicker constantan wire. The current need not be the same as in the first part, but should still remain fixed.

Finally, the experiment is done using thin nickel–chromium wire.

Measure the diameters of the wires.

Questions

a How does the resistance of a wire depend upon its length if it has a fixed cross-sectional area?

b Compare the resistances per metre length for constantan and nickel–chromium wires of the same diameter.

c How does the resistance per metre length for two wires of the same material depend upon their diameters? And therefore upon their cross-sectional areas?

DEMONSTRATION
B6 Effect of temperature on resistance

Three different conductors are connected in turn into a series circuit with an ammeter or milliammeter and a fixed supply of about 2 V. The conductors are: 3 metres of very thin copper wire (about $2\,\Omega$), a carbon resistor ($150\,\Omega$), and a thermistor (a semi-conducting device of normal resistance about $400\,\Omega$).

The copper is screwed into a tight bundle (so that it heats up appreciably when a current is passed); then laid out loosely; and finally put in hot and cold water and cooled by a freezer spray. The other two components are subjected to the same temperature variation.

Since the p.d. is fixed, any changes in current are due to changes in resistance.

Questions

a Do equal changes in current indicate equal changes in resistance?

b Summarize the resistance–temperature behaviour of each component.

c Consider the theoretical relation $I = AvnQ$. You have observed changes in I. Which of the quantities on the righthand side are likely to have changed for each component? Will they have increased or decreased?

EXPERIMENT
B7 Four-terminal boxes

cell holder with four cells
milliammeter, 100 mA
voltmeter, 10 V
four-terminal boxes
leads

Each box has its lower (green) terminals joined by a direct link. There is at least 60 Ω between any other pair of terminals so that a 100 mA meter can be used with a 6 volt supply.

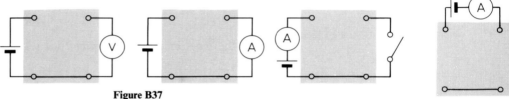

Figure B37
Some circuits which may be used to find out about the boxes.

You should investigate all of boxes **A–D** and at least one from **E–G**. Your teacher might make some harder boxes available to you. Using a voltmeter and ammeter as suggested in figure B37, make appropriate measurements to find out what you can about the circuits inside the boxes.

Question
You might find it useful to work through question 18 in this *Guide* to get an idea of the kind of information you can extract from a measurement, and how the readings can be used to build up a picture of the circuit inside the box.

DEMONSTRATION
B8 Drop in terminal p.d. of source on load

Worcester circuit board kit (3 lamps and 1 cell)
l.t. variable voltage supply
lamp, 12 V, 36 W
lampholder (s.b.c.) on base
e.h.t. power supply
demonstration meter, 10 V d.c., 15 V a.c., and 10 mA d.c.
leads

Using the circuit board, connect the voltmeter across one cell. Then connect first one, then two, and then three lamps in parallel across the cell. Note any changes in p.d. as more current is drawn from the cell.
Repeat the experiment using the 12 V lamp (only use one lamp) both on a.c. and the d.c. supply.
Finally use the e.h.t. supply (under supervision). Set it to 1000 V and connect the milliammeter across the output. Observe both the voltmeter (on the supply) and the milliammeter readings. (Any very high resistance shown as a possible connection on the front panel should not be used.)

Questions
a When you measure the terminal p.d. in the first two parts *with no lamp connected*, what property of the supply are you measuring (at least very nearly)?

b Why does the terminal p.d. drop as more current is drawn?

c From your observations on the e.h.t. supply, roughly what is its internal resistance? Why is it constructed to have this sort of value?

d From casual observations at home (DON'T make any measurements) does the mains supply voltage behave in the same way as the supplies in these experiments? What about the water supply?

EXPERIMENT
B9 Comparison of voltmeters

either
source of e.m.f., 1.5 V, with high internal resistance
or
cell holder with 3 cells
resistor, 10 kΩ *or* resistance substitution box

voltmeter, 5 V or 10 V
voltmeter, same range, low resistance

optional
oscilloscope

leads

You are provided with a source with a fairly high internal resistance. You are going to use different voltmeters to try to measure its e.m.f. List the voltmeters in order of increasing resistance. (For the moving-coil meters you may need to ask what the full-scale current is.)

Connect each voltmeter in turn across the supply, removing one before the next is connected. Start with the meter which you think has the lowest resistance and work your way up. Note each reading.

When the last meter has been connected, add the others again *in reverse order*, but this time do not disconnect one before adding the next (so that at the end you may have three or more meters all connected).

Questions

a Comment on and explain the changing pattern of measured voltage as you increase meter resistance.

b Which of your readings is the best estimate for the e.m.f.?

c Explain the observations in the second part, when you add meters one at a time.

d What other information would you need to be able to estimate the internal resistance of the supply?

DEMONSTRATION
B10 High resistance voltmeter

cell holder with one cell
2 resistors, 220 kΩ
2 clip component holders
voltmeter, 1 V, moving coil
high impedance voltmeter (*e.g.*, digital meter or electrometer/d.c. amplifier)
oscilloscope
leads

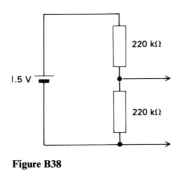

Figure B38

The p.d. across one of the 220 kΩ resistors in the circuit shown in figure B38 is measured using each of the voltmeters in turn: first the moving coil meter, then the oscilloscope, then the digital meter and/or electrometer.

Questions

a Which of the instruments has the biggest loading effect on the circuit? Which has the highest resistance?

b Explain what happens when more than one voltmeter is used at the same time.

c Suggest experiments where it would be important to use a meter which has a negligible loading effect.

EXPERIMENT
B11 Potentiometer balancing an e.m.f.

metre wire potentiometer
l.t. supply (or 2 V accumulator)
cell holder with one cell
voltmeters, 1 V and 10 V
microammeter, 100 µA
leads

Set up the circuit as in figure B39(a). Adjust the position of the sliding contact (do *not* scrape it along the wire) and observe the voltmeter. Compare it with the voltmeter in the second part of experiment B4.
    Now connect a cell into the voltmeter circuit as in figure B39(b). You should be able to find a length, *l*, of the wire so that the meter reads zero.

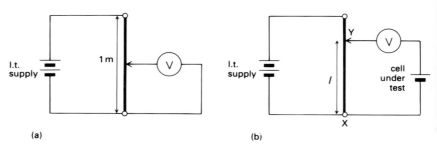

(a)   (b)

Figure B39
Potentiometer (a) without, and (b) with a balancing e.m.f.

(This is the *balance length*.) When you have found the balance point as accurately as you can, replace the voltmeter with the microammeter. You will find that this arrangement is much more sensitive and you can find the balance point very precisely.
    Finally, replace the cell with another apparently similar one. Why does the balance length change slightly?

Questions

a How does the potential difference between points X and Y – figure B39(b) – depend on the length *l*? What is the potential difference per unit length of wire? Is it constant?

b If the potential at point X is taken to be zero, how does the potential at Y depend on the length *l*? In which direction must Y be moved for the potential at Y to increase – in the direction of the current or in the opposite direction?

c At balance, what property of the cell does the length *l* represent?

d State as precisely as you can any differences between the two cells.

EXPERIMENT

B12 Using two potentiometers to make a bridge circuit

Apparatus as for experiment B11, but with a second metre wire potentiometer, and without cell holder and cell

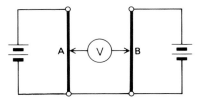

Figure B40
Two potentiometers balanced together.

Join two potentiometers together, as in figure B40. In this experiment the positive terminal of the cell in experiment B11 is replaced by point B on the second potentiometer. Adjust for balance using a voltmeter. Check that a full range of pairs of A–B positions will give balance.

Questions

a State a condition for balance using the word 'potential'.

b When a balanced state has been reached, what might upset it (A and B remain fixed)?

c Suggest a modification to the circuit which would remove the disadvantage of **b**. Make the modification and confirm that you have removed the disadvantage.

d With this change made, replace the voltmeter with a microammeter. Observe what happens to the current through the meter as *small* adjustments are made to the position of B, keeping A fixed. How does the current depend upon the movement of B?

EXPERIMENT
B13 Detection of small resistance changes

cell holder with four cells
thermistor
potentiometer, 1 kΩ or 5 kΩ
resistance substitution box
microammeter, 100 μA
leads

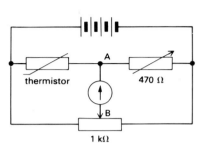

Figure B41
Detecting a small resistance change.

Set up the circuit as in figure B41. It is essentially the same as that for experiment B12, except that there is now only one sliding contact. Adjust the potentiometer until the bridge (the name for this kind of circuit) is balanced.

Demonstration B6 showed that the resistance of the thermistor changes markedly with temperature. This experiment shows how that demonstration can be made more sensitive. Slightly warm or cool the thermistor (gently blow on it, or put on a few drops of a volatile liquid) and observe the bridge move off balance.

Questions

a   When the thermistor is warmed its resistance gets less. What will happen to the potential of the junction point between the thermistor and the 470 Ω resistor?

b   There is now a p.d. across the meter. Which way will current flow through it? Observe the meter connections carefully (normally current into the red terminal produces a deflection to the right), and check your prediction.

c   You have a circuit to detect small resistance changes. Suggest a range of applications for this circuit. How does the out-of-balance current depend upon the change in resistance (see question **d** of experiment B12)?

EXPERIMENT
B14 Capacitors and charge
(a series of 8 short experiments)

cell holder with four cells
3 clip component holders
2 electrolytic capacitors, 500 μF
2 electrolytic capacitors, 1000 μF
2 electrolytic capacitors, 2200 μF
oscilloscope
stopwatch
resistance substitution box
2 milliammeters, 10 mA
resistor, 150 Ω
leads

N.B. These experiments use electrolytic capacitors which can be destroyed by connecting them the wrong way round in the circuit. Always check that the + sign on the capacitor is connected nearest to the + terminal of the supply.

B14a *i* Observe the meter deflections when the capacitor is charged and then discharged (broken line). (See figure B42.)

ii What happens if the capacitor is not discharged between two attempts at charging?

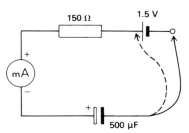

Figure B42

B14b This circuit adds a second meter to the circuit in order to investigate the currents flowing on both sides of the capacitor (figure B43).

What can you deduce about the quantities of electric charge flowing into and out of the capacitor during charge and discharge?

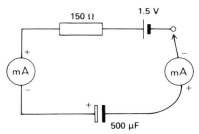

Figure B43

B14c In order to find out precisely how the current in the circuit changes during charge and discharge, an oscilloscope can be used to measure the changing potential difference across a known resistance (figure B44). When using the oscilloscope ensure that:

1 The a.c./d.c. switch is on the d.c. setting.

2 The time-base is initially switched off and the spot is located in the middle of the screen.

3 The *Y*-gain is adjusted so that a p.d. of 1.5 V applied to the oscilloscope causes a vertical deflection of 2 divisions.

i How does the current flowing in the circuit change with time during charge and discharge?

ii Turn the time-base on to the slowest setting and repeat the charge and discharge process.

iii Which quantity on the graph of current against time traced out by the oscilloscope represents the charge flowing around the circuit?

iv How would you calibrate the oscilloscope in order to estimate a value for the charge?

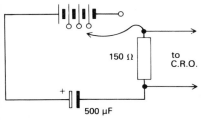

Figure B44

B14d Extra cells can be used to give larger pulses. An important deduction can be made by connecting the charging lead successively to the 1.5, 3.0, 4.5, and 6.0 V terminals of the cell holder without discharging between each stage.

B14e Use capacitors of differing value. The capacitor rating is the charge stored on each plate for a potential difference of 1 V. Check this by using the oscilloscope trace to estimate the charge flowing for each capacitor.

B14f How is the amount of charge flowing around the circuit affected by adding a second capacitor in parallel (figure B45)?

Charging and discharging the capacitors as a pair, and then separately, should show that the charges on the plates of the two capacitors can be added.

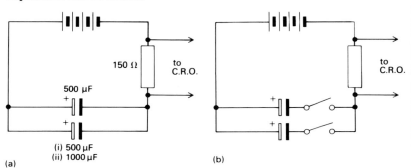

Figure B45 (a) (b)

Try a similar investigation with the two capacitors connected in series (figure B46). Does this arrangement store more, less, or the same amount of charge as a single capacitor?

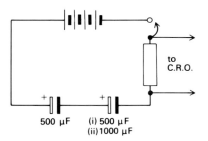

Figure B46

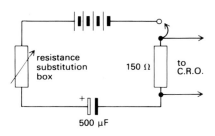

Figure B47

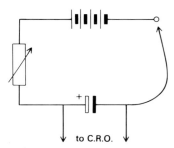

Figure B48

B14g Use a resistance substitution box to investigate the effect of adding extra resistance to the circuit (figure B47).
N.B. The 150 Ω oscilloscope resistor must not be changed, otherwise the constant of proportionality linking trace movement and current would alter.

B14h The oscilloscope can also be used to show how the p.d. across the capacitor changes with time (figure B48).
The 150 Ω resistor is no longer necessary as the current in the circuit is not being measured.

What might you expect to find if the oscilloscope were connected across both the capacitor and resistor? Try it.

DEMONSTRATION
B15 Charging a capacitor at a constant rate

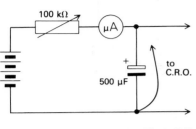

Figure B49
Charging a capacitor at a constant rate.

The capacitor is initially short-circuited and the current adjusted to, say 80 μA. The shorting lead is removed and timing is started. The rheostat is adjusted so as to keep the charging current at its initial value. The progress of the charging can be followed on the oscilloscope as the p.d. across the plates rises. The capacitor is fully charged (*i.e.*, the p.d. across its plates equals the supply p.d.) when the current falls to zero.

Since the current is constant, charge is flowing on to the plates at a steady rate ($80 \, \mu\text{C s}^{-1}$ if the current is 80 μA), and we can find the total charge which has flowed onto one plate and off the other from $Q = It$, where t is the charging time.

If the experiment is repeated for different charging p.d.s V, applied to the same capacitor, we find that $Q \propto V$.

B

Questions

a If the current is to be kept constant during charging, what change has to be made to the rheostat resistance?

b Sketch graphs showing the variation with time of
i the p.d. across the capacitor
ii the p.d. across the rheostat.

DEMONSTRATION
B16 Spooning charge

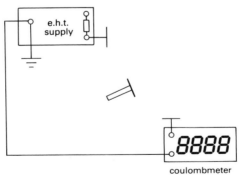

Figure B50
Spooning charge.

*N.B. The e.h.t. supply must **not** be connected directly to the coulombmeter input.*

A metal plate with an insulating handle is used to transfer charge from the e.h.t. supply to the coulombmeter, which measures the total charge transferred to it.

The e.h.t. supply is kept constant at, say, 1000 V.

If the same amount of charge is transferred each time, the coulomb-meter reading should increase by the same amount for each transfer. How does the amount of charge transferred each time depend on the p.d. of the supply? On the size of the 'spoon'?

Questions

a The coulombmeter measures the p.d. across a capacitor, and so gives a reading proportional to charge. What determines how much of the plate's charge transfers to this capacitor? How significant an error might there be in ignoring any charge remaining on the plate?

b This demonstration brings out an important property of electric charge. Some other physical quantities also have this property. From the following list, pick out those which have this property in common with charge: volume, pressure, energy, temperature, potential difference, mass.

DEMONSTRATION

B17 The energy stored in a charged capacitor (motor)

The capacitor is charged and then discharged through a small motor which raises a load.

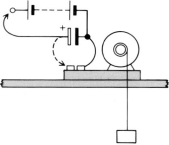

Figure B51
Charged capacitor used to run a motor.

Questions

a Trace through the energy changes in this demonstration, from power supply to load.

b Suppose the demonstration were performed first with a charging voltage of 3 V, and then with 6 V. Ignoring any inefficiency of the motor, what difference would you expect between the two demonstrations? (In fact the efficiency of energy transformation in the motor is low, and this demonstration should be thought of as qualitative only.)

DEMONSTRATION

B18 Energy proportional to V^2 (lighting lamps)

Using the circuit of figure B52(a) the capacitor is charged to 3 V and discharged through the lamp. The brightness and length of flash are noted. The p.d. is doubled to 6 V (circuit (b)) and another lamp added to avoid burning out. The discharge produces brighter and longer flashes than in circuit (a). However, if the capacitor is charged to 6 V and discharged through *four* lamps (circuit (c)), each will flash with the same brightness as the single lamp in circuit (a). At 9 V the capacitor will light three parallel banks of three lamps, each to the same brightness as the single one in circuit (a).

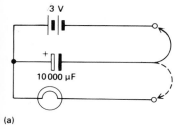

(a)

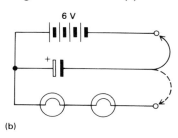

(b)

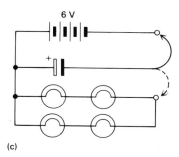

(c)

Figure B52
Discharge of capacitor through lamps.

B

Question

This demonstration shows that the energy stored in a capacitor is proportional to the square of the p.d. across its plates. Using the numerical values above, how many lamps could you light with an 18 V charging p.d. so that each lamp has the same brightness as the original single one?

DEMONSTRATION

B19 Energy proportional to V^2 (heating a coil)

large electrolytic capacitor, 10 000 μF, 30 V
l.t. variable voltage supply *and*
smoothing unit
3 m of insulated constantan wire, 0.4 mm diameter
1 m of copper wire, 0.28 mm diameter
sensitive galvanometer
aluminium block
demonstration meter, 30 V d.c.
leads

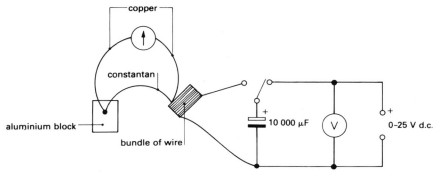

Figure B53
Capacitor discharge through a bundle of wire.

The capacitor is charged to at least 10 V and discharged through the bundle of constantan wire, heating it slightly. The temperature rise is detected by the copper–constantan thermocouple. When a temperature difference is maintained between the two junctions of copper and constantan wires, a small current flows which is proportional to the temperature difference between the junctions. One junction is held in a large aluminium block and may be regarded as being at constant temperature. Thus the galvanometer reading is proportional to the temperature rise of the hot junction, and hence to the energy dissipated in the coil.

The capacitor is charged and then discharged for a range of charging p.d.s between 10 V and 30 V, and the corresponding galvanometer deflections (proportional to energy stored) are noted.

Questions

a Demonstration B18 suggested that the energy stored in the capacitor is proportional to the square of the charging p.d. What graph could you plot to verify this?

b As it stands, this experiment gives a quantity which is *proportional* to the energy dissipated – the galvanometer deflection. What steps *in principle* would be needed to calibrate this deflection in joules?

DEMONSTRATION
B20 Decay of charge

l.t. variable voltage supply
voltmeter, 10 V
capacitor, 500 µF
clip component holder
demonstration meter, 100 µA
resistor, 100 kΩ
leads
stopclock

A 100 µA demonstration meter might not be available. Suitable alternatives are listed here, together with appropriate alternative values for resistance and capacitance.

meter	(0–1 mA)	$R = 5 \text{ k}\Omega$	$C = 10\,000 \,\mu\text{F}$
	(0–5 mA)	$R = 2.2 \text{ k}\Omega$	$C = 25\,000 \,\mu\text{F}$

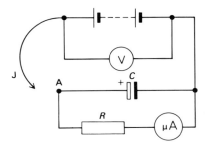

Figure B54
Charge decay circuit.

The capacitor is charged by connecting the flying lead J to A. The discharge is started by removing J from A; timing is started at that instant. The current reading is taken every ten seconds.

Questions

a Suppose the current at any instant is I. What is the p.d. across the resistor? (Either take a numerical value for R or work in symbols.) What must be the p.d. across the capacitor (again, either numerically or algebraically)? Hence what is the charge on the capacitor for a given current I?

b Use your answer to **a** to plot a graph of capacitor charge against time.

c Why does the charge *not* decay at a constant rate with time?

DEMONSTRATION
B21 Electron streams

transformer
e.h.t. power supply
stand for tubes
Maltese cross tube
Perrin tube
deflection tube
pair of coils
gold leaf electroscope
h.t. power supply
battery, 12 V
fine beam tube
Magnadur magnet
rheostat, 10–15 Ω
demonstration meter, 100 mA

Manufacturer's instructions should be followed for the setting up of the tubes.

Safety note: The h.t. power supply is potentially dangerous, so treat it with respect. Always switch it off before making or altering connections to it. It is advisable to use leads having shrouded 4 mm plugs.

A selection of the following demonstrations may be made:

stream of electrons making a shadow of an obstruction (Maltese cross)
stream of electrons coming through a slit to make a splash across a screen
deflection of stream of electrons by electric fields
fine beam tube, raising and lowering gun voltage
deflection of beam (fine beam tube) by electric field
deflection of fine beam by magnetic field
electrons collected to determine the sign of their charge (Perrin tube)

Figure B27 (page 105) shows some of the more important effects to be observed.

Questions

a For each demonstration performed, attempt to answer each of the following questions:

i Is there evidence of a *flow* in the tube?

ii Does it indicate that charge is carried on particles?

iii What does it tell you about the sign of charge carried by the stream?

iv Where there is a deflection, why does the stream bend just the amount that it does?

b (Do this question if you know the relation between the direction of the force on a current in a magnetic field and the directions of current and field.)

Consider the deflection of the beam in a magnetic field. The fact that the magnetic field can move the beam around suggests that the beam is equivalent to a current. Observe the deflection in a field whose direction you know, and decide which is the current direction. What is surprising about this result? How does the result relate to demonstration B1?

OPTIONAL EXPERIMENT

B22 The Millikan experiment (charge on an electron)

Millikan apparatus

either
e.h.t. power supply
or
h.t. power supply
$\left.\right\}$ according to manufacture

multirange meter
leads

Safety note: The h.t. power supply is potentially dangerous, so treat it with respect. Always switch it off before making or altering connections to it. It is advisable to use leads having shrouded 4 mm plugs.

Follow the manufacturer's instructions for the detailed setting up and adjustment. You will need to spend some time familiarizing yourself with the various procedures in this experiment: obtaining oil drops, observing them conveniently, confirming that some are charged, and controlling their fall by varying the applied p.d. so as to balance them.

When a charged drop is balanced the electrical force on it (QV/d) is equal in magnitude to its weight. (Q is the charge on the drop, V the p.d. needed to hold it stationary, and d the spacing of the plates in the instrument.) (See also question **48** on page 158.)

In order to measure the weight of a selected drop, time its fall over a standard distance, with the p.d. turned off. Figure B55 shows a graph of weight against time to fall 1 mm in air, for drops of oil of density $864\,\mathrm{kg\,m^{-3}}$. If the oil you are using has a different density, *multiply* the weight obtained from the graph by $\sqrt{(864/\text{your oil density})}$ to obtain the correct weight.

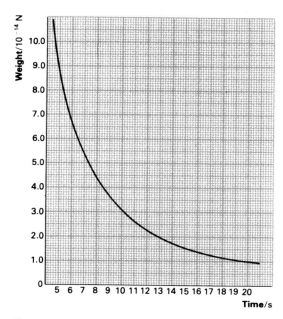

Figure B55
A graph of weight against time for oil drops (density $864\,\mathrm{kg\,m^{-3}}$) to fall 1 mm in air at $23\,°C$.

Question
Calculate the charge on a particular drop and repeat the measurement for as many different drops as possible. To what extent do your results show that the charge is always a small multiple of some basic value? Calculate this value and try to put some limits of experimental uncertainty on it.

HOME EXPERIMENTS

BH1   The resistor

'Off-the-shelf' resistors of predetermined values are very inexpensive today, and easily taken for granted. In this practical task you should try to make a $25\,\Omega$ resistor using only those materials that you can find in, or around, your house. You should not design your resistor on a 'trial and error' basis, but rather using knowledge of the resistivity of the material which you are working with and its dimensions.

If you use a voltmeter and ammeter to check your design certain factors must be taken into account. What are they? What resistance-measuring device(s) could you use to check your resistor?

Check the precision of your resistor and compare it with others in the class.

BH2   The voltaic pile

In your physics course you use a variety of voltage sources with differing characteristics and limitations. In this task you should attempt to make a 'voltaic pile', able to provide the biggest e.m.f. that you can contrive, though your design should have a volume no bigger than that of a large match box.

The e.m.f. of your voltaic pile is found when 'open circuited', where it is often assumed the voltmeter draws negligible current. Find the p.d. of your design when it provides current through a $500\,\Omega$ resistor. Also measure the peak current which flows through the resistor when the circuit is initially closed. You will then be able to find the internal resistance of the pile.

(Warning: Do not try to obtain electrolyte or electrodes by taking a dry cell apart: some modern dry cells contain dangerous materials.)

Tabulate your results and compare them with those obtained by others in the group like this:

Name	e.m.f.	p.d.	Peak current	Internal resistance
B. Mange	1.2 V	0.3 V	0.6×10^{-3} A	$1.5 \times 10^{3}\,\Omega$
...	...	...	...	...
...	...	...	...	...

Table B1

Study the tabulated results. Can you see any patterns? Perhaps plotting a few rough graphs may help you.

QUESTIONS

Flow

1(L) Think of a tube containing liquid which has a cross-sectional area of one square centimetre (10^{-4} m²). Suppose the liquid conducts a current of 10^{-2} ampere. For simplicity, think of the current being carried by a lot of charged particles, all moving along at the same speed, v, which you will be able to calculate if some more assumptions are made about the charged particles.

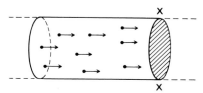

Figure B56

Suppose there is a counting station at XX, which records the number of particles crossing that slice of the tube.

a In 16 seconds, with a current of 10^{-2} ampere, what electric charge passes XX?

b If each particle has a charge of 1.6×10^{-19} coulomb (the magnitude of the charge on an electron), how many particles pass XX in 16 seconds?

c Suppose there are 6×10^{26} charged particles in each cubic metre of liquid. How many charged particles are there in each one-metre length of the tube, if the area of cross-section is 10^{-4} m²?

d From the answer to **b**, the number of particles crossing XX in 16 seconds, and the answer to **c**, find the length of liquid in the tube behind XX from which all the particles cross XX in 16 seconds. (Of course, this length is 'filled up' again with more particles from behind.) You should have an answer much less than one metre. If not, think again.

e The length in the answer to **d** is also the distance a particle travels in 16 seconds at the unknown velocity v, since it is that velocity which carries it past XX. What is the velocity v? Is this result surprisingly large or small? Is there any evidence that helps you to have confidence in it?

f Now go through questions **a** to **e** above but using I for current, t for the time of flow, Q for the charge on each particle, n for the number of particles in each cubic metre, A for the cross-sectional area of the tube, and v for the velocity. You should get $v = \dfrac{I}{AnQ}$. Why is there no time, t, in this answer?

g For a given current in a given sized tube, what is the effect on v of decreasing the density of charged particles, n? What is the effect of decreasing the charge on each particle, Q? What reasons might be given to support a value of around 10^{26} particles per cubic metre for n? Can you think of a liquid (pure? solution?) in which n might be substantially less than this?

2(P) Figure B57 shows a junction of two pieces of copper wire which form part of a simple series circuit around which a current is flowing. Discuss the accuracy of the following statements.

a Charge must be piling up at the junction.

b The conduction electrons in B are moving more slowly than those in A.

c The conduction electrons in B are moving faster than those in A.

d There are more electrons per cubic metre doing the conduction in A than in B.

e The current in B is less than the current in A.

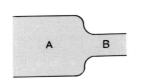

Figure B57

Electrical energy

3(E)a Estimate how much energy is stored in a 1.5 volt 'U2' (D-type) cell.

b Estimate how many joules are needed to set a train in motion.

c Estimate the least time in which a train of mass 500 tonnes could attain a speed of 30 metres per second on level track taking 100 amperes from a 20 000 volt supply.

d An electric train runs at its maximum power of 2.5 MW. What current does it use if it is supplied at
i 2500 V?
ii 25 000 V?
iii If the current in each case runs through supply cables, by what factor would the power wasted in heating the cable be greater for *i* than for *ii*?
iv For the same wastage in the two cases, how much longer might the cable be in case *ii* than in case *i*, assuming the same area of cross-section of cable in each case?

4(R) A lamp B_1 is connected to the terminals X and Y of a battery as shown in figure B58(a).
A second, identical, lamp B_2 is then added to the circuit as shown in figure B58(b). The battery has negligible internal resistance and does not run down.

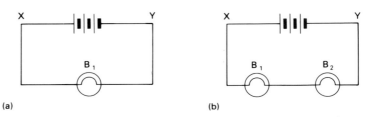

(a)   (b)

Figure B58

The following are statements about the circuits.

1 The number of joules transformed per second is the same in each circuit

2 The number of joules expended per coulomb by the battery is the same in each circuit

3 The number of coulombs flowing out of the battery per second is the same in each circuit.

Which of the above statements is/are correct?

A 1 only   **B** 2 only   **C** 3 only
D 1 and 2 only   **E** 1 and 3 only

(Coded answer paper, 1971)

5(E)a What current should flow in the heating element of a domestic electric kettle in order to boil, in 5 minutes, the water for making a pot of tea? Make sensible estimates of any quantities you need.

(Short answer paper, 1978, part question)

b A bathroom shower heater is rated at 8 kW. It is fed by cold water at a temperature of 15 °C. If the shower temperature is not to exceed 40 °C:
i What mass flow rate does this requirement represent? Is this an upper or lower limit?
ii What sensors and control actions would have to be part of the system to take account of fluctuations in the water inlet temperature and flow rate?
(Specific heat capacity of water $= 4200 \, \mathrm{J \, kg^{-1} \, K^{-1}}$)

6(E) The two passages below were written by someone who knows something about the ideas involved, but who is confused and puzzled about them.

You are asked to write a short explanation about each. Your explanation should first point out any mistakes, or ambiguities, or important features and then go on to make a more accurate and complete statement to help the author to understand.

a You can't really understand electrical potential difference except by way of a model. If you think of a current as like water pumped round a circuit of pipes, then the potential difference is like the pressure of the pump. It's a bit mysterious; for example, the electrical unit, the volt, is nothing like the unit of pressure (newtons per square metre), but this model is the nearest you can get to saying what it means.

(Long answer paper, 1978)

b Some people think that what they pay the Electricity Board for is electricity; that lights, cookers, and heaters use up electricity. They know that a complete circuit is needed, but imagine more electricity coming in than going out. They have the idea that a power station manufactures electricity which is piped like water along wires. They may find it hard to see how an alternating current supply can 'supply' anything if the current is going backwards and forwards all the time.

(Long answer paper, 1979)

Voltage–current relationships

7(P) Figure B59 shows the results of measurements of the current I through and the potential difference V across four different electrical components, each concealed inside a box having only two terminals. What can be said about the contents of the boxes, using only the information conveyed by the graphs? The graphs are all drawn to the same scale, even though no markings are shown.

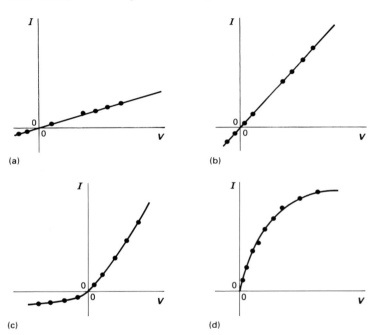

Figure B59

8(P) Here are some values of the potential difference V across a device, and the resulting current I through the device.

V in volts	0	50	100	150	200	250
I in amps	0	0.18	0.25	0.31	0.36	0.40

Which one of the relationships **A** to **E** agrees best with these results?

A $V=kI$ **B** $V=kI^2$ **C** $V=kI^3$
D $V=k/I$ **E** $V=k/I^2$

(Coded answer paper, 1982)

9(E) Part **a** of this question is about interpreting graphs, which is a useful thing to be able to do. Part **b** is about making graphs to test a suggested rule. Part **c** is more philosophical, and considers when it is reasonable to discuss laws and whether they need always be true.

Here is a statement of Ohm's Law.

'The current between two points in a conductor is proportional to the potential difference between these points, provided that physical conditions such as temperature remain constant.'

a Which of the graphs of experimental results in figure B60 could be taken as agreeing with the Law as stated?

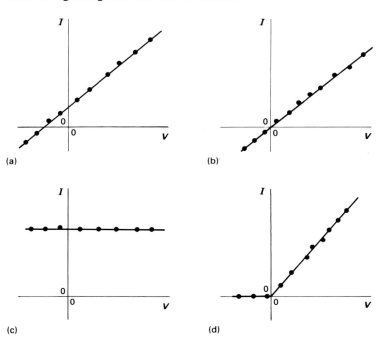

(a)   (b)

(c)   (d)

Figure B60

Current/mA	Potential difference/V
1.0	14.5
1.5	21.8
2.0	29.0
2.5	36.2
3.0	43.5
3.5	50.7
4.0	58.0

Table B2

b Table B2 gives some results taken for a sample of a new material. Does the relation between current and potential difference agree with Ohm's Law?

Can you think of any other ways, apart from the one you used, to make this test? What is the resistance of the sample when the current is 2.0 mA? Is the resistance the same when the current is 4.0 mA?

c There are materials (thyrite, metrosil) for which the current–p.d. graph is curved as in figure B61, even though they do not become hot during the experiment. In the light of this fact and of what you already know, what do you think of the following statements?

1 Ohm's Law, as stated, is a false law, as there are exceptions to it. The Law should be rejected.

2 Ohm's Law is a useful summary of the behaviour of some materials, but not all.

3 There are materials that do not obey Ohm's Law, and it would be best to keep the word 'conductor' only for those that do. Then the Law would always be true.

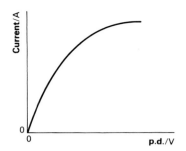

Figure B61

4 Over small enough ranges of current, even a curved current–p.d. graph will be approximately straight. So it would be better to say that, experimentally, Ohm's Law is true over small ranges of current only, to a good approximation. (Look at figures B60(a) and (b) before making up your mind.)

5 I would prefer not to call it a 'law' but to keep that word for statements that always work exactly.

10(L) Thick wires conduct better than thin ones. Because a thick wire could be thought of as several thin ones side by side, it is useful to begin thinking about the effects of thickness by considering resistors connected side by side, in parallel. A p.d. of 12 volts drives current through two resistors in parallel as in figure B62.

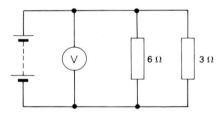

Figure B62

a What current goes through the 6 Ω resistor?

b What current goes through the 3 Ω resistor?

c What is the total current taken from the battery?

d If the two resistors were replaced by a single resistor, which conducts the same current as the total through the 6 Ω and the 3 Ω resistors together, what would be its resistance?

e Now try repeating **a–d** algebraically. Let the p.d. be V and the two resistors R_1 and R_2. If the parallel combination can be replaced by a single resistor, R, you should end up with the useful result

$$\frac{1}{R} = \frac{1}{R_1} + \frac{1}{R_2}$$

11(P) The graph in figure B63 shows the relationship between the current and the potential difference for two different conductors A and B.

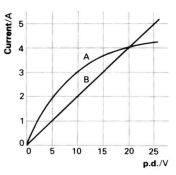

Figure B63

B

a Describe how
  i the resistance of A varies with the potential difference across it.
  ii the resistance of B varies with the potential difference across it.

b If A and B are joined in series and a current of 2 A passes through them, what is the total resistance of the combination? Explain how you arrive at your answer.

c If A and B are joined in parallel and draw a current of 3 A from the supply, what is the potential difference across the combination? Explain how you arrive at your answer.

d Within the range of values shown on the graph in figure B63 will the *lowest* resistance of a *parallel* combination of A and B occur at a high value or at a low value of the potential difference across the combination? Explain your reasoning.

(Short answer paper, 1974)

12(E) This question is to be done as an experiment: detailed instructions on carrying it out will be given to you.
   The circuit shown in figure B64 has been set up using a 3.0 V power supply, in an attempt to measure the resistance of the lamp at different values of potential difference, varying from 0.2 V to 2.8 V as measured by the voltmeter.

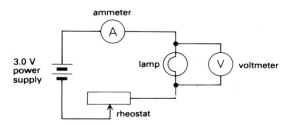

Figure B64

a Switch on the supply, adjust the rheostat, and note that the required range of potential differences *cannot* in fact be achieved.
(Do not make any other changes to the circuit.)
Write down
1 the maximum value, and
2 the minimum value of the potential difference obtainable across the lamp and their corresponding currents and lamp resistances.
i Max. p.d. .... current .... lamp resistance ....
ii Min. p.d. .... current .... lamp resistance ....

b Explain why this circuit, using a power supply which was set to 3.0 V before connecting up the circuit, does not enable the p.d. across the lamp to:
i go down to 0.2 V
ii go up to 2.8 V

c Estimate a value for the maximum resistance of the rheostat and show how you arrive at your estimate.

(Practical problems paper, 1982)

Resistivity

13(L) Figure B65 shows the electrical resistivity of a number of materials, and whether the resistivity rises or falls with a rise in temperature. The height of each bar represents the resistivity.

a Notice that the scale is a peculiar one, with equal *multiples* of resistivity spaced out evenly along it. The difference in the height of the bars for silicon and germanium is roughly the same as the difference between those for germanium and carbon. What can you say about the relative sizes of the resistivities of carbon, germanium, and silicon?

b Suppose the printer had not put 10^{-6}, 10^{-3}, 1, 10^{3}, 10^{6}, etc., along the scale but had printed the powers -6, -3, 0, 3, 6, and so on, instead. What quantity would now be plotted on the vertical scale?

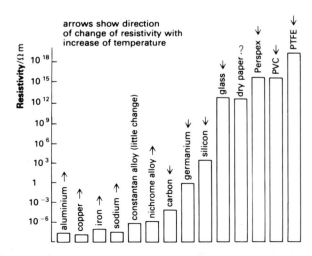

Figure B65

c Estimate roughly how high the bar for PTFE – poly(tetrafluoroethene) – would be if the scale were conventional, with each millimetre representing an increase of resistivity of 10^{-8} ohm metre (so that the bar for copper would be about 1 mm high).

14(E) Table B3 shows three groups of materials, X, Y, and Z. There follow some suggestions about the possible nature or uses of such materials. For each suggestion, pick the group or groups to which it *might* apply, on the information given.

Resistivity at 20°C/Ω m

		Resistivity	
X	silver	1.6×10^{-8}	
	copper	1.7×10^{-8}	all increase with
	aluminium	2.7×10^{-8}	rise in temperature
	mercury	69×10^{-8}	
Y	graphite	350–6500×10^{-8}	all decrease with
	germanium	0.47	rise in temperature
	silicon	2.3×10^{3}	
Z	Pyrex glass	10^{12}	all decrease with
	paraffin wax	10^{14}	rise in temperature
	polystyrene	10^{15}	

Table B3

a They contain many electrons which are free to move.

b Practically no charged particles at all are free to move.

c When it becomes hotter, increased atomic vibrations somehow free more charged particles so that they can move.

d When it becomes hotter, increased atomic vibrations somehow get in the way of moving charge and obstruct its flow.

e The number of charge carriers free to move is much less than for a metal, but is more than for a typical 'insulator'.

f Some might be useful as the wrapping for a submarine cable.

g A piece the size of a pea will pass a current between $10\,\mu A$ and $10\,mA$ under a p.d. of one volt (one of the group might pass more).

h Electrical properties would be significantly affected by surface moisture.

i Some could be used to make high-current shunts for meters.

15(P)a A tube containing a column of mercury passes a current of 0.1 A when connected to a source of p.d. of low resistance. If all of this mercury is now put into a tube which has twice the radius of the first tube, what current will flow when this tube is connected to the same source of p.d.?

b A given length of a conductor, of rectangular cross-section, has a resistance R. If every linear dimension were halved what would the new resistance be? What relevance might this result have to the problem of putting resistors on to microchip circuits?

16(L)   This question is about the way the resistance of a wire changes when it is stretched. If the wire has resistivity ρ, length l, and radius r, then its resistance, R, will be given by

$$R = \frac{\rho l}{\pi r^2}$$

If it is stretched, l will increase, and also r decreases: both of these factors will increase the resistance.

a   This part of the question calculates an approximate expression for the increase in resistance, δR, caused by increases in length, δl, and in radius, δr. From the formula above, we can write the new resistance as

$$R + \delta R = \frac{\rho}{\pi}\left(\frac{l + \delta l}{(r + \delta r)^2}\right)$$

If $\delta r \ll r$ (so that $(\delta r)^2$ can be neglected in comparison with r^2), show that δR is approximately equal to

$$\frac{\rho}{\pi}\left(\frac{r\delta l - 2l\delta r}{r^3}\right)$$

b   This rather complicated result can fortunately be simplified if we think about *proportional* changes. If we divide the expression in **a** by R, we shall obtain the fractional increase in R, $\delta R/R$. Show that

$$\frac{\delta R}{R} = \frac{\delta l}{l} - 2\frac{\delta r}{r}$$

What is the connection between this formula for fractional changes and the original formula for R? (Hint: look at the powers of l and r in the formula.)

c   For many metals the ratio $\left(\dfrac{-\delta r/r}{\delta l/l}\right)$ is constant and equal to about $\frac{1}{3}$. (This fraction is called Poisson's ratio; it varies somewhat from one material to another. Note the negative sign, indicating a reduction in radius when stretched.) Show that for a constantan wire, with a Poisson's ratio of $\frac{1}{3}$,

$$\frac{\delta R}{R} = \frac{5}{3}\left(\frac{\delta l}{l}\right)$$

A constantan wire of resistance 500 Ω is subjected to a strain of 1 %. By how much does its resistance change?

d   A wire used in this way is called a *strain gauge*. What useful property for measurement purposes does the formula in part **c** show? The reading in this *Guide*, pages 111 and 112, describes one important application of strain gauges.

17(L)   In question **1** you showed that if a current I flows through a conductor of cross-section A, in which there are n carriers per cubic metre, the velocity v of the drift of carriers with charge Q is:

$$v = \frac{I}{AnQ} = \frac{I}{A} \times \frac{1}{nQ}$$

(I/A is often called the 'current density'.)

a   If a length of wire has a resistance R, and there is a p.d. V across it, then

$$I = V/R$$

Write an equation for I/A in terms of V, R, and A.

b   If the length of the wire is l, and the resistivity of the material is ρ, then the resistance is

$$R = \frac{\rho l}{A}$$

Write a new expression for I/A without R in it.
(*Check*: A should have vanished.)

c   In a wire of copper, resistivity 1.7×10^{-8} ohm metre, when there is a potential difference of, say, 10 volts across 10 metres of wire (1 volt per metre), what is the current density in amperes per square metre?
   Check: this seems big, doesn't it? What is the current through a wire of cross-sectional area $1\,mm^2$ at this current density?
   Do you think that a piece of copper is often arranged in a circuit with 1 volt per metre across it?

d   As was said above, the carrier velocity v is, on average

$$v = \frac{I}{A} \times \frac{1}{nQ}$$

You now have a value for I/A. Assuming that the carriers in copper are electrons, use $Q = 1.6 \times 10^{-19}$ coulomb, to find the value of the constant in

$$v = \frac{\text{constant}}{n}$$

when the wire has 1 volt per metre across it.

e   Suppose that there is just one conduction electron for each copper atom (a very risky guess). Then n is the same as the number of atoms in a cubic metre of copper.
   Calculate the value of n from these data:
   1 1 cubic metre of copper has a mass of 9.0×10^3 kg.
   2 63.5 kg of copper contain 6.0×10^{26} copper atoms.

f   Now find the average drift velocity, v, of the electrons in metres per second, combining the answers to **d** and **e**.

Circuits and meters

18(P)a–e In the figures below, A, B, C, and D are four terminals of a box containing resistors connected between the terminals. The resistors are not shown, being inside the box and not visible. All the resistors have constant resistance and there are no diodes, lamps, or other components with non-linear characteristics. The terminals C and D are joined by a wire of low resistance, as shown. The battery has a potential difference of 6 V and has negligible resistance.

Say what you can about the contents of the box as a result of each test shown in figures B66(a) to (e).

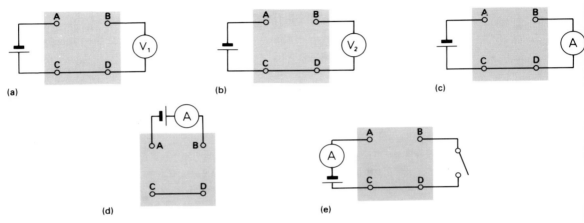

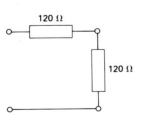

Figure B66

a V_1, a high-resistance voltmeter, reads 6 V. (Figure B66(a).)

b V_2, a voltmeter of resistance 1000 Ω, reads 4.5 V. (Figure B66(b).)

c A is a milliameter, and reads 18 mA. (Figure B66(c).)

d A again reads 18 mA. (Figure B66(d).)

e When the switch is open, A reads zero. (Figure B66(e).) When it is closed, A reads 18 mA again.

Taken together, what do all these tests suggest might be the circuit inside the box?

19(R) Two 120 ohm resistors are connected as shown in figure B67 between four terminals. In the circuits of figure B68 (a) to (e), this network is connected to a 6 volt battery of negligible internal resistance. In circuit (a) and (b), a very high resistance voltmeter is shown connected in two ways to the circuit. In circuits (c), (d), and (e), an ammeter of very low resistance is shown connected in three ways to the circuit.

Beside each meter is written a prediction of its reading, which may or may not be correct.

Figure B67

120 Ω

120 Ω

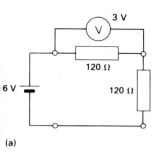

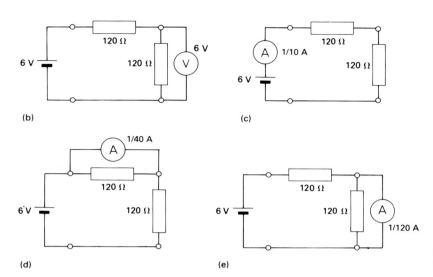

Figure B68

a Which one prediction is correct?

b Which one prediction is just twice as large as it should be?

(Coded answer paper, 1979)

20(R) A student wanted to light a lamp labelled 3 V, 0.2 A, but only had available a 12 V battery of negligible internal resistance. In order to reduce the battery voltage the student connected up the circuit shown in figure B69(a). The student included the voltmeter – using it rather stupidly – so that the voltage could be checked before connecting the lamp between A and B. The maximum value of the resistance of the rheostat at CD was 1000 Ω.

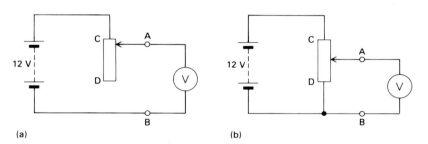

Figure B69

a The student found that, when the sliding contact of the rheostat was moved down from C to D, the voltmeter reading dropped from 12 V to 11 V. What was the resistance of the voltmeter?

b The student modified the circuit as shown in figure B69(b), using the rheostat as a potentiometer, and was now able to adjust the rheostat to give a meter reading of 3 V. What current would now flow through the voltmeter?

c Assuming that this current is negligible compared with the current through the rheostat, how far down from C would the sliding contact have been moved?

d The student then removed the voltmeter and connected the lamp in its place, but it did not light. How would you explain this? (The lamp itself was not defective.)

(Short answer paper, 1979)

21(E) The starter motor of a particular car draws about 100 A from the 12 V battery of the car. If the 100 W, 12 V car lamps become much dimmer, but still stay alight when the starter motor is running, which of the following is the most reasonable rough estimate of the internal resistance of the car battery?

A $0.001\,\Omega$
B $0.05\,\Omega$
C $0.5\,\Omega$
D $5\,\Omega$
E $100\,\Omega$

(Coded answer paper, 1977)

22(P) The p.d. across the output terminals of a source of electrical energy connected to a variable resistor was measured for different values of current as the resistance, R, was varied. The following results were obtained:

p.d./V	10	9	8	7	6	5	4	3	2	1
current/A	0	1	2	3	4	5	6	7	8	9

What is the internal resistance of the source? Plot a graph of the resistance in the external circuit (R) against the output power as ordinate. What general conclusion do you draw from the graph?

Potentiometers and related circuits

23(R) In the circuit shown in figure B70, resistors X and Y are connected to a 6 V battery which has negligible internal resistance. A voltmeter which has resistance R is connected across Y.

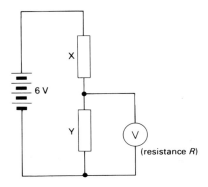

Figure B70

Which one of **A** to **E** below correctly gives the reading of the voltmeter

a if X and Y both have resistance *R*?

b if X has resistance *R* and Y has resistance *R*/2?

    A zero
    B between zero and 3 V
    C 3 V
    D between 3 V and 6 V
    E 6 V

(Coded answer paper, 1981)

24(P) In the circuit in figure B71, what is the current in the low-resistance galvanometer? (The batteries have negligible resistance.)

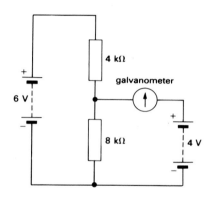

Figure B71

    A 2 mA
    B 1 mA
    C 0.5 mA
    D 0.4 mA
    E zero

(Coded answer paper, 1980)

25(P) If the meter in the circuit (figure B72) reads zero, what is the e.m.f. of battery X? (The 6 V battery has negligible internal resistance.)

    A 2 V
    B $1\frac{13}{17}$ V
    C $1\frac{1}{2}$ V
    D $\frac{4}{5}$ V
    E $\frac{4}{7}$ V

(Coded answer paper, 1979)

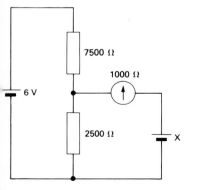

Figure B72

26(R) Two resistance wires, A and B, made of different materials, are connected into a circuit with identical resistors R_1 and R_2 ($R_1 = R_2$), a sensitive high-resistance galvanometer G, a cell C, and a switch S, as shown in figure B73.

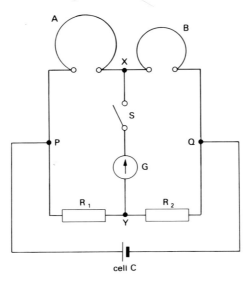

Figure B73

a If A and B have equal resistance, no current will flow through G when the switch S is closed even though the cell is still delivering a current. Explain why this is so.

b The diameter of A is twice that of B and the resistivity of the material of which B is made is 6×10^{-6} $\Omega\,$m. It is found that for zero current through G the length of A has to be three times that of B. Calculate, showing your working, the resistivity of the material of which A is made.

c If the length of wire B is now reduced by a small amount so that the current through G is no longer zero, say which way the current will flow through G and explain why.

(Short answer paper, 1981)

27(L) The circuit (figure B74) shows a 4 kΩ potentiometer which is nearly balanced with a 1.5 V cell. You are asked to apply Kirchhoff's circuit laws to find the current through the galvanometer. Both cells have negligible internal resistance. Draw a large copy of the circuit.

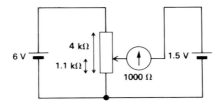

Figure B74

a Decide which way the current flows through the meter (think of the potential at the sliding contact). Mark it in on your circuit.

b Suppose the current drawn from the 6 V supply is I_1, and the galvanometer current is I_2. What current flows in the part of the potentiometer below the sliding contact?

c Kirchhoff's Second Law refers to complete loops in a network. How many loops can you identify in the circuit? How many do you need to find the two unknown currents I_1 and I_2?

d Apply the Second Law to as many loops as you need to produce equations containing I_1 and I_2. Eliminate I_1 from them to find the galvanometer current I_2.

Charge and capacitors

28(I) Suppose a battery supplies a current of two amperes for 10 hours, the change of current with time being shown in figure B75.

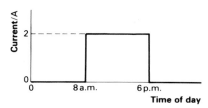

Figure B75

a Describe how the current changes with time.

b What quantity of electric charge flows past each place in the circuit every second during the ten hour period?

c What is the total amount of charge which has flowed around the circuit during the ten hour period?

d Copy figure B75 and show how the charge which has passed from 8 a.m. to some time before 6 p.m. can be represented on the graph.

e A damaged battery produces a graph of current against time like that in figure B76. Show how the charge passed can be represented on the graph as in part **d**.

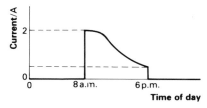

Figure B76

f Estimate the amount of charge which flowed around the circuit.

29(R)   This question is about the circuit in figure B77.

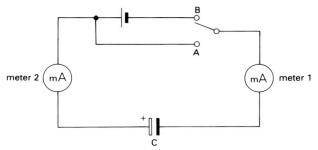

Figure B77

a   When the switch is moved from A to B, meter 1 moves momentarily 5 divisions to the right and returns to zero. Which way and by how much will meter 2 move?

b   When the switch is moved back to A, how do the pointers on the meters move?

c   What evidence is there, from the meter readings, that no charge is conducted through the insulation of the capacitor?

d   Mark on copies of the diagram the direction of the current flow around the circuit when the capacitor is charging (switch in position B) and discharging (switch in position A).

e   What is the evidence that the charges on the plates of the capacitor are equal in size but opposite in sign?

30(P)   When the flying lead is moved from point A to point B in figure B78, the trace on the oscilloscope is as shown in figure B79.

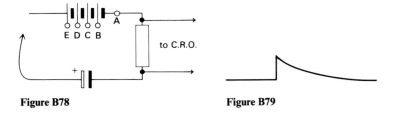

Figure B78   **Figure B79**

What will be seen on the screen of the oscilloscope in the following sequences of changes? Sketches of the traces indicating the height of the pulse and the direction are required for each change. The capacitor is initially discharged.

a   *i* from A to B
ii from B to A
iii from A to C
iv from C to A
v from A to E
vi from E to A

b   *i* from A to B
    ii from B to C
    iii from C to D
    iv from D to E
    v from E to A

c   How do the observations made during sequence **b** lead to the conclusion that the charge on the capacitor, Q, is proportional to the p.d., V, across the plates?

31(L)   A constant current of 1 mA flows for 100 s onto one plate of an uncharged capacitor connected to a battery in a circuit. At the end of this time the p.d. across the capacitor is found to be 10 V.

a   What current flows off the other plate of the capacitor?

b   What is the charge on one plate after 100 s?

c   How many coulombs are needed on one plate to give a p.d. of 1 V?

d   What is the capacitance of the capacitor?

e   If the plates of the fully charged capacitor were allowed to touch, what would now be the total charge on the plates?

32(P)   A 100 μF capacitor is charged to 10 V.

a   How much charge is there on one plate of the capacitor?

b   The charging of the capacitor could be thought of as the removal of electrons from the positive plate and the transfer of the same number of electrons by the battery to the negatively charged plate. How many electrons have moved around the circuit in charging the capacitor? (Charge on the electron, $e = -1.60 \times 10^{-19}$ C.)

c   The capacitor is then discharged through a variable resistor – figure B80(a). Describe how you would have to vary its resistance in order to obtain a discharging graph like the one in figure B80(b).

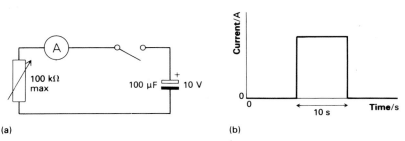

(a)   (b)

Figure B80

d   If you did succeed, what would be the value of the discharging current if the capacitor ended up uncharged?

e   Sketch a graph of how the potential difference across the plates of the capacitor changes during the ten seconds discharge time.

33(R) A teacher is given a box of unmarked capacitors. In order to try to find their approximate values, the teacher constructs the circuit shown in figure B81(a).

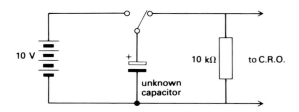

(a)

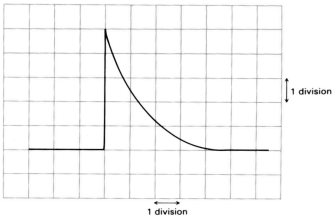

1 division

1 division

(b)

Figure B81

When the capacitor is charged up to 10 V and discharged through the 10 kΩ resistor, the resulting oscilloscope trace is as illustrated in figure B81(b).

The time-base is set at 0.1 s per division, and the voltage sensitivity at 2 volts per division.

a What is the maximum discharge current?

b Estimate the quantity of charge which flows around the circuit during discharge.

c Use the equation $C = Q/V$ to estimate the value of the unmarked capacitor.

d Using a similar scale to figure B81(b), sketch the traces which would be observed on the oscilloscope if the capacitor had:
i half the capacitance of the original;
ii twice the capacitance of the original.

34(L)   *Capacitors in parallel*
This question explains how capacitors in parallel add up.

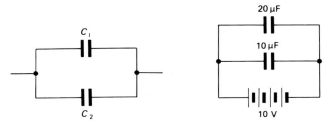

Figure B82   Figure B83

a   What charge is stored on a 10 µF capacitor connected to a 10 V supply?

b   What charge is stored on a 20 µF capacitor connected to a 10 V supply?

c   In figure B83 what is the potential difference across the 20 µF capacitor when it is fully charged?

d   What is the potential difference across the 10 µF capacitor when it is fully charged?

e   What is the total charge stored in the arrangement in figure B83?

f   What single capacitor would store the same charge as the two together?

g   Is the following statement correct? 'Capacitances in parallel add up because their charges add up and the p.d. across each capacitor is the same.'

35(L)   *Capacitors in series*
The following question explains how to calculate the total capacitance of the combination in figure B84.

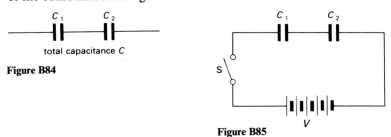

Figure B84

Figure B85

a   When switch S is closed both capacitors start to charge.
i In a series circuit can the current flowing in one part of the circuit be larger than in another part?
ii Can the current flow in one part of the circuit for a longer time than in another part? (Think of Kirchhoff's Laws.)

b   Can the capacitors have different charges on their plates? Explain your answer.

c If the charge that flows round the circuit is Q, what is the potential difference, V_1, across the capacitor C_1 in terms of C_1 and Q?

d What is the potential difference V_2 across capacitor C_2?

e By Kirchhoff's Second Law the sum of the potential differences around the circuit must equal the battery e.m.f. Show that

$$V = \frac{Q}{C_1} + \frac{Q}{C_2}$$

and hence that

$$\frac{V}{Q} = \frac{1}{C_1} + \frac{1}{C_2}$$

f If V is the p.d. across the whole circuit and Q is the charge flowing around the circuit, what is the capacitance, C, of the whole circuit?

g You should now be able to show that

$$\frac{1}{C} = \frac{1}{C_1} + \frac{1}{C_2}$$

h Discuss the following:
Capacitors connected in series must have the same charge on their plates, and the p.d.s across both capacitors must add up to the supply voltage, therefore

$$\frac{1}{C} = \frac{1}{C_1} + \frac{1}{C_2}$$

The result that the total capacitance of two capacitors connected in series is less than either capacitance is often useful in constructing electronic circuits.

36(P) This question is about the circuit in figure B86.

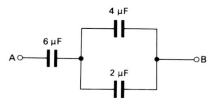

Figure B86

a What is the total capacitance between points A and B?

b It is found that as the p.d. across AB is increased, the insulating material between the plates of the 4 µF capacitor suffers electrical breakdown (it starts to conduct and its capacitance falls to zero). What is the equivalent capacitance now?

Energy in capacitors

37(L) Electric charge is stored on a capacitor, and as indicated in figure B87, at 20 V the charge stored is 0.2 C.

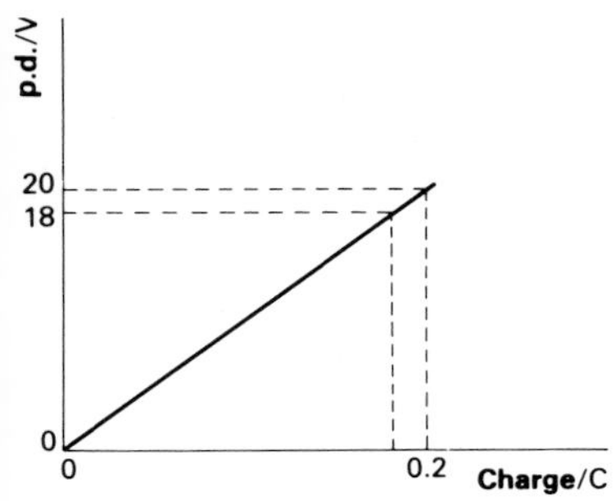

a What is the capacitance?

b If the p.d. across the capacitor dropped to 18 V, how much charge would have flowed off the capacitor?

c During this discharge the p.d. was first a little above 19 V, and then a little below. How much energy was transformed? Shade, on a copy of figure B87, an area that represents this energy.

d If the p.d. drops from 18 V to 16 V the same charge will flow. Will the energy transformed be the same? What will it be? (Take an average p.d. again.)

e On your copy of figure B87, shade in an area which represents the total energy transformed when the p.d. drops from 20 V to 16 V.

f By considering the energy transformed every time the p.d. drops by 2 V, calculate the total energy transformed when the capacitor discharges.

g The total energy can also be calculated using the formula $\frac{1}{2}QV$. Why are the two answers the same? What area on figure B87 represents this total energy?

38(P)a Calculate the energy that can be stored in a 10 000 µF capacitor capable of being charged to a p.d. of 30 V.

b This capacitor is then discharged through a photographic flash tube. If the power output of the flash tube is 500 W, for how long does the flash last?

39(E)a A student is designing an electric car. He has heard that capacitors store electrical energy and starts some calculations on the size of capacitance and charging voltage required. He jots down some data:

Mass of car   500 kg
Top speed   30 m s^{-1}
Largest capacitance available   100 000 µF

    Calculate the charging voltage required if the capacitor is to supply enough energy to accelerate the car from rest to its maximum speed.

b Comment on your answer and the feasibility of using capacitors to store large amounts of electrical energy.

40(E) Figure B88 shows a plan view of a model intended to represent a capacitor in a circuit.

Figure B87

B

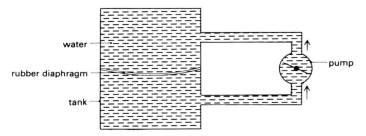

Figure B88

The pump pushes water into the top of the tank, which is divided into two compartments by a leakproof rubber diaphragm. The whole model is filled with water. When the pump is running the diaphragm bulges as shown.

a A student says that the device doesn't store water but that the capacitor does store electricity, so it is a poor model. Comment.

b Does the model store energy?

c How would you modify the model to represent an increase of capacitance?

d For a capacitor, charge is proportional to potential difference. What quantities would you measure to investigate whether an analogous relationship holds for this model?

e A small leak occurs in the rubber diaphragm. What would be the analogous situation for a capacitor?

Exponential changes

41(P) When a 100 µF capacitor is charged though R_1, as shown in figure B89(a), the graph of current against time looks like the one in figure B89(b).

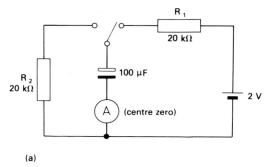

(a)

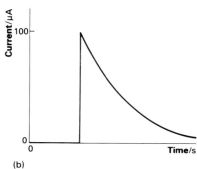

(b)

Figure B89

a Sketch the graph of current against time when the switch is now connected to R_2 (assume the resistance of the meter is small).

b Now sketch the graphs you would expect if the process were repeated with $R_1 = R_2 = 40\,k\Omega$. Remember that the area under the graph represents the total charge, which must be the same as before, but that the currents will be different.

42(P) A 400 μF capacitor is discharged through an unmarked resistor. Table B4 shows readings of current against time.

Time/s	Current/μA
0	50
10	39
20	30
30	23
40	18
50	14
60	8
70	6
80	5

Table B4

a Plot a graph of current against time and estimate the value of the resistance.

b To what p.d. was the capacitor originally charged?

c Estimate the quantity of charge on the capacitor using the area under the current–time graph. How does this compare with the value obtained by using $Q = CV$?

d Sketch the graph which would have been obtained if the capacitor had been initially charged to 15 V.

43(E) Table B5 shows the monthly figures for sales and rent of video recorders during 12 consecutive months.

	January	February	March	April	May	June	July	August	September	October	November	December
1000s of sets	1.5	1.64	1.8	1.96	2.19	2.36	2.57	2.8	3.06	3.33	3.8	4.3

Table B5

a Plot a graph of monthly sales and rentals against time.

b How many sets were rented or sold during the year?

c Find the ratios of sales in one month to the sales in the previous month. For example, June/May, $\dfrac{2.36}{2.19} = 1.08$.
What do you notice about these ratios?

d How do you account for the anomalies in November and December?

e Why do you think sales of video recorders have increased exponentially?

f Plot a graph of lg (monthly sales or rental) against time. Use the graph to predict how many months would pass before the monthly totals top the 100 000 mark. Is this prediction likely to be accurate, and if not, what other factors need to be taken into account?

44(E) A relay is a magnetic switch. Two contacts are closed when the coil creates a sufficiently strong magnetic field (figure B90). The smallest current through the coil which will close the contacts is 0.1 mA. The contacts can be used to switch a current of up to 4 A.

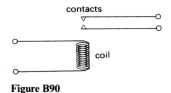

Figure B90

B

Invent a circuit based on a capacitor, variable resistance, and a relay for use as a photographic enlarger timer. For making exposures it is required to switch on the enlarger lamp for times ranging from 5 to 15 seconds. (Give appropriate values and state any assumptions you make.)

Electrons

45(P) The beam current in a cathode ray tube is 1 mA. How many electrons hit the screen in

    a one second?

    b one microsecond?

    c 10^{-12} s?

    d Could the tube be used to test the suggestion that the charge is in fact carried by particles with a finite charge?

46(P) How fast is an electron going if it has an energy of

    a one electronvolt?

    b 100 electronvolts?

47(E) Estimate the energy emitted from the screen of a television set. (Compare it with the brightness of a lamp of known power – preferably a fluorescent lamp.) If the electrons in the tube are accelerated by a p.d. of 20 kV, estimate the minimum current required to maintain the picture. How many electrons per second per square millimetre of the tube face does this give?

48(L) The results shown in table B6 were obtained with a Millikan apparatus, using oil drops. For each of several drops, the experimenter measured the potential difference across the plates (which were 4.42 mm apart) at which each charged drop was just held poised against the gravitational pull of the Earth. The weight of each drop was found by calculating it from the observed steady speed of fall in the air between the plates.

    a Suppose an object with charge Q lies between the plates and the plates have a potential difference V across them. How much energy would be transformed if the charge Q moved from one plate to the other?

    b Suppose the plates are a distance d apart, and a steady force F is exerted on the charge. How much energy would be transformed if an object were moved under force F from one plate to the other?

    c If the pull, F, of the charged plates on the charge, Q, balances the weight, W, of the drop, the drop will be held at rest. Express Q in terms of W, V, and d. Check that C (coulomb) is the unit of measurement for your expression for Q.

    d Find the charge on each drop in the results in table B6.

e Try to identify a basic charge e, such that each charge Q is some whole number multiple n of e ($Q=ne$).

f Given that other experimenters have shown that Q is always accurately a whole number multiple of $e=1.6 \times 10^{-19}$ C, what do you think about the accuracy with which this experimenter was determining the balancing p.d.s and drop weights?

V p.d. to balance drop/V	W Weight of drop/N	Q Charge on drop/C	n Multiple
470	5.05×10^{-14}		
820	5.90×10^{-14}		
230	3.35×10^{-14}		
770	2.85×10^{-14}		
1030	3.65×10^{-14}		
395	7.00×10^{-14}		

Table B6

B

Unit C
DIGITAL ELECTRONIC SYSTEMS

Mark Ellse
Emanuel School, London

David Grace
Eaglesfield School, London

SUMMARY OF THE UNIT

'If the aircraft industry had evolved as spectacularly as the electronics industry over the past 25 years, a jumbo jet would cost £300 and it would circle the globe in 20 minutes on 20 litres of fuel.'

Adapted from: Hoo-min D. Toong and Amar Gupta 'Personal computers'.
Copyright © 1982 by *Scientific American, Inc*. All rights reserved.

C

SUMMARY OF THE UNIT

INTRODUCTION

Electronics provides a very powerful tool for the physicist and the engineer. Today, few physicists work without the aid of a microcomputer to control experiments and to read, store, process, and analyse data from experiments. Computers provide a means of developing and testing theoretical models, and to use them effectively requires some understanding of the basic processes that take place within them.

The circuits studied in this Unit are at the heart of the microprocessor (the integrated circuit which has made the microcomputer possible). Together with linear circuits (which you will learn more about in a later Unit), the microprocessor has made a big impact on our lives. It has made the industrial robot a reality; it is revolutionizing communications systems; in the home it controls the modern washing machine.

In the experiments in this Unit the logic gates that you will investigate are treated as black boxes. It does not matter whether you use circuits based on single transistors or integrated circuits. We will be concerned with how the output of a box depends on the input, not with what goes on inside the box.

You are asked to study the behaviour of each gate and then to make systems that will solve various problems. You will need some basic electric circuit theory, for example, the use of potential dividers and the action of capacitors in circuits.

When you attempt to solve the problems set in this Unit you are being asked to behave like an engineer. The task of engineers is to solve real problems, inventively using the resources available.

To begin with your resources may be rather limited, but they will grow as you design 'new' circuits.

When tackling a problem, engineers rarely start with equipment in front of them. The solution begins as an idea, perhaps inspired by knowledge and experience gained from earlier problems. These ideas may develop in discussion with other engineers, and will probably soon reach expression on paper. Only when a clear plan has been produced will engineers put their ideas into action. You are encouraged to follow that example.

Section C1   COMBINATIONAL LOGIC

Electronic systems

Everyone knows how to buy a hi-fi system. You either go for a 'music centre' which contains all the parts you need, or for 'separates'. Examining electrical circuits is a bit like that. You either look at the system as a whole or you look at bits of the system. Designing in electronics is a matter of putting parts together to produce a system which does the required job.

QUESTIONS 1 to 5

The parts of a music system are shown in figure C1.

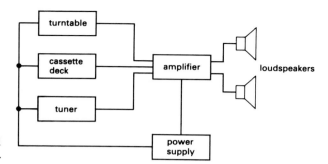

Figure C1
A music system.

When deciding on a hi-fi system it sometimes helps to consider each part individually. In the same way this can be helpful in designing electronic circuits.

A digital clock (figure C2) is a more complicated system. It uses the 50 Hz a.c. mains as a frequency standard. After being reduced to a suitable size (not shown in figure C2) the alternating voltage is changed from a sine wave to a square wave because square pulses are easier to count. Dividing by 50 gives us one pulse per second. The counter registers each pulse and the associated decoder and display circuits give a readout as a decimal number. The first counter counts from 0 to 9 and then starts again, having sent on a pulse to the next counter. The next counter counts up to 5, and then resets and passes on a pulse, and so on. Each counter passes its count at any instant to the display.

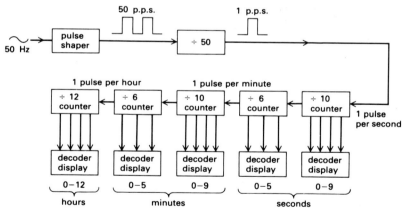

Figure C2
A digital clock.

This circuit could be made out of many separate transistors or a few integrated circuits. Nowadays it would be much more common to find all the components on a single integrated circuit (IC or 'chip'). All you need is a power supply, the 'clock' chip, a display, and a board to mount them on and you've got a clock.

To make a digital watch we need to use an internal oscillator to provide the reference frequency instead of the mains. The oscillator frequency can be very precisely controlled by a quartz crystal.

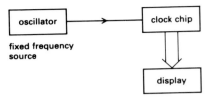

Figure C3
A block diagram of a digital watch.

Single integrated circuits that perform complicated tasks are very much the order of the day in modern electronics. Engineers look at what the circuit can do and can then fit it into a system. Few people beyond the designer of the chip, know or care very much about how it works in any detail.

Digital systems

QUESTIONS 6 and 7

Binary arithmetic is the language of computers and of many electronic systems. Each 'bit' (BInary digiT) can either be 1 or 0, and can be stored as such in the computer's memory. 1 and 0 are the only states possible for a binary digital system. Groups of bits can represent data or instructions upon which the computer operates.

A wide variety of information can be represented by a 'two state' signal: whether a circuit is open or closed, on or off, whether a signal is absent or present, whether an event has occurred or not, and so on. In digital electronics we represent the digits 0 and 1 by low and high voltage levels (figure C4). If we examine the signals that are passed between various parts of a microcomputer, we would see a pattern of '1's and '0's. We might not be able to see the transitions between these states as these often occur extremely quickly, in times measured in nanoseconds ($1 \text{ ns} = 10^{-9}$ s).

DEMONSTRATION C1
Digital signals

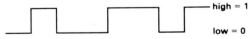

Figure C4

In digital electronics we generate digital outputs from circuits usually in response to digital inputs.

Unit I,
'Linear electronics,
feedback and control'

Much of the real world is not digital. Temperature variations are usually continuous, as are changes in light intensity and the alternating voltage output from a microphone. When dealing with such signals you can either process them as continuously variable voltage signals, or convert them to digital signals and then process them in this form. One group of circuits that is of particular interest is those that convert linear signals to digital signals and vice versa. 'Analogue to digital converters' and 'digital to analogue converters' are found in many microcomputer systems.

Of the two systems we examined earlier, the hi-fi system is linear. However, digital methods are being used more and more as superior techniques of processing digital signals are developed – for instance, digitally encoded discs that are read by a laser. The digital clock is of course digital.

Another example of a digital signal involves no integrated circuits. The nervous system of the body is 'digital'. For example the 'rod' cells on the retina of the eye send digital messages to the brain. In response to the stimulus of light, the rod cells produce electrical impulses which pass along nerve fibres to the part of the brain that interprets vision. These electrical impulses are all the same size and shape. There are no small impulses and large impulses. If light of greater intensity falls on the rods they respond by producing pulses more rapidly, and vice versa.

The lower limit to the amount of light that the eye can see is the amount of light that stimulates one nerve impulse. Less light will not produce an impulse at all.

This picture is a little simplified; there is more than one rod in the eye. In fact one pulse from a single rod is not enough for the brain to 'see'. Some recent experiments have shown the brain 'sees' when about ten rods each send a pulse to the brain.

Our understanding of computer systems suggests new ways of thinking about the brain. It is not unusual to find that understanding of systems of one kind, electronic systems say, can be applied to systems in apparently unrelated fields.

C

Gates and truth tables

$$
\begin{array}{r}
10110101 \\
+01100001 \\
\hline
1\ 00010110 \\
\end{array}
$$

carry sum

EXPERIMENT C2
The digital electronics kit

In the heart of a microcomputer is a microprocessor which can perform binary arithmetic. For example, it can add two eight-bit binary numbers and generate an eight-bit sum and a carry. Another task may be to compare two numbers to see which is the larger; another to take a number in binary form and to display it as a decimal number on an L.E.D. (light-emitting diode) display. All these tasks can be achieved by combinations of simple devices called 'gates'. Gates are basic components in digital electronics.

The inverter

EXPERIMENT C3
Investigating a single-input gate

A very simple gate is the inverter or NOT gate. It has one input and one output. When the input is low, the output is high; when the input is high, the output is low. The diagram of the inverter and a table that describes its behaviour (the truth table) are given in figure C5.

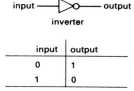

input ——▷o—— output

inverter

input	output
0	1
1	0

Figure C5

The NOR gate

EXPERIMENT C4
Investigating a NOR gate

The output of a two-input NOR gate is high when neither one inpu NOR the other input is high; that is, the output is only high when bot inputs are low (figure C6).

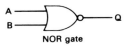

NOR gate

inputs		output
A	B	Q
0	0	1
1	0	0
0	1	0
1	1	0

Figure C6

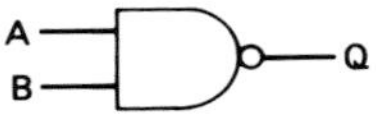

NAND gate

inputs		output
A	B	Q
0	0	1
1	0	1
0	1	1
1	1	0

Figure C7

The NAND gate

EXPERIMENT C5
The behaviour of the NAND gate

QUESTIONS 8 to 12 and 31 and 32

A two-input NAND gate and its truth table are shown in figure C7.

NAND means 'NOT AND'; the output Q is NOT high when input A AND B are high.

Both the NOR and the NAND gates are inverting gates – hig outputs only occur with low inputs. It is convenient and very commo practice to make inverters out of NOR or NAND gates. If the tw inputs of each of these gates are joined together (figure C8) their trut tables become the same as that of the inverter.

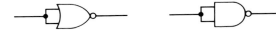

Figure C8
Making an inverter from a NOR or a NAND gate.

Other gates

EXPERIMENT C6
Designing more gates

There are several other useful gates with two inputs. The symbols an truth tables of some are shown in figures C9 to C11.

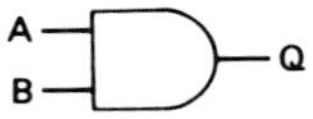

AND gate

inputs		output
A	B	Q
0	0	0
1	0	0
0	1	0
1	1	1

Figure C9

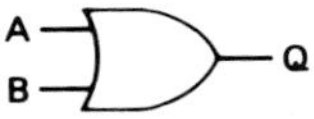

OR gate

inputs		output
A	B	Q
0	0	0
1	0	1
0	1	1
1	1	1

Figure C10

Exclusive OR gate

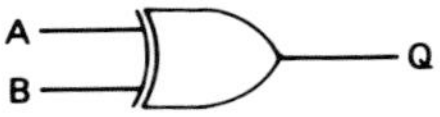

inputs		output
A	B	Q
0	0	0
1	0	1
0	1	1
1	1	0

Figure C11

The non-inverting gate (figure C12) can be used to isolate one part of a system from another.

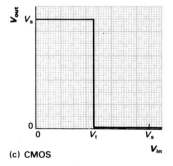

Non-inverting gate

input	output
0	0
1	1

Figure C12

Hardware and software

Although it seems a large jump from these two-input logic gates to microprocessors, this jump is not as large as you may think. Among the many other functions that it can perform, a microprocessor can do the same job as simple logic gates.

EXPERIMENT C7
Using a microcomputer to carry out logic functions

A microprocessor has to be programmed: the program is a list of instructions and data. The instructions are written in code ('machine code'); the set of possible coded instructions for a particular type of microprocessor (such as the Z80 or the 6502) is called its 'instruction set'. Among the instructions that most have are those that instruct the processor to perform OR, Exclusive OR, and AND functions on two pieces of data, and NOT (or inverse) on one piece of data. The data may be in the program or it may be somewhere in the memory of the computer.

QUESTION 18

A programming language such as BASIC can be used to instruct a computer to carry out simple logic functions. These functions are part of the microprocessor's instruction set.

Logic gates: a more careful look at input and output voltages

QUESTION 19

So far we have described voltages as being either 'high' or 'low'. Sometimes it is useful to know in more detail how the output voltage depends on the input voltage to the gate. This is easiest to investigate with an inverter, since it has only one input. These 'characteristics' depend on the type of gate used. Three types are likely to be encountered, based on:

i discrete components, for example, the 'basic unit'
ii TTL (Transistor Transistor Logic) integrated circuits
iii CMOS (Complementary Metal Oxide Semiconductor) integrated circuits.

EXPERIMENTS C8a and b
Measuring the characteristics of an inverter

Figure C13 shows approximately how each type of inverter behaves. V_s is the supply voltage.

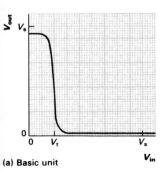

(a) Basic unit

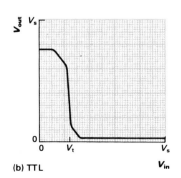

(b) TTL

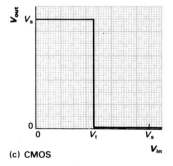

(c) CMOS

Figure C13
Characteristics of inverters.

The similarities between the graphs are more important than their differences. For all the inverters, an input lower than a certain value produces a high output and vice versa. There is a transition value of the input V_t, around which the output changes between high and low. This value is avoided in most logic circuits: the input voltage is fixed well below or well above V_t.

QUESTION 19

EXPERIMENT C9

Making a light-operated switch

QUESTIONS 20, 21, 22, 31, 34

If we use one light-sensitive resistor and one fixed resistor in a potential divider that controls a gate, we can use a light level to control a logic circuit – we can make a light-dependent switch.

The system shown in figure C14 gives a high output when the light-dependent resistor (L.D.R.) is dark. The output could be made to go high in the light either by exchanging the positions of the L.D.R. and the other resistor, or by replacing the inverter with a non-inverting gate.

The gate can be controlled by some other physical parameter (for example, temperature) if the L.D.R. is replaced by some other sensor (for example, a thermistor – a resistor sensitive to temperature).

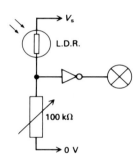

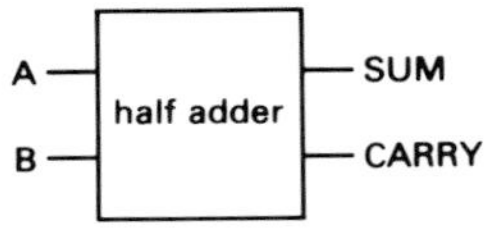

inputs

A	B	SUM	CARRY
0	0	0	0
0	1	1	0
1	0	1	0
1	1	0	1

Figure C14
Light-dependent switch (for simplicity the 0 V connection to the indicating lamp is not shown).

Figure C15
A half adder.

Binary addition

A microprocessor performs arithmetical and logical functions in a section called the Arithmetic and Logic Unit (ALU).

The rules of binary addition are:

0 plus 0 equals 0
0 plus 1 equals 1
1 plus 0 equals 1
1 plus 1 equals 0, carry 1.

Figure C15 shows a circuit that can add two binary digits together, and its truth table.

EXPERIMENT C10

Making a half adder

Apart from the 'CARRY' column this table is the same as the truth table for the Exclusive OR gate. The circuit in figure C16 is one way of producing this result. It is called a 'half adder'.

The half adder is all right for adding two digits together, but most calculations are more complicated!

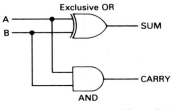

Figure C16
A half adder circuit.

Suppose we wish to add two numbers, A_1A_2 and B_1B_2 (for example 01 and 11).

$$A_2A_1 \quad A_1 + B_1 = S_1 \text{ with the CARRY } C_1$$
$$B_2B_1 \quad A_2 + B_2 + C_1 = S_2 \text{ with the CARRY } C_2$$

$$\overline{C_2S_2S_1}$$

$$\overline{}$$

$$\begin{array}{c} \swarrow\!\!\!-1+1=0, \text{ CARRY } 1 \\ 0\ 1 \\ 1\ 1 \\ \nwarrow\!\!\!-0+1+1=0, \text{ CARRY } 1 \end{array}$$

$$\overline{}$$

$$1\ 0\ 0$$

$$\overline{}$$

QUESTIONS 23 and 24

EXPERIMENT C11
Making a full adder

One half adder can add A_1 and B_1; but then we have to consider the CARRY from the first addition, C_1, as well as adding A_2 and B_2. A circuit which overcomes this problem is called a 'full adder' and is shown in figure C17.

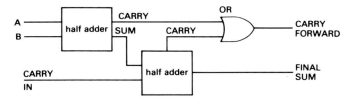

Figure C17
A full adder.

EXPERIMENT C12
Using a microcomputer to perform binary addition

Physics Reader
Microcomputer circuits and processes

In order to carry out the addition $A_2A_1 + B_2B_1$ we would need the circuit shown in figure C18.

All that is needed to add two binary numbers with more bits is an extra full adder for each extra bit. The addition is always started from the right (the least significant digit) as in ordinary arithmetic.

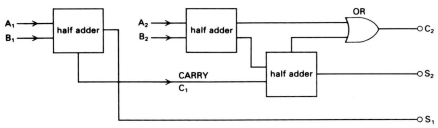

Figure C18
A circuit to add two two-bit numbers ($A_2A_1 + B_2B_1$).

Section C2 **SEQUENTIAL LOGIC**

Making pulses

EXPERIMENT C13
Making pulses

If the switch in the circuit shown in figure C19 is moved from 0 V to V_s the input to the gate is made high and then falls exponentially at a rate determined by the values of C and R. The output of the gate will change state when the input passes from low to high or from high to low. When an inverter is used as in the diagram, the pulse produced is 'negative going' – figure C20(a) – that is, the output is normally high but goes low for the period during which the input is high (more than V_t, the transition voltage).

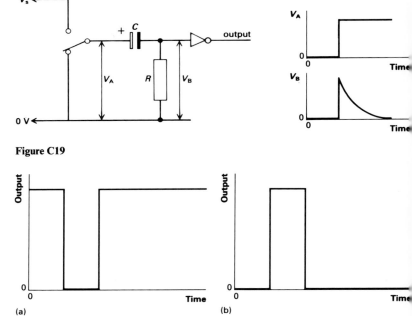

Figure C19

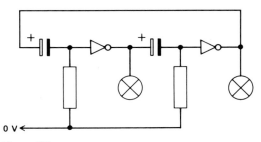

Figure C20
(a) Negative going pulse.
(b) Positive going pulse.

A positive going pulse – figure C20(b) – may be made by replacing the inverter with a non-inverting gate.

The astable circuit

QUESTIONS 25 and 26

EXPERIMENT C14
The astable circuit

If two pulse producers are connected together, one after the other with the output of each stage fed to the input of the other (figure C21), then each stage will send a pulse to the other stage which sends it back to the first. The circuit is said to be astable (unstable), and the frequency with which it oscillates depends on the values of the resistors and capacitors.

Figure C21
An astable circuit.

In practice, other astable circuits are normally used but they have their frequency controlled by an *RC* circuit in a similar way. An astable circuit may be used for many forms of timing if the pulses from it are counted, or for generating audible signals.

The bistable circuit

This circuit, as its name implies, has two stable states. It is made by connecting a pair of inverters together as shown in figure C22. If the input to the first is high, its output and therefore the input to the second will be low. Hence the output of that will be high, keeping the input to the first high. The circuit is also stable with the input to the first low.

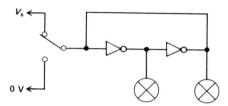

Figure C22
A bistable circuit.

This circuit can be used as a memory. One of the states represents the binary digit 0, the other representing binary 1. Large numbers of bistable circuits are used in the memories of computers.

A clocked bistable module, as provided in the electronics kit, can be used to divide by 2. Each pulse going into the module changes it from one state to the other. So *two* pulses are needed to get the bistable module from one state to the other and back to the first, that is, to get one pulse out.

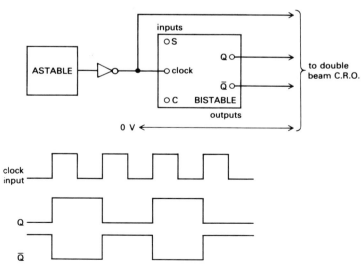

Figure C23
Investigating the bistable module.

Section C3 DESIGNING DIGITAL SYSTEMS

EXPERIMENT C18
Designing digital systems

The basic devices and principles of digital electronics are the basis of many modern communication and control systems, from turning on a lamp through the programming of a domestic washing machine to the automatic control of complex manufacturing processes. Most computers and industrial robots use the same digital ideas.

You can start exploring some of these possibilities with the digital electronics kit.

READING

DIGITAL SOUND

(Adapted from J. Borwick *The Gramophone guide to hi-fi*, David & Charles, 1982.)

Introduction

The future of every aspect of sound recording and reproduction is dependent on digital techniques.

Put simply, digitalization consists of replacing the audio signal waveform with a stream of pulses encoded to represent numbers. These numbers each correspond to the amplitude of the signal waveform sampled instant-by-instant during the recording process. For playback, the pulse stream is transformed back to instantaneous signal amplitude values, so reconstituting the original waveform. The considerable circuit complexity which this technique requires at the recording and reproducing end of any audio chain has become a practicable reality only because of the accelerating development of integrated circuits and microprocessors.

Binary numbers

At the heart of these developments is the use of a binary scale of numbers. Table C1 shows how the binary scale works in the particular case of a 'word' of four binary digits ('bits').

Decimal	Powers of 2	Binary	Decimal	Powers of 2	Binary
0		0000	8	2^3	1000
1		0001	9		1001
2	2^1	0010	10		1010
3		0011	11		1011
4	2^2	0100	12		1100
5		0101	13		1101
6		0110	14		1110
7		0111	15	(2^4-1)	1111

Table C1
Decimal and binary scales compared.

Our four-bit word can have sixteen different values (of which one is 0). In general a word-length of n bits provides 2^n values as shown on table C2. This has a special significance in sound recording, as we shall see.

Analogue-to-digital conversion

In digital sound recording, the signal to be recorded starts off as an electrical imitation or 'analogue' of the original sound waveform (see

Number of bits, n	Range of values, 2^n	Signal-to-noise ratio/dB*
1	2	6
2	4	12
3	8	18
4	16	24
5	32	30
6	64	36
7	128	42
8	256	48
9	512	54
10	1 024	60
11	2 048	66
12	4 096	72
13	8 192	78
14	16 384	84
15	32 768	90
16	65 536	96

Table C2
Range of values provided by different word lengths.

figure C24). To convert this to digital form, the amplitude of the waveform is sampled at regular intervals of time and each sample is encoded and stored as a binary number corresponding to its value on a fixed amplitude scale. In the four-bit words used, only sixteen different binary code numbers are possible (as we have seen from table C2).

The digitally encoded signal is made up of pulses for the symbol 1 and blank spaces for 0, the system being referred to as pulse code modulation (p.c.m.). The process of allocating each sample a binary code number is called quantizing, and clearly involves some approximation since the analogue waveform will usually be at some intermediary

*Note on the decibel scale

The decibel scale is a method of expressing the ratio of two powers.

If voltages are known one calculates $20 \log_{10} (V_1/V_2)$.

For example, the ratio between 12 V and 6 V is

$$\frac{12\,\text{V}}{6\,\text{V}} = 2$$

In decibels

$$20 \log_{10} \left(\frac{12\,\text{V}}{6\,\text{V}} \right) = 20 \log_{10} 2 \, \text{dB}$$
$$= 20 \times 0.30 \, \text{dB}$$
$$= 6 \, \text{dB}$$

The signal-to-noise ratio of a digital signal is the ratio between the number of bits that represent a signal level and the maximum error (noise) that could be present (1 bit).

So for a 16-bit system, the signal-to-noise ratio is

$$\frac{2^{16}}{1}$$

or

$$20 \log_{10} \frac{2^{16}}{1} \, \text{dB}$$

$$= 96 \, \text{dB}$$

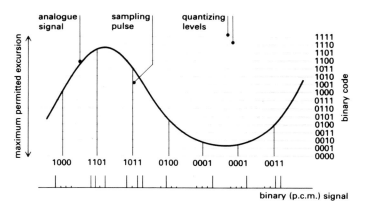

Figure C24

In digital recording, the analogue waveform is sampled at fixed time intervals and each sampled level is expressed or 'quantized' in binary code.

amplitude when sampled. This approximation leads to 'quantization noise' but this can be reduced to very acceptable proportions by using more bits. This is confirmed by the last column of table C2 which lists the signal-to-noise ratio obtainable for p.c.m. systems using up to sixteen bits.

Provided the sampling frequency is more than twice the highest sound frequency to be recorded, it has been found that the waveform can be recorded and reproduced to high standards of accuracy. If an audio bandwidth of 15 kHz is considered adequate, for example, a sampling frequency of 32 kHz might be chosen. Where fidelity up to 20 kHz is required, the sampling frequency might be 44 kHz or 50 kHz. To record or transmit p.c.m. digital signals, the system requires a bandwidth at least equal to the sampling frequency multiplied by the word-length in bits. Thus a sixteen-bit system with a sampling frequency of 50 kHz would need a bandwidth of 800 000 Hz per channel – well beyond the capabilities of ordinary sound recording equipment.

Digital evolution

One of the first important applications of digital audio techniques was the B.B.C.'s construction of a p.c.m. network for distributing radio programmes to its various transmitting stations in the late 1960s. Though this was only a thirteen-bit system with a 32 kHz sampling frequency (audio bandwidth is in any case restricted to 15 kHz on VHF/FM radio) it produced a dramatic improvement in broadcast quality. The important advantages of digital sound which appeared in this application included the following:

Reduced noise Whereas noise is a continual nuisance in analogue distribution networks, becoming worse with each mile of landline or radio link travelled, it can effectively be ignored in the digital system. The digital-to-analogue converter at the receiving end of the line has only to recognize the presence or absence of pulses, not gauge their relative amplitudes, to be able to reconstitute the sound signal almost perfectly. The B.B.C.'s thirteen-bit system gives a signal-to-noise ratio of 78 dB regardless of network distance (see table C2) – much better than analogue.

Reduced distortion   Normal analogue networks introduce considerable non-linear and phase distortion, whereas the digital process restores the original waveform practically undistorted.

Wide frequency response   The full 15 kHz range is preserved in the digital system, whereas high-frequency losses were almost inevitable in analogue lines.

Around 1972, the Nippon Columbia Company of Japan (Denon) began p.c.m. sound recording, using professional videotape recorders to provide the necessary bandwidth. The conventional analogue LP records which they produced from these digital master recordings showed a degree of clarity compared with discs manufactured from analogue tape masters. In addition to the three advantages of the digital process mentioned above, the new master tapes possessed the following

No wow-and-flutter   Because the replay system relied on an accurate quartz clock to synchronize the sampling rate on replay with that recorded, the usual short-term pitch fluctuations normally associated with tape (and disc) systems were eliminated.

No tape modulation noise   Analogue tape introduces several types of noise such as granular hiss, modulation noise, and print-through These are all avoided in digital recording.

Improved channel separation   As the left and right stereo signals are encoded separately, very little interchannel crosstalk takes place.

With such a formidable array of advantages, it is not surprising that most of the major record companies and independent studios soon followed the Nippon Columbia lead and installed digital machines for master recording. The designs varied, some using videotape equipment and others data storage units. This made it difficult to exchange recordings between studios, since each digital master could only be played back on the same type of machine on which it had been recorded. However, the in-house benefits were considerable. Masters did not deteriorate in store and copies could be virtually identical to the original.

By 1980, many of the LP records coming on to the market bore the word 'Digital', and the companies stressed the advantages of the new recording technique in their advertisements and sleeve-notes. It was however, perfectly obvious – and often reiterated by the critics – that the different digital mastering machines gave inconsistent sound quality that indeed these early digital recorders had limitations as well as advantages. These limitations included:

Increased distortion at low levels   While distortion in analogue recording increases at high signal levels, the opposite occurs in digital. Resolution deteriorates at low quantizing levels leading to quite severe distortion. Techniques exist for disguising this effect, so that it is barely audible under most conditions. Nevertheless, degrees of acoustic dryness and tonal hardness were noticed in many early digital records ings, suggesting that the reverberant 'tail' of musical sounds, and the harmonics of certain instruments, notably strings, were not being reproduced cleanly.

Bandwidth restrictions All professional digital recording systems used a sampling frequency in excess of 40 kHz, thus permitting an audio bandwidth of 20 kHz. This might seem to meet all hi-fi requirements but it is a feature of p.c.m. encoding that a very sharp filter must be included to cut off all frequencies above the chosen upper limit. Such filters must be designed very carefully.

Other practical limitations of the early digital equipment affected the way in which it was used. Editing was more difficult, for example, so that musicians were encouraged to make longer 'takes' with fewer stoppages. Ironically, this gave rise to some favourable comment from critics, who felt that editing had previously become too prevalent in the recording studios. Again, digital masters were mainly two-track only, so that balance engineers often felt obliged to go back to a simpler microphone technique and mix-down straight to two-track stereo. This was in contrast to the multi-microphone multi-track procedures evolved over recent years, yet, as with the simpler editing, it frequently produced a more natural, cleaner sound balance which drew critical acclaim for many of the new 'digital' LP records, even though it was an indirect, rather than a direct, consequence of the adoption of a digital mastering process.

Digital discs

While the so-called 'digital' LP records could, in the best examples, be identified as possessing at least some of the benefits inherent in the original studio master recordings, they were nevertheless cut and pressed in analogue form and by the old traditional methods. They were therefore subject to all the same limitations and just as easily damaged or invaded by dust and static as conventional LPs.

It was obvious that, if all the benefits of digital recording were to be enjoyed in the home, the sounds must be retained in digital form right up to the moment of playback. Research into truly digital disc systems was intensified and soon about half a dozen competing types were being demonstrated at engineering conventions and trade shows. This diversity of incompatible domestic systems, with each disc unplayable except on its own machine type, was much more serious in commercial terms than the similar lack of standardization amongst professional studio equipment. The general public would like to be able to buy records in the shops without having to select a particular format suited to their own player unit. A great deal of confusion already exists in the videotape market, for example, where three quite different incompatible tape cassette formats are in general use at the time of writing – VHS, Beta, and V2000.

The Compact Disc

Pride of place must be given to the laser-scanned Compact Disc developed by Philips in Holland, with subsequent collaboration from Sony in Japan. This is the most technically advanced of the systems so far on offer and has a number of important practical, as well as high quality, features.

As with the earlier Philips invention, the Compact Cassette, the small size of the Compact Disc is already an attractive feature. The disc is only 120 mm (4.7 inches) in diameter and consists of clear PVC 1.1 mm thick. The digital sound signals are pressed into the PVC, on one side only, in the form of tiny pits or indentations measuring a mere 0.6 μm across. The binary code is represented by pits for the 1 symbol and blank spaces for 0. The track spirals outwards from the disc centre and the microscopically small track spacing of 1.66 μm gives up to 60 minutes of stereo music per disc. During manufacture, the stamped out PVC surface is given a thin coating of reflective aluminium and this in turn is overlaid with a transparent 0.15 μm coating. A fixed linear track speed of about $125 \, \mathrm{cm \, s^{-1}}$ is employed and so the rotational speed decreases as the record plays, being about 500 r.p.m. at the centre and 215 r.p.m. at the outer edge.

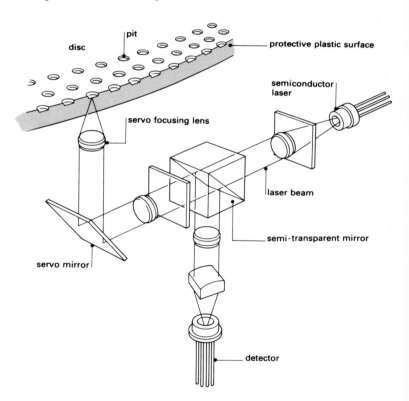

Figure C25
Playback system for the Compact Disc, showing the semiconductor laser light source, the system of lenses and mirrors, and the detector which receives the pulses reflected from the disc surface.

As shown diagrammatically in figure C25, the optical playback system consists of a laser light-source, a series of lenses and mirrors, and a light-sensitive detector. The laser beam is reflected and focused so as to scan the helical track of pits and blank spaces from underneath the disc. As pits and blank spaces are scanned, the laser beam is either scattered or strongly reflected back downwards to be picked up by the photo-electric detector and converted into pulses of electric current. Thus the original digital signals have been reproduced and can be changed back into analogue form by a digital-to-analogue converter built into the player. The stereo electrical output of a Compact Disc player is therefore similar to that from a conventional record-player, tuner, or cassette deck and can be connected to any hi-fi amplifier or domestic system.

Though the mass-production techniques for the Compact Disc are basically similar to those used for vinyl LPs, the small dimensions and close tolerances needed demand extremely high standards of air cleanliness and quality control measures. For the user, however, the Compact Disc is an extremely robust medium. Dust or finger marks are ignored by the laser beam and the normal problems of stylus cleaning and careful pick-up alignment are a thing of the past. Also, since there is no physical contact, there is no wear of the recorded track or the scanning mechanism. The player itself can be quite small, yet the constant linear speed feature allows more flexible access and cueing facilities than the conventional LP. In technical terms, the Compact Disc uses a sixteen-bit format with a sampling frequency of 44.1 kHz. This gives a claimed frequency response up to 20 kHz, immeasurable wow-and-flutter, with dynamic range and channel separation of up to 90 dB.

While the Philips Compact Disc uses the same basic principle of a pitted surface and laser-beam scanning as their 300 mm diameter Laser Vision video disc system, the two media are in all other ways incompatible. Philips and their associated companies are convinced that separate players and different standards are preferable for digital audio and television playback.

Questions

a A flute plays the note A (frequency, $f = 440$ Hz) into a microphone. The output signal of the microphone amplifier, V, varies with time according to $V = V_0 \sin 2\pi f t$. V_0, the maximum value of the output is 1.5 V.

i What is the value of the signal at the following times: 0, 0.1 ms, 0.2 ms, 0.3 ms, 0.4 ms, 0.5 ms, 0.6 ms?

ii Using the four-bit binary number 1000 to represent 0.8 V, express each of the answers to *i* in similar form.

iii In parts *i* and *ii* you have been digitalizing an analogue signal using four bits and a sampling frequency of 10 kHz. Explain, with reference to your results, why more bits and higher sampling frequencies are used in practical systems.

b A microphone amplifier needs an input of 0.5 mV for an output of 1.5 V.
What is its gain
i expressed as a ratio
ii expressed in dB?
Power levels are compared on the dB scale using the formula

$$10 \log_{10}\left(\frac{P_1}{P_2}\right)$$

Current and voltage levels use the formulae

$$20 \log_{10}\left(\frac{I_1}{I_2}\right) \text{ and } 20 \log_{10}\left(\frac{V_1}{V_2}\right)$$

Explain why the figure 10 occurs in the power formula and 20 in voltage and current ones.

c *i* Explain the term 'hi-fi' (high fidelity).
ii Why is digital transmission more faithful than the transmission of an analogue signal?

d *i* Using the figures given in the passage, calculate the maximum length of the recording track on a Compact Disc.
ii Estimate the separation between adjacent tracks on the disc and check that it is consistent with the quoted figure of 1.66 μm.

LABORATORY NOTES

DEMONSTRATION

C1 Digital signals

C1a Simple circuit having two states

mounted bell push
lamp, *e.g.*, 2.5 V m.e.s.
holder
cell holder with two cells
leads

N.B. The lamp must, of course, have a second connection to the power supply, but to simplify the drawing we omit it here and elsewhere. But don't forget it is always included.

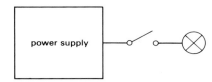

Figure C26

How many possible states does this circuit have?
What are these states?

C1b Slow electronic flashing circuit

digital electronics kit with power supply
leads

In what way is this circuit similar to figure C26?

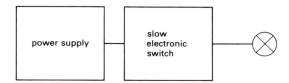

Figure C27

C1c Fast electronic flashing circuit

digital electronics kit with power supply
oscilloscope
leads

How many states does this circuit (figure C28) have?
How can you tell?

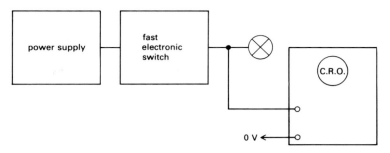

Figure C28

C1d Other sources of digital signals

Many familiar pieces of equipment depend on digital signals. For example, when a single digit, say 6, is dialled on a mechanical telephone dial, a mechanical switch opens and closes six times. A signal consisting of 6 pulses is sent. It could be used to control a lamp as in demonstration C1a above.

Most school scaler timers have sockets at which their internal clock pulses are available.

A microcomputer running a program has digital signals at many parts of the circuit.

EXPERIMENT
C2 Introducing the digital electronics kit

digital electronics kit with power supply
voltmeter appropriate to the power supply
leads

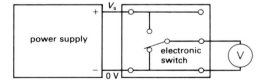

Figure C29

Examine the electronics kit or the manufacturer's instructions to determine the voltage of the power supply required (V_s). Then connect the circuit as shown in figure C29.

The circuit gives two outputs, one that is low (near 0 V) representing the binary digit 0, the other being high (near V_s) representing binary digit 1. Use the voltmeter to get a more accurate value for the 'high' and 'low' voltages.

If your electronics kit has a separate indicator unit, connect this in place of the voltmeter. What happens? What advantage does this have over a voltmeter?

EXPERIMENT

C3 Investigating a single-input gate

digital electronics kit with power supply
leads

Take a NOR gate (which will probably be labelled with the symbol shown in figure C30) and connect an appropriate power supply. Identify the inputs to one gate (usually on the left), and the output. Connect the inputs together as shown to make a system with one input and one output. If your kit has separate indicators, connect one to show the state of the output.

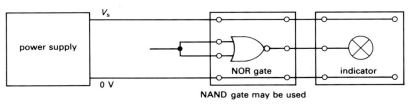

Figure C30
Investigating a single-input gate.

What is the state of the output when the input is low?
What is the state of the output when the input is high?
What happens when the input is unconnected?
Draw a table to record your results.
Suggest an appropriate name for this circuit.

EXPERIMENT

C4 Investigating a NOR gate

digital electronics kit with power supply
leads

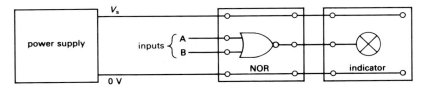

Figure C31
Investigating a NOR gate.

Using the two inputs to your gate, make a 'truth table' that describes the behaviour of the gate. Start by making both inputs low – is the output high or low? Make one of the inputs high. What state is the output now? Continue until you have tried all four combinations of inputs.
What could these gates be used for?
What happens if an input to the gate is not wired directly high or low but left 'floating'?
Why are they called 'gates'?

Figure C32
Simplified version of figure C31.

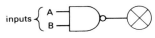

Figure C33
NAND gate.

From now on we shall use the simplified form of circuit diagram shown in figure C32 for all circuits. The power supply lines V_s and 0 V must be connected up, but in general they will not be included in the diagrams. The symbol used for the NOR gate is a standard one. There is a summary of standard symbols on page 498.

EXPERIMENT
C5 The behaviour of the NAND gate

digital electronics kit with power supply
leads

Repeat experiment C4 with a NAND gate to establish its truth table.

EXPERIMENT
C6 Designing more gates

digital electronics kit with power supply
leads

Try to work out solutions to these problems before you attempt to make the circuits. Remember that you can use NAND or NOR gates as inverters.

Suggest a use for each gate that you design.

C6a Design a two-input AND gate whose output is high when one input AND the other are high. This is very easy with NAND gates but harder with NOR gates. Try to solve the problem using both types of gate.

C6b Design a two-input OR gate whose output is high when one input OR the other OR both are high. This time it's easy with NOR and harder with NAND. Again, try both methods.

C6c Design a non-inverting gate whose output is the same as the input. What might this be useful for?

C6d Design a two-input NOR gate using NAND gates. The solution can be achieved with four gates.
or:
Design a two-input NAND gate using NOR gates. Again, it can be done with four gates.

C6e Design a three-input NOR gate using two-input NOR gates. This task can be done with three gates.
or:
Design a three-input NAND gate using two-input NAND gates. Again, this can be done with three gates.

C6f Design a two-input Exclusive OR gate: a gate whose output is high when one input OR the other is high but not when both are. The best solution to this uses two NOR gates and a single AND gate.

EXPERIMENT

C7 Using a microcomputer to carry out logic functions

microcomputer with BASIC language

The following programs demonstrate how the logic functions NOT, OR, and AND work on a microcomputer.

```
10  INPUT A
20  PRINT NOT A
30  GOTO 10
```

```
10  INPUT A
20  INPUT B
30  PRINT A OR B
40  GOTO 10
```

```
10  INPUT A
20  INPUT B
30  PRINT A AND B
40  GOTO 10
```

Try the programs and see what happens. For some microcomputers (for example, Sinclair, Apple), the inputs should be either 0 or 1. For other microcomputers (for example, Research Machines, B.B.C.), the inputs should be 0 or -1.

How can you use the functions NOT, OR, and AND to write a program that will perform **a** the NOR function, **b** the NAND function? Test your solutions.

Suggest a reason why the functions NOR and NAND are not standard in BASIC.

EXPERIMENT

C8a Measuring the characteristics of an inverter

digital electronics kit with power supply
2 voltmeters appropriate to the supply
potentiometer, 1 kΩ
leads

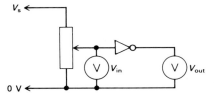

Figure C34
Circuit to measure the characteristics of an inverter.

Set up the circuit in figure C34. Remember that an inverter may be made from a NAND gate or a NOR gate with both inputs connected together as shown in figure C35.

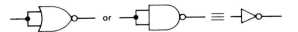

Figure C35
Making an inverter from NOR or NAND gates.

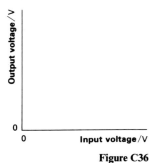

Figure C36

Use the potentiometer to vary the input voltage. For each value of input voltage, measure and record the output voltage. You may wish to change the input voltage in small steps when the output voltage changes. Draw a graph of output voltage against input voltage.

From your graph:

i When the input voltage is high, is the output voltage high or low?

ii Is there a distinct input voltage at which the output suddenly changes? If so, record that voltage.

ii Does the output change from high to low over a small range of input voltages? If so, what is the smallest change in input voltage that causes the output to go from high to low?

DEMONSTRATION
C8b Plotting the characteristics of an inverter on an oscilloscope

digital electronics kit with power supply
signal generator
oscilloscope
diode, *e.g.*, 1N4001
leads

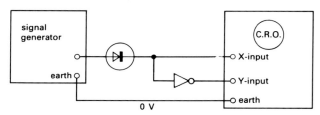

Figure C37
Using an oscilloscope to display characteristics.

The oscilloscope time-base is turned off and the oscilloscope used with the a.c./d.c. switch set to d.c. The signal generator should be set to give a sine wave output of about 100 Hz. The output of the signal generator is connected to the gate input as shown in figure C37. The diode ensures that only inputs of the correct polarity are applied to the inverter. This input is also connected to the X-plates of the oscilloscope. The output from the gate is connected to the Y-plates.

The output of the signal generator is increased to about 3 or 4 V until the voltage characteristic is plotted on the screen. It may be necessary to adjust the gain of the X- and Y-amplifiers and the position of the trace on the screen.

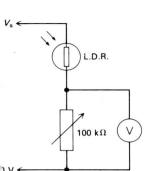

Figure C38
Potential divider with light-dependent resistor.

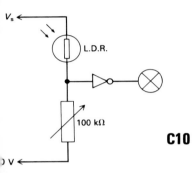

Figure C39
A circuit which turns a lamp on in the dark.

EXPERIMENT
C9 Making a light-operated switch

digital electronics kit with power supply
light-dependent resistor (L.D.R.) *e.g.*, ORP 12
resistance substitution box
voltmeter appropriate to the supply
leads

From experiment C8, determine the potential required at the input of an inverter to change the state of the output. Set up the circuit in figure C38 and investigate briefly how its output varies as the light incident on the L.D.R. varies.

What would you expect to happen if you connected this circuit to the input of the logic gate as shown in figure C39? When you have decided what you expect to happen, try it out.

Some discrepancy between the predicted and observed behaviour may occur. Explain.

The circuit you have made should make the indicator light when the illumination of the L.D.R. falls below a certain level. How would you change this circuit so that the opposite occurs, that is the indicator goes out when the light level falls?

EXPERIMENT
C10 Making a half adder

digital electronics kit with power supply
leads

Make a system to add the two digits A and B together to give a SUM and CARRY digit in binary arithmetic. This is called a 'half adder'.

A + B = SUM and CARRY

$0 + 0 = 0$ and 0
$0 + 1 = 1$ and 0
$1 + 0 = 1$ and 0
$1 + 1 = 0$ and 1

EXPERIMENT
C11 Making a full adder

digital electronics kit with power supply
leads

In experiment C10 you constructed a half adder which will add together two digits. A computer adding two numbers, like 1001110 and 1101101, will add them in pairs of digits starting at the right as usual. Except for the addition of the first pair of digits, any of the subsequent additions may also have to include a CARRY digit from the previous addition.

The block diagram (figure C40) describes what we wish to achieve together with the truth table.

Devise one stage of such a binary number adder, using two half adders and a small extra item.

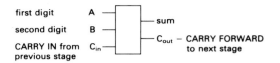

	inputs			outputs	
A	B	C_{in}		sum	C_{out}
0	0	0		0	0
0	0	1		1	0
1	0	0		1	0
1	0	1		0	1
0	1	0		1	0
0	1	1		0	1
1	1	0		0	1
1	1	1		1	1

Figure C40
Block diagram and truth table of a full adder.

EXPERIMENT
C12 Using a microcomputer to perform binary addition

microcomputer with BASIC language

Write a program that will give outputs like the half adder, using only the logic functions you have on your microcomputer (probably only OR, AND, and NOT).

Extension: make a full adder.

EXPERIMENT
C13 Making pulses

digital electronics kit with power supply
capacitor } (1000 µF and 560 Ω are suitable for TTL;
resistor } 47 µF and 10 kΩ for other gates)
leads

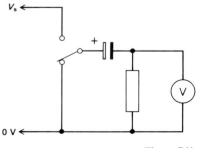

Figure C41

Sketch a graph of the way in which V varies with time when the switch in the circuit shown in figure C41 is moved from 0 V to V_s.

How would you expect a NOT gate to behave when connected to this circuit? Test your prediction. The circuit you need is shown in figure C42.

The circuit shown in figure C42 made a lamp that was normally on go off for a short period. Devise a circuit that will make a lamp that is off go on for a short period.

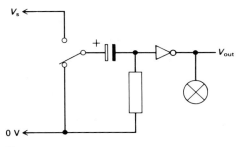

Figure C42

EXPERIMENT
C14 The astable circuit

digital electronics kit with power supply
capacitors ⎫
resistors ⎬ see note to experiment C13
leads

C14a What happens in the circuit of figure C43 when the input is switched from 0 V to V_s? Can you explain why this happens? What will happen if the output from the last stage is sent back to the input of the first?

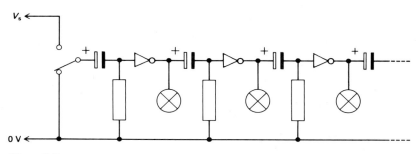

Figure C43

C14b Set up the circuit shown in figure C44. If nothing happens initially, try shorting one of the capacitors momentarily with a flying lead.

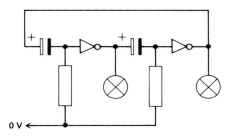

Figure C44

EXPERIMENT

C15 Investigating the astable module

digital electronics kit with power supply
high-impedance earpiece
oscilloscope
leads

Your electronics kit has at least one astable module. Its internal circuit may be very different from your astable circuit, but it is controlled in the same way with an *RC* circuit. The modules vary in the way in which they are controlled, so you will need to consult the manufacturer's data or ask your teacher to find out how to set up your module.

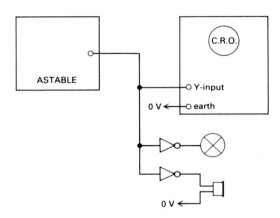

Figure C45
Investigating the astable module.

The astable frequency may be varied over a wide range and used to drive many different circuits. You may be able to get a frequency low enough to see the indicator flashing, and you should certainly be able to get the frequency high enough to produce an audible note in the earpiece. It is good practice to connect a gate between the astable module and anything that you use it to drive, so that the load does not affect the astable module (figure C45).

EXPERIMENT

C16 The bistable circuit

digital electronics kit with power supply
leads

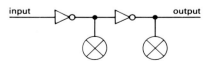

Figure C46

Connect the circuit shown in figure C46. What happens when the input is low? What happens when the input is high?

Connect the output to the input. What happens? How can you change what has happened?

EXPERIMENT
C17 The bistable module

digital electronics kit with power supply
double beam oscilloscope
2 high-impedance earpieces
leads

C17a Check that the module can be made to behave in the same way as the bistable circuit in experiment C16 by connecting each input in turn momentarily to V_s (figure C47). For what could you use this circuit?

C17b What happens to the state of the outputs when the clock input is moved between $0\,V$ and V_s? It helps to use an electronic switch to make these connections (figure C48).

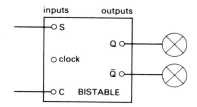

Figure C47
The bistable module.

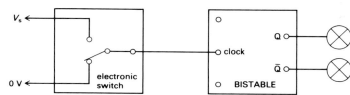

Figure C48
Investigating the clock input.

Alternatively, you can feed in a series of pulses from a slow astable circuit into the clock input (figure C49).

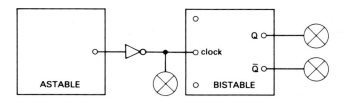

Figure C49
Investigating the behaviour of the bistable module.

Compare the astable module's output with the output from the bistable module. What do you notice?

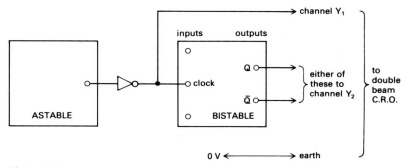

Figure C50
Monitoring input and output of a bistable module.

C17c A double beam oscilloscope can be used to monitor the input and the output of the bistable module as shown in figure C50.

Think of another use for the bistable module.

Why does figure C50 show a gate between the astable module and the input to the bistable module?

C17d Set up the circuit shown in figure C51. Listen to the audible outputs produced in the two earpieces. What do you notice? Give an explanation for what you hear.

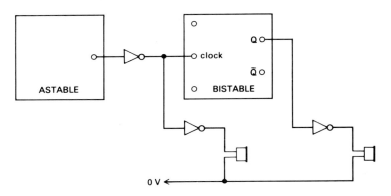

Figure C51
Investigating the effect of the bistable module at an audio frequency.

EXPERIMENT
C18 Designing digital systems

Using the digital electronics kit it is possible to design a very large number of systems to do a wide variety of jobs. The suggestions below will, we hope, start you off. No doubt you will think of others. The idea is for you to do some electronic systems engineering of your own, trying to build a system that does some interesting and potentially useful job.

Of course it is important for you to have a plan for your solution before you start connecting up any circuit.

Some or all of the following apparatus may be required:
digital electronics kit with power supply
various capacitors (between $10 \,\mu F$ and $10\,000 \,\mu F$)
resistance substitution boxes
light-dependent resistor
high-impedance earpieces
thermistor
aerosol freezer
leads

C18a Make a lamp go on for half a second.

C18b Make a lamp flash from an input falling from V_s to $0 \,V$.

C18c Make a Morse code sender.

C18d Make a bleeper that emits an audible pulse of sound when a button is pushed.

C18e   Make a lamp give six flashes in a row.

C18f   Make a device that emits six audible pips in a row.

C18g   Make a device that emits a warning tone when a thermistor's temperature becomes high. Then try a warning of low temperature.

C18h   Use a light-dependent resistor to produce a warning tone when the light level is low. Then try high.

C18i   Make a system that can be used with a counter to time the interval between two successive pushes of a button.

C18j   Make a human response timer to measure the time a person takes to respond to a lamp coming on (or a tone sounding), by pressing a button.

C18k   Make a monostable circuit. Halfway between a bistable and an astable circuit is one that has one stable state but may be momentarily switched to another state for a time. This time is dependent on the values of C and R, not on the time for which a button is pressed.

C18l   Make a binary counter. Use three bistable modules to make a counter to count from 0 to 7 pulses. Extension: modify the first circuit so that if all the lamps are on to start with, incoming pulses will turn them off in the binary sequence from 7 down to 0.

C18m   Make a digital frequency meter that displays the frequency of a supply in digital form (binary form using the electronics kit, in scale of ten using a counter/scaler).

C18n   Make a safety interlock system for a furnace which is to have a start button and a stop button; but the heating is to be turned off if the temperature is too high or if the safety door is open, and must not come on if the start button is pressed under these circumstances.

C18o   Design a circuit that compares two one-bit numbers and gives a high output when they are both the same. This circuit is called a 'comparator', and is usually found within the Arithmetic and Logic Unit (ALU) of a microprocessor.

C18p   Extend your solution to C18o to compare two two-bit numbers.

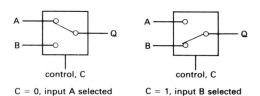

Figure C52

C18q   Design a circuit that will pass one of two digital inputs to an output depending on the state of a single control input (figure C52). This circuit is called a 'data selector' or 'multiplexer'. A multiplexer allows signals from several lines to share a single output line; for example, it allows the

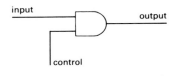

if control = 0, then output = 0
if control = 1, then output = input

Figure C53
AND gate used as a switch.

28 segments of a four-digit display to be controlled with 8 rather than 29 lines (one for each digit and one common to all digits).

Hint: an AND gate may be used as a switch as shown in figure C53.

The 74LS157 is a multiplexer which has four pairs of inputs and four outputs (figure C54).

When SELECT is low the A inputs are passed to the output. SELECT high passes the B inputs to the output. This particular integrated circuit has another input, ENABLE, that allows the user to disable the device totally if necessary. When the ENABLE is high all outputs are forced low.

Notice that only one control line is needed to select one of two sets of inputs.

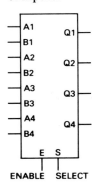

Figure C54
The 74LS157 multiplexer.

The CMOS 4051B is an eight-channel multiplexer. One of the eight one-bit inputs is selected and connected to the output. Three select lines are now required. Each can be high or low, giving a total of $2^3 = 8$ combinations as the truth table in figure C55 shows.

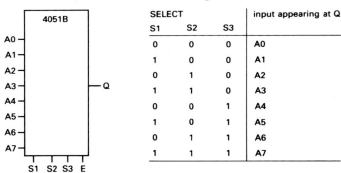

SELECT			input appearing at Q
S1	S2	S3	
0	0	0	A0
1	0	0	A1
0	1	0	A2
1	1	0	A3
0	0	1	A4
1	0	1	A5
0	1	1	A6
1	1	1	A7

Figure C55
The 4051B multiplexer and its truth table.

The ENABLE, E, on this chip performs the same function as the ENABLE on the 74LS157. When it is high the output is held low.

C18r Design a circuit with two inputs and four outputs that obeys the truth table in figure C56. Notice that only one of the outputs is high at any time.

The inputs 00, 01, 10, 11 are the binary equivalents of 0, 1, 2, 3; that is, the circuit is decoding the binary input.

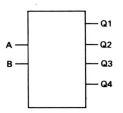

Figure C56

inputs		outputs			
A	B	Q1	Q2	Q3	Q4
0	0	1	0	0	0
1	0	0	1	0	0
0	1	0	0	1	0
1	1	0	0	0	1

Hint: Note that Q4 = A AND B which means you can start your design as shown in figure C57. Look for similar relationships between the other outputs and A and B.

This circuit is called a 'decoder'. It is used in microcomputers and in many other digital circuits. A very common application is activating a 7-segment LED display. An integrated circuit commonly used to do this is the TTL 7447A. The 7447A decodes four input lines to seven outputs to produce the pattern required to activate the display. With four input lines, $2^4 = 16$ different combinations of four inputs are possible, but to represent the decimal numbers 0 to 9, only ten combinations need to be decoded.

The circuit of a 7-segment display and the layout of the segments are shown in figure C58.

Q4

to other gates

Figure C57

C

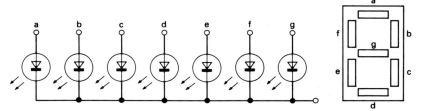

Figure C58
A seven-segment display.

7447A

A
B
C
D

a
b
c
d
e
f
g

Figure C59
The 7447A decoder.

So, if we wished to show the figure '3' for example, segments a, b, c, d, and g would have to be activated.

The inputs to the 7447A would be the 'binary coded decimal' numbers representing the decimal numbers 0 to 9, that is 0000 to 1001.

To display '3' the input would be 0011 and the output would be 1111001, so that all the segments of the display except e and f would be on.

Decoders are also used in microcomputers to select blocks of memory.

C18s   The opposite of a decoder is an 'encoder'. The problem below illustrates the idea. Design a circuit which produces the correct binary equivalent of a decimal number when one of the keys 1, 2, or 3 is pressed. A block diagram and truth table are shown in figure C60.

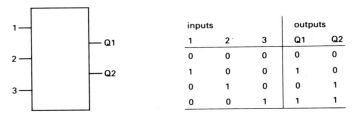

inputs			outputs	
1	2	3	Q1	Q2
0	0	0	0	0
1	0	0	1	0
0	1	0	0	1
0	0	1	1	1

Figure C60
An encoder.

C18t   Make a coincidence detector. In an experiment in nuclear physics, pairs of counts from decays of particles may occur together in time if they have a common origin. Devise a system to indicate if two pulses fall within a fixed time of each other.

C18u   Make a pulse delay system. Start by extending a square pulse by a fixed time. Then add to the system to remove the front part of the extended pulse and so produce a pulse like the original one but delayed by a fixed time.

C18v   Make a traffic lights system with red, amber, and green flashing in the correct sequence.

C18w   Make a ring counter. A way of counting pulses is to make them light lamps in rotation. Ten lamps numbered 0 to 9 make a decade counter. Light three lamps in rotation. Then try four (which can be done with only two bistable modules, since two bistable modules will count up to four in binary arithmetic).

C18x   Playing tunes. The digital electronics kit can be used to play tunes. One can either make a simple organ with a keyboard (switches or push buttons) controlling a series of notes, or one can arrange that a system itself selects notes of the proper pitch, duration, and sequence to play a tune. Try something like the first bars of 'Three blind mice'.

QUESTIONS

Systems

These questions are about systems, and the usefulness of thinking about complicated electronic circuits as systems made up of interconnected parts.

1(P) A system is a collection of parts, in which each part has a definite job to do, all the parts acting together to do whatever task is required of the system.

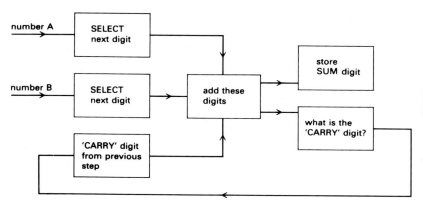

Figure C61
A system for adding pairs of numbers.

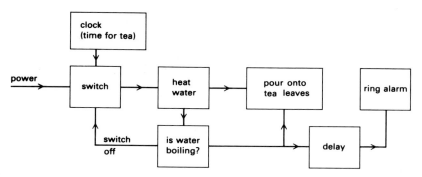

Figure C62
A system for making tea.

To illustrate the idea, figure C61 shows a system for adding up numbers, while figure C62 shows one for making tea automatically.

a Explain how the system for adding numbers works. That is, say what happens step by step. You need not say how a machine might achieve each step: that is another problem.

b Explain how the tea-making system works.

c Draw a similar system for one or more of the following things:
Controlling cars at a pedestrian crossing.
A 'pop-up' toaster.
An automatic washing machine with control of water temperature, washing time, and spinning time.

2(I) Suppose you are listening to a song on a record player; trace the series of things which have happened to the song signal on its way to you.

3(E) Write an essay about 'systems'. What are they, what do they do? What does the word system mean? How can we use the concept of a system?

4(E) Most electronic systems have an input, an output, and a power supply. Why do they need a power supply? What about other systems?

5(I) Almost all digital electronic systems use the binary system. What are the reasons for this?

Digital signals

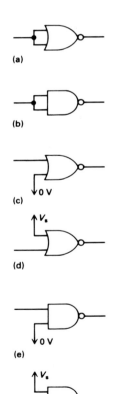

(a)

(b)

(c)

(d)

(e)

(f)

Figure C63

6(P) The following are all types of information you might want to communicate:
i whether a heater is on or off,
ii the temperature of a room,
iii the direction in which a sign points at a junction,
iv the magnitude and direction of an electric current,
v the number of pages in this book,
vi the area of this page,
vii the number of letters on this page.
　Which of these pieces of information could easily be transmitted by a digital signal? Give reasons for your answers.

7(R) A 'bit' is a single BInary digiT (either 0 or 1).

a At a crossroads, there is a set of traffic lights controlled by a central computer. How many bits of information need to be communicated?

b 1 'byte' = 8 bits. 1 kilobyte = 1024 bytes. Why is it 1024 and not 1000?

c Two electronic systems communicate with each other in chunks of one byte. How many different characters can they communicate?

d How many bytes of information are there on this page?

Logic gates

8(P) Figure C63 shows some simple applications of NOR and NAND gates. Draw a truth table for each circuit and describe what each will do.

9(L) Suppose two inverters and a NOR gate are connected as shown in figure C64. Copy and complete the table shown in figure C65, showing what the outputs are for various inputs. What might such a system be used for? Test your answer in the laboratory.

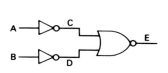

Figure C64

inputs		outputs		
A	B	C	D	E
0	0	1	1	0
1	0	0		
0	1			
1	1			

Figure C65

10(P) Draw up your own tables to describe the behaviour of the circuits shown in figure C66. Some of the circuits are useful. Suggest applications for them.

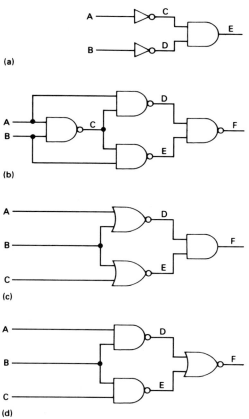

(a)

(b)

(c)

(d)

Figure C66

11(P) Design a digital circuit that gives a high output when two signals are the same (that is, both high or both low).

12(P) Figure C67 shows several two-input gates. In each case a pair of signals is fed into the inputs of the gate. For each gate sketch the signal you would expect to see at the output of the gate.

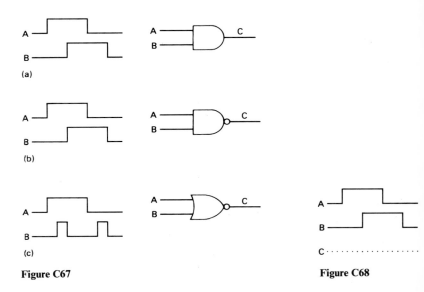

(a)

(b)

(c)

Figure C67

Figure C68

Hint: Copy the input signals shown in figure C67, and draw the output directly below, as illustrated in figure C68.

13(P) A grocer buys a new deep freeze. He wishes to make a battery-powered alarm system that will do the following:
i Light a red light-emitting diode (L.E.D.) if the compressor is running but the temperature is too high.
ii Light a green L.E.D. if the temperature is low and the lid is open.
iii Light a yellow L.E.D. if the mains fails.
Four sensors are used.
Sensor A gives a logic '1' if the compressor is running.
Sensor B gives a logic '1' if the temperature is too high to store groceries.
Sensor C gives a logic '1' if the mains is on.
Sensor D gives a logic '1' if the lid is shut.
Design a logic system that will solve the grocer's problem using

a any gates;

b NAND gates only.

14(R) Baked beans are produced in cans of three different sizes: small, medium, and large. The cans have to be sorted ready for packaging. Three photodetectors are available that give a logic '1' output when a light shines on them; otherwise they are at logic '0'. They could be arranged on either side of a conveyer belt, as in figure C69, so that the cans would cut the light incident on one or more photodetectors.
Design a system that will light an indicator if a medium size can passes but not a small or large can.

15(P) Suppose the customs officials at an airport decide to allow travellers through without inspection as long as they say that they have nothing

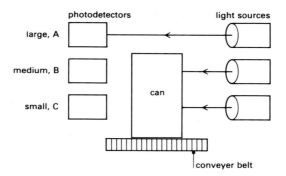

large, A — photodetectors | light sources
medium, B
small, C
can
conveyer belt

Figure C69

to declare, and that they have not flown in from Paris. The decision is to depend on how the travellers press two switches. One is labelled 'Have you arrived from Paris?', the other 'Have you anything to declare?'. Draw a circuit using a logic gate and two switches to light a lamp behind a sign saying 'Proceed'. Must the switch position that corresponds to V_s be marked 'yes' or 'no'?

16(P) The system shown in the answer to question 15 has the defect that the lamp saying 'Proceed' is lit as the passengers walk up to the switches if these are at rest in the 0 V position. Add a third switch to the system so that, after the two questions have been set, the sign only lights up as the passenger walks over the third switch (hidden in the floor), closing it, on his or her way to the sign.

    You have then made a system, 'NEITHER from Paris, NOR with anything to declare, AND has passed the questioning point'.

17(P) Show how you could use an Exclusive OR gate as an 'optional inverter'; that is, to invert a signal or not at the control of another input.

18(P) Look at question **10**. The following BASIC program has the same behaviour as the circuit in figure C66(a).

```
10 INPUT A
20 INPUT B
30 LET C = NOT A
40 LET D = NOT B
50 LET E = C AND D
60 PRINT
70 PRINT E
80 PRINT
90 GOTO 10
```

a Enter the program into a microcomputer and check that it gives the answer you expect.

b Write similar programs to behave in the same way as the circuits in figures C66(b) to (d).

19(P) The following table summarizes some of the most important specifications for three types of logic kit.

	Supply voltage, V_s	Input voltage 'low'	Input voltage 'high'	Output voltage 'low'	Output voltage 'high'
'basic unit'	5 to 6 V	<0.5 V	>1.5 V	≈0.1 V	≈4.7 V
TTL	4.75 to 5.25 V	<0.8 V	>2 V	<0.2 V	>3 V
CMOS	3 to 15 V	<0.2 V_s	>0.8 V_s	<0.01 V_s	>0.99 V_s

Table C3

a You have an unmarked electronics kit. What power supply voltage would you use at first to try it? Explain your answer.

b Why is it important that all the figures in column 4 (output voltage 'low') are less than the corresponding figures in column 2 (input voltage levels 'low')? There is a similar relationship between another pair of columns. Which pair?

c The figures in the last two columns are specified in different ways for each of the systems. Explain why this is so.

20(P) The circuit shown in figure C70 is a rain detector. The indicator lights up when rain falls on the bare wires.

a Explain why the circuit works.

b Why is the logic gate necessary? Why is it not adequate to use a circuit such as that shown in figure C71?

c When pure water is used instead of rain water, the indicator fails to light. Why is this so?

d The system is to be modified to control the pump filling a water tank. The pump should switch off when the water level gets high enough. How would you modify the circuit? What extra components would you need?

21(P) Figure C72 shows two arrangements for monitoring the temperature of a chemical processor. Both thermistors have a resistance that decreases with temperature.

a Which circuit warns when temperatures are too high? Which protects against temperatures that are too low? Explain your answer.

b The circuit shown in figure C73 combines both functions. Explain how it works. If an OR gate were used instead of the NOR gate what difference would it make to the operation of the circuit?

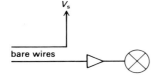

Figure C70
A rain detector.

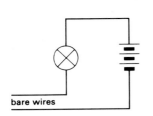

Figure C71

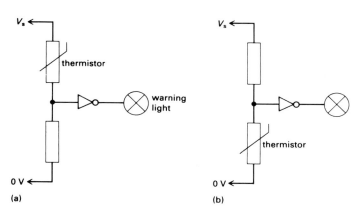

Figure C72
Circuits for monitoring temperature.

(a)

(b)

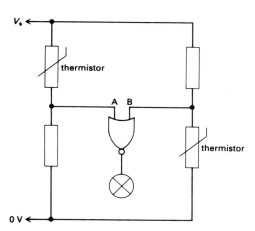

Figure C73
A temperature warning circuit.

22(R) To answer this question and the following one, you need to apply what you know about potential dividers, that is, arrangements of resistors that are used to fix a voltage at a particular value at a point in a circuit (Unit B, 'Currents, circuits, and charge').

The circuit diagram in figure C74 shows two resistors being used to control the input to an inverter. The voltage transfer characteristic for the gate is also shown.

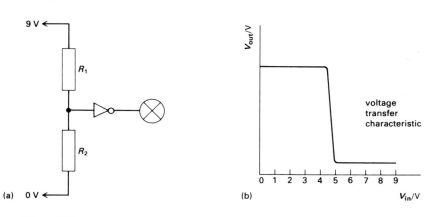

(a) 0 V

(b)

Figure C74

a According to the characteristic curve, what is the smallest input voltage that will keep the output of the gate low?

b If R_2 is $1\,k\Omega$, what is the largest value that R_1 can have to keep the output of the inverter low? Assume that the current drawn by the input of the inverter is negligible.

c R_1 is now replaced with a light-dependent resistor (L.D.R.) which has a resistance of $500\,\Omega$ in daylight and $10\,k\Omega$ in darkness.

What resistance must R_2 have so that the indicator will be off during daylight but will come on when the resistance of the L.D.R. rises above $1\,k\Omega$? Again, ignore the effect of any current drawn by the gate.

d If you set about building this circuit using the value of resistor that you have calculated in part **c** you would probably have some difficulty in finding a resistor of exactly that value: it would be quite expensive to buy. Resistors are sold in *preferred values* to a specified *tolerance*. Commonly used tolerance values are 5 % and 10 %. Find out what the preferred values are for these tolerances. (A good source of information for such data is an electronics components catalogue such as the RS Components catalogue.)

e Suppose the current drawn by the gate itself is not negligible. How would it affect your calculations? Assume that the gate draws a current of 1 mA as a worst case.

f Why might it be a good idea to use a variable resistor for R_2?

Arithmetic with logic gates

23(R)a Draw the truth table for a 'half adder'.

b Draw the truth table for a 'full adder'.

c What is meant by the 'SUM' and 'CARRY' terms?

24(R)a Draw a 'half adder' using NOR gates.

b Show how two half adders can be connected together to make a 'full adder'.

c In a full adder, why can there not be a 'CARRY' from both half adders at the same time?

Circuits with feedback

25(L) An electronic circuit can be made which behaves as follows. When the input goes from low to high, the output goes from high to low, stays there for a short time, and then goes high again. Figure C75 illustrates the idea.

a What circuit could you use to get this result?

b Two such circuits are put in a row as in figure C76.

What does output 2 do when input 1 goes from low to high?

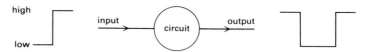

Figure C75

Figure C76

c If output 2 is joined to input 1, what happens?

26(P) Suppose you have a signal source whose output goes from 0 V to V_s for five seconds, returns to 0 V for five seconds, and so on, indefinitely. How could you use logic gates with such a supply to drive a lamp, perhaps a buoy at sea which

a flashes, going on for one second every ten seconds?

b occults, going out for one second every ten seconds?

27(L) Figure C77 shows an inverter connected to a NOR gate. Complete the table giving the output of the combination.

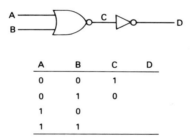

A	B	C	D
0	0	1	
0	1	0	
1	0		
1	1		

Figure C77

Write down in words the conditions that D is high (1) in terms of whether or not A and B are high. Why is this called an OR gate?

28(L) Look at question **27**. Join D to A in figure C77.
Let us consider what happens if B is low.

a If A is low as well, what is C?

b If A is low, what is D? If A is low and D is fed back to A, does A change?

c If A is high, what is D? If D is fed back to A, does A change?

Now suppose A is low and B is low, with D fed back to A. The system will sit there, with A kept low by D just because A is already low. Now think of B being made high. C goes low.

d What happens to D? What happens to A? What happens to the circuit?

e How would you get it back to the condition with A low?

f Why is this system called bistable?

29(L) Look back at questions **25** and **28**.

a If two inverters are connected in a row, the first feeding the second, what happens to the output of the second if the input of the first starts to go from low to high?

b Now suppose that the output is fed back to be the input of the first. If the input of the first starts to go high, what happens to it next? What happens later? What happens in the end?
  This is called *positive feedback*; a change to the system drives the system further in the same direction.

30(E) Suppose the small local shop looks like running out of chocolate, and there isn't another shop nearby. Naturally, some people hasten to buy what they can before it is too late. Seeing this happen, others do the same. What happens? What has this to do with question **29**?
  Many economic systems are like this. What might happen if one market trader cut the price of his apples one day? How might his customers react? How might his competitors react? What will happen in the short term? What will happen in the long term?

31(R) Figure C78 shows a crude burglar alarm system. If any of the door or window switches are closed, the alarm sounds.

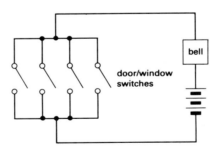

Figure C78

  Figure C79 shows an improved system. In this circuit, the alarm will sound if one of the door or window switches is opened.

a Explain how this arrangement works.

b Why is it an improvement on the first arrangement?

c A practical arrangement uses the circuit shown in figure C80. In what way is this a further improvement?

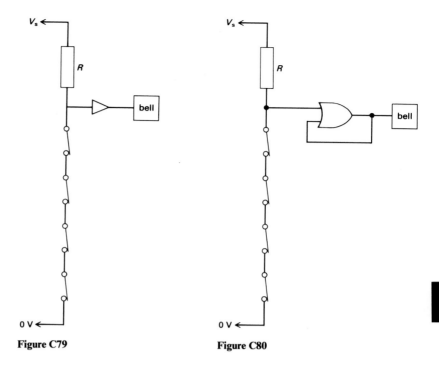

Figure C79   **Figure C80**

Logic gates

32(R)   A NOR gate has two inputs, p and q. The output of the NOR gate and a third input, r, are the inputs to an AND gate (figure C81).

   Which one of the following sets of values of the inputs p, q, and r will result in the output of the AND gate being 1?

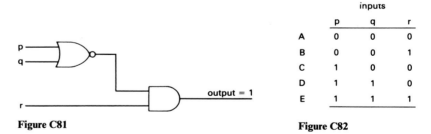

Figure C81   **Figure C82**

	inputs		
	p	q	r
A	0	0	0
B	0	0	1
C	1	0	0
D	1	1	0
E	1	1	1

(Coded answer paper, 1979)

33(R)   The four inputs, 1 = high and 0 = low, to a pair of two-input NOR gates are as shown in figure C83.

   Which of **A** to **E** is the correct combination of the values X and Y at the input to the third NOR gate, together with the output Z of the third gate?

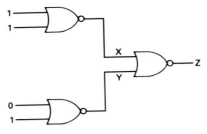

Figure C83

	X	Y	Z
A	0	0	0
B	0	0	1
C	0	1	0
D	0	1	1
E	1	0	1

Figure C84

(Coded answer paper, 1981)

34(R) The graph in figure C85 shows how, for a self-contained inverter, the output voltage V_{out} changes when the input voltage V_{in} is varied over a range from -4 V to $+4$ V.

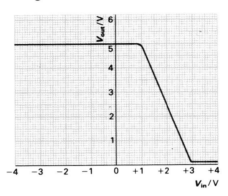

Figure C85

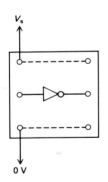

Figure C86

a The inverter is shown in figure C86. Copy the drawing and add the input and output circuits which could have been used to obtain the values of V_{in} and V_{out} used in plotting the graph in figure C85.
Label your added circuit components.

b Two different inputs to the inverter are represented on the pair of axes shown in figure C87. Carefully draw graphs, on appropriate axes, representing the corresponding outputs from the inverter.

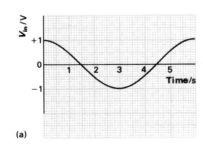

(a)

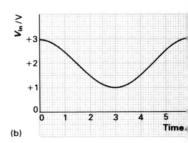

(b)

Figure C87

(Short answer paper, 1984)

Unit D
OSCILLATIONS
AND WAVES

Charles Milward
Royal Grammar School, Worcester

April Bueno de Mesquita
St Paul's Girls' School, London

Susan Ross
Godolphin and Latymer School, London

D

SUMMARY OF THE UNIT

INTRODUCTION

This Unit deals with oscillations in mechanical systems and the waves they set up. Engineers are especially concerned with such mechanical vibrations, which have important implications for the stability and even the safety of the structures they build. For example, the designer of a new car must consider how the parts of the vehicle might oscillate and the effect such oscillations would have on passenger comfort, road holding, and so on.

The study of mechanical oscillations and waves is important too as a preparation for understanding electrical oscillations (Unit H, 'Magnetic fields and a.c.') and electromagnetic waves (Unit J, 'Electromagnetic waves'). Analogies between mechanical and electrical oscillations and waves are easy to draw, and the mathematical models developed in this Unit to describe mechanical situations transfer directly to equivalent electrical situations.

You will use several of the ideas developed in Unit A, 'Materials and mechanics', including those of interatomic forces and spacings, the Young modulus, and the spring constant. And it will be helpful if you can recall some of the important features of wave behaviour from earlier science courses.

Section D1 | ## INTRODUCTION TO OSCILLATIONS

What are oscillators and why study them?

Anything that exhibits a rhythmic, repetitive, or to-and-fro motion may be considered as an oscillator. Although the cause of the oscillations and the nature of the oscillator differ from example to example, we can describe common properties.

Unit H, 'Magnetic fields and a.c.'

Unit I, 'Linear electronics, feedback and control'

Unit J, 'Electromagnetic waves'

READING
Quartz and atomic clocks (page 237)

'Buildings, bridges, and wind' in the Reader *Physics in engineering and technology*

Almost every object in the universe, large or small, can oscillate in some way or another; and if oscillating electric and magnetic fields, currents, and potential differences are considered, then the study of oscillations is a major theme in physics, as the list below indicates.
Quartz crystals as used in clocks and watches.
Atoms oscillating in solids.
Metal structures oscillating (leading to fatigue) as, for example, bridges aircraft wings.
A car bouncing on its suspension.
An oil platform oscillating in rough seas.
A boat pitching and rolling.
The Earth's atmosphere after an explosion such as that of Krakatoa.
The larynx in the human voice-box.

Any real system, electrical or mechanical, subject to a sudden change will begin to oscillate, unless damping (for example, by friction) is very large.

As well as the fact that oscillations can occur in so many different systems, as the list above suggests, there are four particular reasons why they are worth studying:

i some oscillators have a constant period and so can be used for timekeeping;

ii some oscillations can be destructive if uncontrolled;

iii mechanical waves may originate from an oscillating body if the body can cause particles or other objects within the surrounding medium to oscillate and so transmit the wave;

iv electromagnetic waves are radiated into free space by oscillating charged particles; the same oscillators are also able to absorb electromagnetic radiation.

Time traces of oscillators

EXPERIMENT D1
How do oscillators move?

You should be familiar with the words:
displacement
amplitude
period
frequency

QUESTION 1

Some oscillators are isochronous; that is, the period of the oscillation is constant (from the Greek *iso* = equal, *khronos* = time). Some are not. Some have smooth graphs of displacement against time and some have not.

A typical near-isochronous time trace is given by the loaded lath oscillator (see figure D1). Note that:

i one cycle of the graph resembles a cosine graph;

ii the amplitude dies away due to damping – the oscillator loses energy to the surroundings;

iii the period is independent of the amplitude.

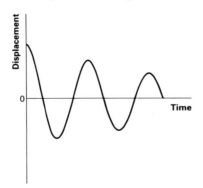

Figure D1

Oscillations and clocks

HOME EXPERIMENT DH1
Making a chronometer

READING
Quartz and atomic clocks (page 237)

Isochronous oscillators from the pendulum to the precise oscillations of caesium atoms in the atomic clock obviously have a use in measuring time. But is our desire to measure time more and more accurately the end of the story? Such questions arise as: 'What is time?', 'Does time run steadily?', 'Could time run backwards?', and 'Would we know if it was doing so?'. These questions are not just amusing, they have great importance to physicists studying fundamental particles, and influenced

both Newton and Einstein. So it is well worth reading about time ar
its measurement, the history of timekeeping, the use of timekeeping
navigation, and the problems that have arisen with our ideas of time.

Oscillations and circular motion

QUESTIONS 3 to 5

DEMONSTRATION D2
Oscillators and circular motion

If a swinging pendulum and an object rotating on a turntable a
viewed from the side, the two motions can appear identical. They see
to have a great deal in common, and the relationship between them is
very useful one. This demonstration leads to two important quantiti
associated with oscillations: phase, ϕ, and angular velocity, ω.

Phase (ϕ) If a pendulum is the right length for its natural frequency
equal the frequency of rotation of the turntable, then the shadows ca
on a screen by the pendulum and the rotating object will move togeth
and be 'in step' (providing of course that the pendulum is released at t
right moment). We say that they are *in phase* – figure D2(a). If t
pendulum is released at some other instant, there will be a constant tir
interval between one shadow reaching the outermost limit of its swi
and the other reaching that position. The fraction of a comple
oscillation by which one is ahead of the other is known as the *pha*
difference. It can be expressed as a fraction of a revolution or oscillatio

QUESTION 2

or, more usually, as an angle. See figure D2(b). Such an angle is usual
measured in radians (see below). If the pendulum is too long or t
short, the two movements will not stay in step, and the phase differen
will alter continuously.

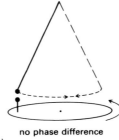

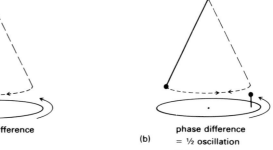

Figure D2 (a) no phase difference   (b) phase difference
= ½ oscillation
= π

Angular velocity (ω) is the change in angle per unit time. It is usual
measured in radians per second, rather than degrees per second.

Definition of the radian

QUESTION 6

One radian is the angle at the centre of a circle of radius r subtended
an arc of length r.

  If an arc of length r subtends an angle of 1 radian, then the who
circumference (length $2\pi r$) will subtend an angle of 2π radians. That is

2π radians are equivalent to $360°$

therefore $1 \text{ radian} = \dfrac{360°}{2\pi} = 57.3°$

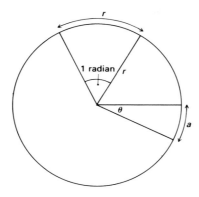

Figure D3
Radian measure of angles.

In general for any angle θ, if θ is measured in radians, then

$$\theta = \text{arc/radius} = \frac{a}{r}$$

QUESTIONS 7, 8 For small enough angles $\tan \theta \approx \sin \theta \approx a/r$. So $\tan \theta = \sin \theta = \theta$ (in radians) is quite a good approximation when θ is small.

Section D2 MECHANICAL WAVES AND SUPERPOSITION

READING
Applications of ultrasonics (page 232)

All waves are produced by some sort of oscillator; the wave transfers energy from the oscillator to other points. Sometimes this is useful, for example the oscillations of a loudspeaker producing musical sound waves; sometimes a nuisance, as when a loose oscillating panel in a bus produces a non-musical rattle; and sometimes it can be dangerous – an earthquake wave causing buildings to crack or collapse. The enormous variety of waves means that they are of great practical importance: to people trying to insulate buildings against noise; to designers of musical instruments and hi-fi systems; to architects and builders of high-rise flats and suspension bridges; to installers and designers of any equipment that vibrates or rotates; to geophysicists and many others.

Since the same ideas apply to all types of wave, we can learn about them all by studying a few simple systems.

Basic words and ideas about waves

DEMONSTRATION D3
Basic ideas about waves

HOME EXPERIMENT DH2
Make your own wave machine

The basic ideas can be demonstrated using mechanical waves along strings or springs, ripples on water, or special wave machines. The system carrying the waves is called the *medium*. A *displacement against distance graph*, or *wave profile*, shows the displacement of points along the medium at one instant of time.

In figure D4, waves on the spring are being started by oscillations of the end P_1. At the instant shown, P_1 has completed two oscillations. All other points along the medium perform the same oscillations as P_1, only later in time. P_5 is just starting to oscillate; P_3 has performed one complete oscillation; P_2 and P_4 are also oscillating, half a cycle (π) out of phase with P_1. The distance between any two adjacent points which are in phase is one *wavelength* (λ).

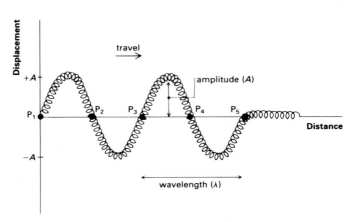

Figure D4
Wave profile: displacement–distance.

Another graph, of *displacement against time*, could be plotted for any particular point along the medium. From this graph the *period*, T, and frequency, f, could be obtained. Figure D5 shows a displacement against time graph for the point P_5, assuming that figure D4 shows the profile at $t = 0$.

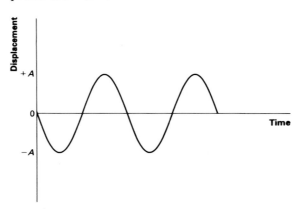

Figure D5
Displacement–time for P_5.

QUESTIONS 9, 12   The speed of travel of the wave, c, is given by $c = f \times \lambda$. If P_1 oscillates continuously, a *continuous wave* travels along the spring. If however, P_1 is just displaced once and then remains at its equilibrium position, a *single pulse* travels along the spring.

Pulses on springs: experimental results

Experiments show that:

EXPERIMENT D4
Properties of mechanical waves   The shape of a pulse on a spring is determined by the nature of the flick creating it: a quick flick gives a short pulse, whereas a slow flick gives long pulse.

Friction makes a pulse grow smaller in amplitude as it travels – its energy spreads out to its surroundings.

The speed of a pulse is not determined by its shape, nor on how you flick the spring to create it.

The speed does depend on the spring – and on the tension with which it is held. The speed increases as the tension is increased.

When pulses meet they superpose – the displacements that each pulse alone would cause on the spring add together; but when the pulses pass beyond each other they continue with their original shape (figure D6).

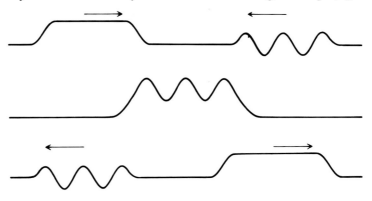

Figure D6
Superposition of pulses.

When a series of pulses is reflected, the returning pulses form a stationary pattern as they superpose with those pulses still moving outward (figure D7).

QUESTIONS 10, 11

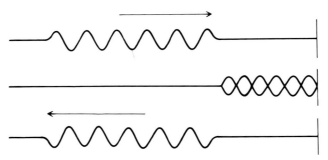

Figure D7
Stationary pattern produced when waves superpose.

EXPERIMENT D4c   A pulse reflected at a fixed end suffers a phase change of π – it turns upside down. If reflected at an open end, it suffers no phase change.

How does a mechanical wave travel?

Section D3 considers the speeds of waves like this in more detail

When a wave travels along a trolley-and-spring model each trolley moves in turn, because of a force on it from the preceding spring. The speed of the wave depends on how long it takes the trolley to acquire enough displacement to exert force on the next spring. This depends on the mass, m, of the trolley and the stiffness, k, of the springs, and it can be shown that the speed, c, is proportional to $\sqrt{k/m}$ for this wave.

Longitudinal waves

DEMONSTRATION D5
Longitudinal waves on a Slinky spring

As a wave travels through a medium the individual particles of th medium oscillate about their rest positions.

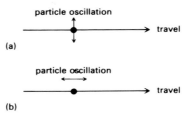

Figure D8

In *transverse waves* the particles oscillate at right angles to th direction of travel of the wave (figure D8(a)). In *longitudinal waves* th particles oscillate along the direction of travel of the wave (figure D8(b This can be demonstrated on a Slinky spring; each part of the sprin oscillates back and forth about its rest position as the wave passe Sound waves are also longitudinal: here pressure variations in the ga cause the molecules to oscillate about their mean positions.

Longitudinal waves behave in much the same way as transver waves.

One difficulty arises in drawing the longitudinal wave: often it represented in the same way as a transverse wave, and this can b misleading. Remember that the displacement is really in the direction travel of the wave (see figure D9).

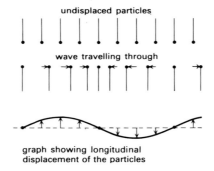

graph showing longitudinal
displacement of the particles

Figure D9

Superposition of waves

DEMONSTRATION D6
What happens when waves meet?

QUESTION 16

In figure D10, waves from both S_1 and S_2 are arriving at P. Th principle of superposition is that at any moment the displacement at is the sum of the separate displacements that the waves from S_1 and S would cause individually.

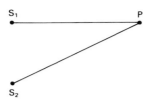

Figure D10

here will be a maximum disturbance at P
if $S_2P - S_1P = n\lambda$ (providing S_1 and S_2 are
emitting in phase)

On the ripple tank S_1 and S_2 vibrate in phase, with the same amplitude. The amplitude of the oscillation at P then depends on the phase difference between the two arriving waves; this in turn depends on the *path difference* $(S_2P - S_1P)$ and on λ. The phase difference in cycles is $(S_2P - S_1P)/\lambda$; if this is an integer then we have oscillations of maximum amplitude, or an antinode at P. That is the familiar condition for a maximum $(S_2P - S_1P) = n\lambda$.

If waves of identical frequency superpose, and if their sources have a constant phase difference (or none), a stationary pattern of nodes and antinodes (or maxima and minima) results. This is known as an interference pattern and may be used to determine the wavelength of the waves.

DEMONSTRATION D7
Path differences and phase difference

D

In the simplest case, a path difference of a whole number of wavelengths means that the two waves arrive in phase, giving a maximum, or antinode; but this only applies if the waves are emitted in phase with each other and if no other factors, such as reflection, affect the phase of either wave. Reflections in some cases result in the wave changing phase by half a cycle, or π. Such phase changes must be taken into account in the calculation of wavelength. If the two waves are exactly out of phase, a minimum, or node, results.

If either the frequency or the speed is already known, the other may now be calculated using $c = f\lambda$.

Superposition at a point can occur with waves from two (or more) different sources – though a stationary pattern will only result if they have a fixed phase relationship; it can also occur with waves from a single source which have travelled by different paths to the point.

EXPERIMENTS D8
Superposition of waves

QUESTIONS 14, 15, 17 to 19

As an example, the 3 cm radio receiver, R, in figure D11 will receive *minimum* radiation if both T and R are very close to M. This is because the difference in path between TMR (reflected path) and TR (direct path) is almost zero; but the reflection at M changes the phase of the reflected wave by π: hence the waves arrive out of phase by π. To obtain a phase difference of one cycle, M must be moved away from T and R until the path difference is 1.5 cm (half a wavelength). See figure D12.

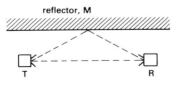

Figure D11

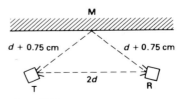

Figure D12

Superposition effects demonstrate the wave nature of radiation such as light, radio, and X-rays where the waves themselves cannot be seen. Superposition effects also show that electrons (and other 'particles') have wave-like properties.

Unit L,
'Waves, particles, and atoms'

Examples and applications of superposition

QUESTIONS 20 to 22

Familiar examples include the colours seen in 'rainbow bubbles' and on oily puddles, and the colours seen on the surface of a long-playing record when it is tilted around in sunlight. The abrupt occurrence and subsequent disappearance of gigantic ocean waves up to 30 metres high are also due to superposition.

Section D3

MECHANICAL OSCILLATIONS

EXPERIMENT D1
How do oscillators move?

DEMONSTRATION D2
Oscillators and circular motion

This Section develops the link between oscillatory and circular motion and uses it to derive a mathematical description which can be applied more or less exactly, to many different oscillators. The reason why many oscillators are isochronous is also dealt with.

Features common to all mechanical oscillators

Every mechanical oscillator, isochronous or not, has these features:
i it is displaced successively to one side then the other of an equilibrium position;
ii it is accelerated towards the equilibrium position by a force; the force is related to its displacement in some way;
iii it has inertia, which means that it continues through the equilibrium position, rather than coming to rest there;
iv it possesses kinetic energy as it passes through the equilibrium position, potential energy at the extreme ends of its motion, and usually a combination of both at points in between;

QUESTIONS 26, 27

v there are resistive forces against which it must do work; as a result the oscillator loses energy.

The mass and spring oscillator

k is the force which would displace the mass m a distance of 1 m

The rest of this Section deals solely with one particular type of oscillator; a mass, m, which oscillates horizontally or vertically, and is attached to springs which provide the restoring force. By Hooke's Law $F = -ks$, where F and s are force and displacement (positive when measured to the right or downward), and k is the spring constant of the whole assembly of springs. This is illustrated in figure D13.

This system is chosen for detailed study because many other systems are analogous to it: if quantities corresponding to m and k can be identified in another system, then the results obtained for the mass and spring system can be used to describe this other system.

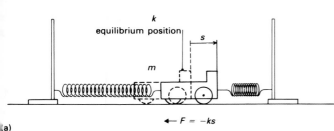

(a)

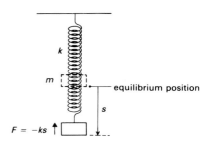

(b)

Figure D13
Mass and spring oscillators.

Damping is ignored, partly to simplify the mathematics, and partly because it does not affect perhaps the most important property of the oscillator, its period.

Periodic time, T

EXPERIMENT D9
Factors affecting the period of an oscillator

Experiments show that for this system
i T does not depend on A, the amplitude;
ii T is proportional to $\sqrt{m}$;
iii T is proportional to $\sqrt{1/k}$.

A qualitative argument explaining why T does not depend on A runs as follows:

HOME EXPERIMENT DH3
A mechanical oscillator

Consider one quarter of an oscillation, first with one amplitude, then with double that amplitude.

In the second case, the object starts with double the displacement;
⇒ twice the force acts on it
⇒ it has twice as much acceleration
⇒ velocity it gains in a given short time doubles
⇒ it covers twice the distance in a given short time
⇒ new amplitude is covered in the same time as the old amplitude.

QUESTIONS 23, 24

Similar qualitative arguments can be used to explain the dependence of T on m and k.

Simple harmonic motion

Simple harmonic motion (S.H.M.) is the name given to the motion of objects moving in such a way that:

restoring acceleration

$a \propto$ displacement s or, with our sign convention,

QUESTION 25

$a \propto -s$.

The mass in the mass-and-spring system obeys this rule, since

$a \propto F$ (Newton's Second Law) and

$F \propto -s$ (Hooke's Law)

EXPERIMENT D10
Oscillation of a tethered trolley

As shown above, the resulting oscillation has a period independent of its amplitude.

Few, if any, oscillators obey this rule exactly; but many obey it approximately, particularly if their amplitude of oscillation is small relative to the dimensions of the system.

Analysis of the oscillation

A ticker-tape obtained from half an oscillation of a trolley tethered between springs shows a time trace like the one in figure D14.

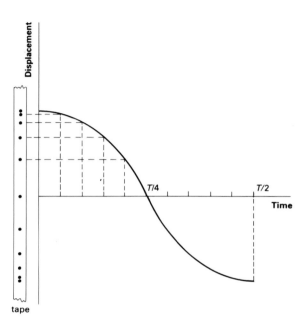

Figure D14
Time trace for a tethered trolley.

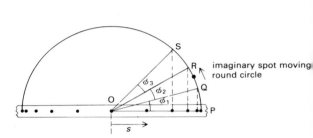

Figure D15
Projection of time trace onto a semicircle.

$\phi = \omega t$
if $t = 0$ when oscillator is at P

The procedure illustrated in figure D15 shows that the motion can be generated by the shadow (projection) of a spot moving round a circle at constant speed. In the time interval between the ticker-tape dots shown, the spot always moves through the same angle round the circle, since measurements show that

$$\phi_1 = \phi_2 = \phi_3 \,(= \Delta\phi)$$

By definition, the angular velocity of the spot $\omega = \dfrac{\Delta\phi}{\Delta t}$.

Also,

$$\omega = \frac{2\pi}{T} \qquad \text{(if } \phi \text{ is in radians; } T = \text{period)}$$

ω can be measured from figure D15; if ω is also calculated from the directly measured T, agreement should be good.

QUESTION 28   It can be seen from figure D16 that the displacement of the oscillator $s = A \cos \phi = A \cos \omega t$.

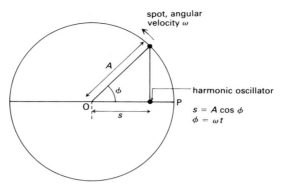

Figure D16
Projection of circular motion onto diameter.

Thus $A \cos \omega t$ gives the complete detail of where the oscillator will be at any given time t after it starts from P.

An alternative definition of S.H.M. is to say that it is a motion which is described by $s = A \cos \omega t$.

Velocity and acceleration of the oscillator

Since velocity is the rate of change of displacement, the velocity at any time is the gradient of the displacement against time curve. Similarly, the acceleration can be obtained from the graph of velocity against time (see figure D17).

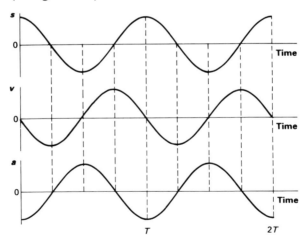

Figure D17
Displacement, velocity, and acceleration of an oscillator.

This result can be obtained more formally by differentiation:

QUESTION 31

$$s = A \cos \omega t$$

$$\Rightarrow v = \mathrm{d}s/\mathrm{d}t = - A\omega \sin \omega t$$

$$\Rightarrow a = \mathrm{d}v/\mathrm{d}t = - A\omega^2 \cos \omega t = - \omega^2 s$$

The dynamics of the oscillator

If the oscillator is displaced distance s to the right, the unbalanced force on it (to the left) is $-ks$.

$$a = F/m$$

$$\Rightarrow a = -(k/m)s \qquad\qquad \text{equation [1]}$$

Now $s = A \cos \omega t$

$$\Rightarrow a = -\omega^2 A \cos \omega t \quad \text{and} \quad a = -\omega^2 s \text{ (see above)} \qquad \text{equation [2]}$$

Equations [1] and [2] are the same, provided

$$\omega^2 = \frac{k}{m}$$

ω, previously calculated, should compare (within experimental error) with the value of k/m measured during the experiment.

Note also that

$$T = \frac{2\pi}{\omega} = 2\pi \sqrt{\frac{m}{k}}$$

This ties up with earlier experiments and qualitative reasoning.

The formula $T = 2\pi \sqrt{\dfrac{m}{k}}$ has many applications. For example, atoms in solids vibrate as though they were masses held by springs (later in this Section this idea is used to derive the speed of sound in a solid); and the vibrations of buildings, bridges, and almost any mechanical oscillator can be analysed by reference to this mass-and-spring model.

Numerical solution of $a = -\dfrac{k}{m}s$

This piece of work illustrates a method widely used in science and engineering for solving difficult mathematical problems. The equation $a = -\dfrac{k}{m}s$ is relatively simple, and can be solved exactly by integration: we already know a solution, $s = A \cos \sqrt{\dfrac{k}{m}}t$. So the numerical method of solution is not really necessary for this problem, but provides a good illustration of how a more difficult problem, impossible by integration, could be solved. Such problems are common in more complex fields like engineering. There is an example from physics in Unit L, 'Waves, particles, and atoms'.

The problem: To obtain a detailed graph of how s varies with t, knowing the constants of the system (k and m), the initial values of s and v, and that $a = -\dfrac{k}{m}s$.

The principle: The position of the oscillator is calculated at successive moments, which are separated by short time intervals Δt

The oscillator's speed is assumed to remain constant during each of these short intervals. Each calculation is approximate, but can be made as accurate as we like by taking a small enough value for Δt.

The steps in calculating each successive displacement value, s, are as follows:

Knowing the old value of displacement, s_0

$$F = -ks_0$$

$$\Rightarrow a = -\frac{k}{m}s_0$$

new velocity, $v_1 = v_0 + a\Delta t$

new displacement $s_1 = s_0 + \Delta s = s_0 + v_1\Delta t$

QUESTION 29 The new value of s is now used to work out a at the new position, hence the new v, etc.

This so-called iterative method, in which the same steps are repeated (re-iterated) many times, is of course ideally suited for programming on a digital computer. Furthermore, in a computer program one could easily include other factors, such as damping, or a QUESTION 30 regularly applied driving force.

D

Energy of an oscillator

The potential energy stored in the springs at any position is $\frac{1}{2}ks^2$. When the oscillator is at maximum displacement (and stationary), this is equal to $\frac{1}{2}kA^2$, which must, therefore, be the total energy of the system. Note particularly that the total energy is proportional to A^2.

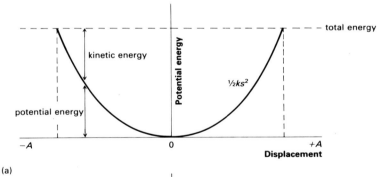

(a)

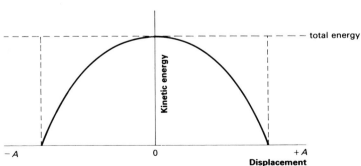

Figure D18
Energy of an oscillator.   (b)

If no energy spreads from the oscillator to the surroundings,

total energy = P.E. + K.E.

$$\Rightarrow K.E. = \tfrac{1}{2}kA^2 - \tfrac{1}{2}ks^2$$

Thus P.E. and K.E. vary with s, as in figures D18(a) and D18(b). P.E. and K.E. vary with time like this:

$$P.E. = \tfrac{1}{2}ks^2 = \tfrac{1}{2}kA^2\cos^2\omega t$$

$$K.E. = \tfrac{1}{2}mv^2 = \tfrac{1}{2}m\omega^2 A^2\sin^2\omega t$$

(See figure D19.)

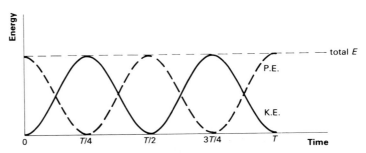

Figure D19
K.E., P.E., and total E of a harmonic oscillator.

Oscillations and the speeds of waves

DEMONSTRATION D11
Longitudinal wave on a trolleys-and-springs model

The passage of a mechanical wave involves energy passing through a medium from one oscillator to the next. In demonstration D11, for example, each trolley is a tethered-trolley oscillator; as a single pulse moves along the line of trolleys, it sets each one into motion in turn.

Figure D20 shows the line of trolleys at three successive short time intervals. In figure D20(b), trolley Q is just being set into motion. By the instant of diagram D20(c), the pulse has advanced by one section-length (x), and R is now in the same situation as Q was previously. The speed of advance of the pulse is thus the distance x divided by the time interval between diagrams D20(b) and D20(c).

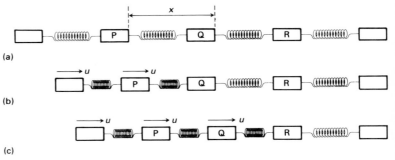

Figure D20

QUESTION 40

If Q is considered as a tethered-trolley oscillator, then in figure D20(b) it is at the lefthand extreme of an oscillation; by figure D20(c) it has reached its equilibrium position between trolleys P and R. This is one-quarter of an oscillation; the time interval for the pulse to be handed on to the next oscillator thus appears to be one-quarter of the period of the oscillation. In fact, a more thorough analysis shows that the fraction is not $\frac{1}{4}$ but $\dfrac{1}{2\pi}$ of an oscillation.

c is the speed of the pulse

$$c = \frac{x}{\text{time interval}} = \frac{x}{\dfrac{1}{2\pi} \times \text{period}}$$

period $\quad T = 2\pi \sqrt{\dfrac{m}{k}}$

$$\Rightarrow c = \frac{x}{\dfrac{1}{2\pi} \times 2\pi \sqrt{\dfrac{m}{k}}}$$

$$= x \sqrt{\frac{k}{m}}$$

The speed of sound in a solid

In Unit A, 'Materials and mechanics', a solid was pictured as consisting of atoms connected by springy bonds. This model was used to calculate the Young modulus for steel. Now it can be used to calculate the speed of a wave in a solid. The atoms and bonds are modelled by trolleys connected by springs. From Unit A, the Young modulus, $E = k/x$, where k is the spring constant of the interatomic bonds.

So

$$c = x \sqrt{\frac{k}{m}} = x \sqrt{\frac{Ex}{m}} = \sqrt{\frac{Ex^3}{m}} = \sqrt{\frac{E}{m/x^3}}$$

But x^3 is the volume occupied by one atom; so $m/x^3 = \rho$, the density of the material

$$\Rightarrow c = \sqrt{\frac{E}{\rho}}$$

QUESTIONS 41, 42

DEMONSTRATION D12
Speed of sound in a metal rod

The same sort of argument applies to all mechanical waves. In each case, the speed of the wave depends on the properties of the substance through which the wave is travelling. Formulae for the speeds of many mechanical waves have the form $\sqrt{\dfrac{\text{force constant}}{\text{mass}}}$, or something equivalent.

Section D4   FORCED VIBRATIONS AND RESONANCE

Free and forced oscillations

A system which can oscillate may be set into oscillation in many ways. Two of these are particularly important.

Section D3 dealt with 'free' or natural oscillations. In this case the oscillator is given an initial displacement or velocity, and then released.

The second important situation is when a periodic repetitive driving force is applied to the oscillator in some way. This in general causes 'forced' oscillations. When the frequency of the driving force equals the oscillator's natural frequency, then the amplitude of the oscillation may build up to a large value. This special situation is called *resonance*.

Resonance

DEMONSTRATION D13
Forced vibrations of a mass on a spring

'Buildings, bridges, and wind' in the Reader *Physics in engineering and technology*

Resonance has wide ranging practical applications. Any machine or structure is likely to be subjected to periodic forces, either as a result of its own operation (*e.g.*, the motor in any vehicle imposes an oscillation or vibration on every part of the vehicle) or through the action of some external agent (*e.g.*, wind exerts a periodic force on buildings and structures through vortex shedding). If you keep your eyes and ears open you will notice countless examples of forced oscillations.

Forced oscillations can prevent machines operating efficiently, as when an unevenly loaded spin drier cannot achieve its normal working speed because much of its energy is being diverted into a violent wobbling. More seriously, forced oscillations can result in fatigue failure of metal components at stresses well below the tensile strength of the metal, simply as a result of repeated flexing (like breaking a piece of wire by bending it to and fro). If resonance occurs, forced oscillations can be violent and may have catastrophic results (as in the Tacoma Narrows bridge collapse). An understanding of forced oscillations is clearly essential to engineering.

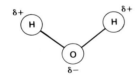

Figure D21

Forced oscillations are not always destructive; sometimes engineers and scientists can make positive use of them. Nor is the phenomenon confined to mechanical oscillations. Microwave ovens heat food as a result of a forced oscillation of the molecules within the food, particularly water molecules, which are polar (they are permanently charged positive at one end and negative at the other, see figure D21). Infra-red absorption spectroscopy, which is an important technique for chemists, involves the forced oscillation of atoms or groups of atoms within a molecule. The conversion of radio waves to electric currents in an aerial is an example of a forced oscillation, and the operation of a tuning circuit in a radio relies on resonance.

READING
Spectroscopy (page 236)

Many more examples are given in the books recommended for this Unit.

EXPERIMENT D14
Investigations of resonance

DEMONSTRATION D15
Barton's pendulums

An investigation of a resonant system reveals the following points.
i When a driving force (driver) acts on something which can vibrate, the initial *transient oscillations* are irregular, with varying amplitude.
ii These transient oscillations give way, in a time which depends on the degree of damping, to a *steady state*, in which the driven oscillator oscillates at the forcing frequency, regardless of its own natural frequency. (Damping is the result of friction-type forces which always act against the motion of an oscillator.)
iii The amplitude of the driven oscillation depends on the forcing frequency and rises to a maximum if the forcing frequency is equal to

the natural frequency of the driven oscillator. These large amplitude vibrations are called resonant oscillations. See figure D22.

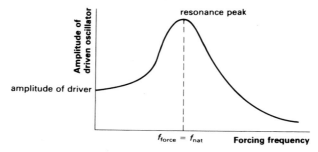

Figure D22
Resonance.

iv At resonance the driver and the driven oscillator are not in phase. The driver leads by one quarter of a cycle.

The photograph of Barton's pendulums in figure D23(c) was taken when the driver was at its maximum displacement to the left. The resonating pendulum is just passing through the centre of its oscillation, and moving to the left. It is one quarter of a cycle, or $\pi/2$, behind the driver. The shorter pendulums at the top of the picture, with higher natural frequency, are moving approximately in phase with the driver. The long pendulums with lower natural frequencies, are approximately in antiphase with the driver (phase difference of π). Notice how the pendulums which have natural frequencies close to the forcing frequency (that is, pendulums of similar length to the driver), oscillate with larger amplitude than the others.

The amplitude of the forced vibrations also depends on the degree of damping. The photographs in figures D23(a) and (b) illustrate how the amplitude of resonant vibrations is reduced by damping.

Figure D23
Photographs of Barton's pendulums.
(a) Time exposure (damped).
(b) Time exposure (less damped).
(c) Instantaneous.
A. W. Trotter.

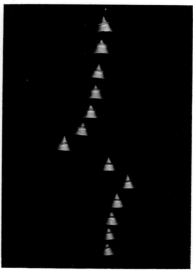

(a)　　　　　　　　　(b)　　　　　　　　　(c)

Resonance curves (figure D24) reveal the effects of damping in more detail. Damping reduces the amplitude at all frequencies. It also makes the resonance peak broader (reduces the sharpness of resonance).

It very slightly reduces the resonant frequency of the driven oscillator.

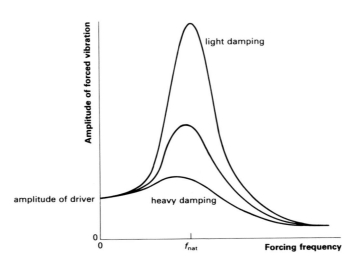

Figure D24
Resonance curves.

Energy in forced oscillation

The driver delivers energy to the forced oscillator during each cycle of oscillation. This energy may be:

i stored in the oscillator, increasing the amplitude (energy stored $\propto$ amplitude^2);

ii used to overcome the resistive forces which cause damping;

iii returned to the driver, later in the cycle (but this does not happen at resonance).

The amplitude of a forced oscillator goes on increasing until energy loss per cycle = energy provided by driver per cycle.

The quality factor, *Q*

The quality factor, Q, of an oscillator can be formally defined like this:

$$Q = 2\pi \times \frac{\text{energy stored}}{\text{energy lost per cycle}}$$

However, there is a much more useful, though non-rigorous, description of Q: it is approximately equal to the number of free oscillations which occur before all the oscillator's energy is gone.

QUESTIONS 43 to 53

Q is related to the degree of damping of the oscillator, and to the sharpness of its resonance peak. Low values of Q are associated with heavily damped oscillations which do not resonate violently and which die away quickly if they are not forced. High values of Q are associated with light damping and sharp resonance.

Some typical values of Q are:

Car suspension	1
Tethered trolley	10
Simple pendulum	1000
Guitar string	1000
Quartz crystal of watch	10^5
Excited atom	10^7
Excited nucleus	10^{12}

QUESTIONS 54, 55   Consider the guitar string, for example. The energy is emitted as sound waves, with a fundamental frequency of, say, 512 Hz (the C above middle C). If $Q = 1000$, then roughly 1000 oscillations occur before all the energy is gone. Thus the plucked string will cease to oscillate after $1000/512 \approx 2$ seconds: which agrees roughly with experience.

Standing waves and resonance

A standing wave is formed when identical waves travelling in opposite directions superpose.

In figure D25, P and Q represent points along a rope. At the instant shown, two wave trains travelling in opposite directions are just about to overlap at point P. Points L_1 to L_4, R_1 to R_4 represent peaks or troughs along the wave train.

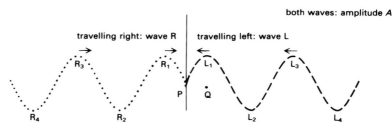

Figure D25
Formation of a standing wave.

Superposition at point P causes P to oscillate with amplitude $2A$, since peaks L_1 and R_1 arrive there simultaneously, followed half a cycle later by troughs L_2 and R_2, etc. Careful inspection shows that superposition at Q will result in Q remaining stationary in space all the time.

Points such as P are called *antinodes* (A); points such as Q are called nodes (N). Adjacent nodes are distance $\lambda/2$ apart. The motion of a section of rope on which a standing wave is occurring can be represented as in figure D26.

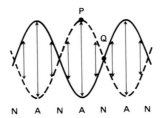

Figure D26
Representation of a standing wave.

All points between adjacent nodes oscillate in phase with each other; they are in antiphase with all points in the next half-wavelength section.

Resonance of a string with both ends fixed

DEMONSTRATION D16
Standing waves on a rubber cord

QUESTIONS 56, 59

A wave in a stretched string cannot escape beyond either end; it must be reflected.

If a string is made to vibrate near one end, waves travel to and fro along the string, being reflected each time they reach an end. If the length of the return trip for these waves is a whole number of wavelengths, that is, $2L = n\lambda$, where L is the length of the string, they will always pass the vibrator in phase with the wave which it is producing, even after several return trips. A standing wave of large amplitude therefore develops. Figure D27 illustrates some of the modes of vibration of a string.

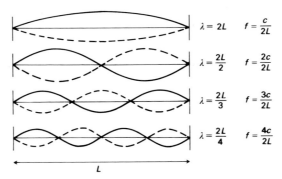

Figure D27

Large amplitude standing waves only occur for these well-defined wavelengths.

Standing waves in bounded systems

DEMONSTRATIONS D17
More complicated standing waves

HOME EXPERIMENTS DH4 to DH6

QUESTIONS 57, 58

Unit L, 'Waves, particles, and atoms'

The edges of any solid object act as boundaries to waves. Superposition of waves travelling towards the boundary with those reflected from it can lead to standing waves, if the object is vibrated at an appropriate frequency (unless the vibrations are damped). In a similar way, standing waves can be set up in fluids if they are contained (air in a trumpet, water in the bath). The same ideas are used to explain the energy levels of atoms.

These more complex standing waves have the following features in common with waves on a string:

i There is a series of definite modes of oscillation, corresponding to different frequencies, at each of which the response is large (resonance).

ii The patterns developed depend on the frequency, there being more nodes at higher frequencies.

iii The standing waves must 'fit' into the system.

READINGS

APPLICATIONS OF ULTRASONICS

Ultrasonic waves are compression waves travelling through a medium with frequencies higher than those of audible sound waves. Their existence may be demonstrated using an ordinary signal generator and loudspeaker to create them, and a microphone and C.R.O. to detect them. However, most practical ultrasonic systems use transducers which depend on either the piezoelectric effect or the magnetostrictive effect. (A 'transducer' converts energy, in this case from electrical energy to ultrasonic wave energy, or vice versa.)

The piezoelectric effect occurs in certain crystals, such as natural quartz. When a p.d. from an external supply is applied across it, the crystal will alter its shape slightly; an alternating p.d. at high frequency will thus cause it to act like a miniature tuning fork, generating ultrasonic waves. Conversely, if the crystal is compressed or stretched by a mechanical force, such as occurs repeatedly when an ultrasonic compression wave arrives, a p.d. appears across it; hence it can also be used to detect the waves.

Magnetostriction involves the change of shape of certain metal alloys when they are magnetized, or the complementary effect of a change in their magnetization when they are stressed.

A number of applications involve an echo-sounding technique. A transducer sends out a pulse of ultrasonic wave energy; the same transducer is then used to detect any returning energy reflected from discontinuities in the medium. The time between transmission of the pulse and its returning echo, together with a knowledge of the wave speed in the medium, allows the distance to the discontinuity to be computed.

Ultrasonic flaw detection   This is an example of a non-destructive testing method. Such methods are used where the material to be tested must not be cut up, broken down chemically, or even removed from its working position. A transducer using a frequency in the MHz range is placed in contact with the material to be inspected. The pulses are reflected from the rear surface of the material; they are also reflected from any cracks or flaws within it, including any which may be invisible from the outside. If the returning signals are displayed on a C.R.O., the position of a flaw and its approximate size can be judged from the trace. This technique is particularly valuable in the inspection of railway tracks and welded pipes.

Ultrasonic foetal scanning   Many expectant mothers now receive an ultrasonic scan as part of their routine check-up on the progress of their as-yet-unborn baby. Ultrasonic waves penetrate the mother's skin, but are reflected back selectively by discontinuities in the tissues beneath. By examining the whole area of the foetus bit by bit, a complete 'picture' of it can be built up. (See figures D28(a) and (b).) This is another example of non-destructive testing: ultrasonic waves apparently cause no harm to mother or baby, whereas X-rays, which could be used to obtain a similar picture, would harm them.

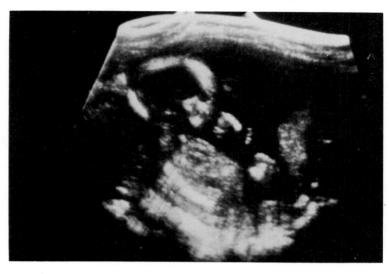

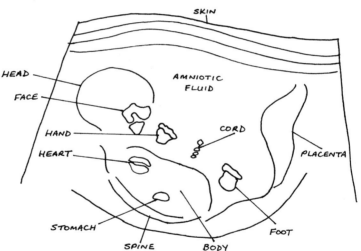

Figure D28
Ultrasound scan of 18-week foetus and explanatory diagram. *Department of Medical Illustration, St. Bartholomew's Hospital, London.*

Ultrasonic flow measurement   If the pulse is projected into a stream of liquid flowing in a pipe, then energy reflected back from minute discontinuities in the liquid will show a Doppler frequency shift depending on the flow rate. This is used as the basis of a 'non-invasive' flowmeter, so-called because a transducer can send and receive sound through the wall of the pipe, and there is no need to obstruct the flow as in many meters which, for instance, measure the rotation of a paddle-wheel inserted in the pipe.

The last example, ultrasonic flow measurement, is quoted from B. Jolly (Ed.) Hobsons Science Support Series, *Waves and sound*. CRAC Publications, Hobsons Press (Cambridge) Ltd, 1982.

Questions

a An ultrasonic transducer converts alternating p.d.s into ultrasonic wave energy (or vice versa). How might this be done using the phenomenon of magnetostriction? (What would be the essential parts of the transducer?)

b The speed of compression waves in a metal is of the order of $5000\,\mathrm{m\,s^{-1}}$. If your best laboratory C.R.O. is to be used for flaw-detection in a metal sample, estimate the length of the smallest sample that could satisfactorily be tested.

THE EFFECTS OF VIBRATION ON PEOPLE

We know that the slow oscillations of a rolling ship can produce sea-sickness, although the origin of car-sickness is less clear. Machine operators are subject to more rapid vibrations; the pneumatic road drill is an extreme example.

Some effects of oscillations of various frequencies are shown in figure D29.

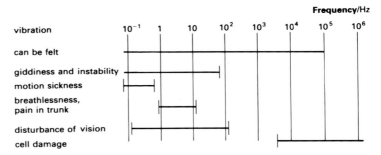

Figure D29

Most serious effects are due to *resonance* – when the natural frequency of oscillation of some part of the body is the same as the frequency with which it is driven. This has been studied by sitting a person on a vibrating platform. Figure D30 shows the motion of the abdomen wall at various frequencies.

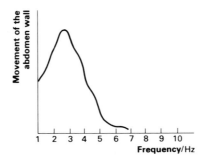

Figure D30

The tolerance of human beings to vibration varies with frequency. You can see this in figure D31, which is a graph showing the results of a study of human vibration tolerance. Such studies are of especial importance in designing aircraft and space probes.

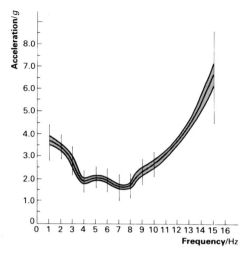

Figure D31
Human vibration tolerance. The curves show the value, and the range, of the limit of tolerable acceleration at various frequencies.
From MAGID, E. B., COERMANN, R. R., and ZIEGENRUECKER, G. H. (1960) Aerospace Medicine, **31**, *page 915.*

Human vibration engineering is also important in designing hand-operated machine tools. The use of such a tool for intricate work would be very difficult if it vibrated at a resonant frequency of the hand–arm system.

Questions

a Estimate the natural frequencies at which various sections of your hand–arm system might vibrate (try swinging or shaking the various sections and the whole arm). Suggest ranges of frequencies which should be avoided for hand-operated machines.

b Figure D31 shows that humans are very intolerant of acceleration when they are subjected to vibrations between 3 to 9 Hz. Use information from figure D29 to suggest what discomforts might be experienced under these conditions.

c Figure D31 shows the maximum tolerable acceleration, during one cycle of an oscillation, plotted against oscillation frequency. Using the formula $a_{max} = -\omega^2 A$, compute the maximum tolerable *amplitudes* of oscillations at 1 Hz, 4 Hz, 8 Hz, and 15 Hz. Hence sketch a graph of maximum tolerable amplitude of oscillation against frequency (use logarithmic scales).

d Comment on figure D30 (showing the movement of the abdomen wall against frequency) in the light of your knowledge of resonance.

SPECTROSCOPY

If one could see the atoms in a molecule vibrating, and time their oscillations, one could obtain useful information about the stiffness, of the bonds between them, using $2\pi f = \sqrt{k/m}$. Although the vibrating atoms cannot be seen, the frequency at which they absorb radiation can be found. Spectroscopy is thus a valuable tool for studying the vibrations of electrons, atoms, molecules, or ions.

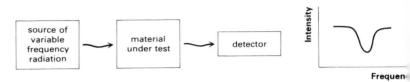

Figure D32

At its simplest, a source sends radiation, at a range of controlled frequencies, to a detector through the material under test (figure D32). Such methods are appropriate if the frequency of vibration is relatively slow, so that the wavelength of the electromagnetic radiation is more than a few millimetres.

Many interesting vibrations are faster, but we cannot vary continuously the frequency of sources of infra-red or visible light.

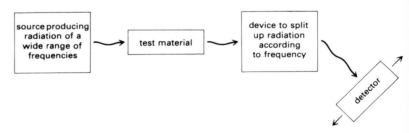

Figure D33

In figure D33, radiation of a wide range of frequencies is shone on the test material, and a device (a grating or prism, for example) sends the radiation of each frequency off in a different direction. Alternatively, the detector could in principle be tuned to each frequency in succession, although this is less useful in practice.

Uses of spectroscopic information

The stiffness of bonds in molecules, or in solids, may be found. For example, the bond stiffness of H_2 is $5.2 \times 10^2 \, N \, m^{-1}$.

The analysis of complex organic compounds is assisted by studying their infra-red absorption spectra, for many types of bond tend to absorb at much the same frequency even though the atoms form part of different molecules. The spectrum can then be used as a means of indicating which bonds are present. For instance, aliphatic C—C bonds oscillate in an in-and-out (stretching) manner at a little below $10^{14} \, Hz$.

Dyes, whose function is to colour, must absorb visible radiation strongly at selected frequencies. It is possible to design molecules which will absorb at a desired frequency.

At microwave frequencies, the spinning motion of molecules can be studied, and information about the length of bonds and the masses of the atoms obtained.

Questions

a The frequencies which molecules absorb are the frequencies which the same molecules emit when they return to their unexcited state. But even though the absorbed radiation is re-emitted, there is a detectable decrease in the intensity of the radiation reaching the detector at this frequency. Why is this?

b The frequencies which are characteristically absorbed by certain groups are listed below:

Functional group	$C-C$	$C=C$	$C\equiv C$	$C-H$
Approximate frequency/Hz	4×10^{14}	5.3×10^{14}	6.6×10^{14}	9.0×10^{14}

i Compare the stiffnesses of the single, double, and triple bonds between carbon atoms (*i.e.*, calculate the ratios of the stiffnesses).
ii Why does the C—H group have the highest frequency of oscillation among those tested?

D

QUARTZ AND ATOMIC CLOCKS

(from G. W. Dorling, Longman Physics Topics, *Time*. Longman, 1973.)

The quartz clock

In the 1930s a new type of clock started to replace the most accurate pendulum clocks as a standard for measuring time. This was the quartz crystal clock. The time keeper in this case is a quartz crystal instead of a pendulum. A quartz crystal will vibrate elastically with a natural period of its own, just like a tuning fork. In this case, however, electrical charges constantly build up and die away on its surface in time with the vibrations. It is this effect, the piezoelectric effect, which makes it so easy both to keep the crystal vibrating and to use the vibrations to control the frequency of electrical oscillations in other circuits.

It is these electrical oscillations, accurately controlled by the vibrations of the quartz crystal, that drive the hands of the clock, or control its display.

The frequency of the quartz crystal vibrations is sharply defined by the dimensions of the crystal. It is much less affected by variations in external conditions than the pendulum.

Why do we believe that these clocks are so much more reliable than pendulum clocks? There is an important test we can do. We can ask how well these clocks keep time with each other. Suppose two quartz clocks are adjusted to read exactly the same time and then left to run without adjustment. Comparisons of their time readings at various times later have indicated a difference of no more than 0.0005 second per day over a period of a week or so. This suggests that a quartz clock will measure a time interval of 1 day, or 86 400 seconds, to within 0.0005 seconds; an accuracy of better than 1 in 10^8. This is ten times better than that which could be obtained with the best pendulum clock.

Such comparisons are made continuously as quartz clocks are usually run in groups of three. This is because the likeliest disturbance to a quartz clock's time keeping is a failure of one of the electronic components. Simultaneous failure of all three clocks is most unlikely.

Quartz clocks were initially developed in response to the demand from scientists and engineers for more and more precise time standards for the purposes of radio communication, navigation, and pure research. It was also in response to this demand that the atomic clock was developed in 1954.

The atomic clock

Atoms can emit and absorb energy only at very sharply defined frequencies. Provided a suitable atom is chosen, they can be used to control the frequency of radio waves from an electronic oscillator.

In 1958 a clock, based on a beam of caesium atoms, was successfully constructed on this principle. The electronic oscillator is controlled by a quartz crystal whose vibrations are in turn controlled by the effect on the beam of caesium atoms of radio waves produced by the oscillator. As in the case of the quartz clock, it is these accurately maintained electrical oscillations which ultimately drive the clock.

Comparison of the timekeeping of two of these clocks showed that they could be relied upon to an accuracy of 1 part in 10^{11} over an apparently indefinite period of time. To put this in a slightly different way, this meant that they could be relied upon to within 1 second in 3 000 years!

A new time-scale

The development of a clock of such high reliability highlighted the strain that this increasing demand for precision had thrown on the astronomical unit of time. Both quartz and now atomic clocks showed greater consistency amongst themselves than they did with the sidereal day. Again, appeal to the laws of physics showed good reason why the Earth's period of rotation should vary from day to day and year to year. At first variations which could be calculated were incorporated into adjustments of the astronomical time-scale to make it more uniform. Atomic time showed, however, that there were some important irregular variations in the Earth's rotation as well.

In 1964 general agreement was reached for a new time-scale based on the atomic clock. The frequency of the energy level transition of the caesium atom involved in the atomic beam clock was determined a precisely as possible in terms of the then accepted value of the second and was found to be

9 192 631 770 ± 20 Hz

The ± 20 Hz represented the uncertainty in the value of the astronomical second rather than the uncertainty in the value of the frequency.

The atomic second was then defined as exactly 9 192 631 770 periods of the oscillation associated with the caesium atom for this particular transition.

In this way the atomic second is the same time interval as the previously defined second based on astronomical time. This is essential, for we must still be able to use the new scale to tell the time of day.

Because the length of the mean solar day and the year vary by small amounts when viewed against the atomic time-scale, occasional adjustments are made to the atomic scale to keep it in step with the year, in much the same way that the number of days in the year is occasionally adjusted to keep *them* in step with the year. Astronomical observations must remain the basis for determining the time of day, and they are constantly compared with the atomic time-scale. In this way the demand for a time-scale of precisely repeated equal intervals is brought into line with the need for a time-scale to tell the time of day and season of the year.

D

Questions

a Show that the frequency of the energy level transition of caesium used to define the second is consistent with a radio wavelength. What is the wavelength?

b A similar redefinition to that which befell the second has occurred with another fundamental unit, the metre. What is now the definition of one metre?

c *i* Explain the meaning of this sentence more fully: 'The ± 20 Hz represented the uncertainty in the value of the astronomical second rather than the uncertainty in the value of the frequency.'
ii Roughly what uncertainty, in Hz, would be attributable to uncertainty in the value of the frequency? (Information which enables you to answer this question is given in the text.)

LABORATORY NOTES

GROUP OF EXPERIMENTS

D1 How do oscillators move?

You will be asked to find some method of showing how the displacement of the oscillators listed below varies with time. A graph of displacement against time is called a time trace. The basic equipment for the oscillator will be available, but you may need to ask for other equipment when you have decided how you are going to produce its time trace.

Whichever oscillator you work with, consider these questions:

i Does the method of obtaining the time trace affect the period?

ii Does the method of obtaining the time trace affect the damping (resistance to motion)?

iii If you are relying on several different 'runs' to take, for example, a series of times, can you be sure that all runs are carried out under the same conditions?

iv What assumptions are you making if you interpolate between the points you have obtained (for example, by joining them to form a continuous graph)?

v Is the oscillation isochronous?

vi Whether or not it is isochronous, on what factors does the period of the oscillation depend? Try some experiments to confirm your ideas. Can you find any quantitative rules?

Here are some suggestions for setting up the oscillators and measuring the time traces.

D1a Pendulum

There are many ways of producing its time trace, from a simple sand pendulum to the electrical methods shown in figures D34(a) and (b). But remember these are not the only possibilities.

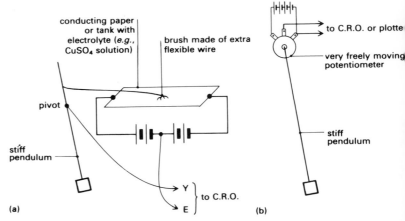

Figure D34
Obtaining the time trace of a pendulum.

D1b Torsion pendulum

string
2 retort stand bases
3 retort stand rods and bosses
G-clamp, large

Hang a retort stand rod horizontally on two parallel lengths of string. The ends of the rod may be loaded with bosses. It might be possible to adapt the potentiometer method illustrated in figure D34(b) to obtain this time trace.

D1c Lath with load

either
metre rule
or
long lath

2 G-clamps, large
clean smooth paper (*e.g.*, computer print-out paper)
felt-tip pen or brush with ink
2 rubber bands
2 masses, 1 kg

Clamp the metre rule to a stool or table so that it oscillates in a horizontal plane.

Figure D35 shows one possible way of obtaining the time trace.

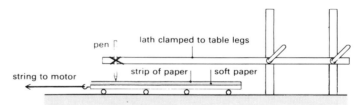

Figure D35
Lath with load.

D1d Inertia balance (wig-wag)

inertia balance
2 G-clamps, small

D1e Ball rolling on curved tracks

3 lengths of curtain rail
large ball-bearing, 1 or 2 cm diameter

Three curvatures to investigate are shown in figure D36.

Figure D36
Ball rolling on curved tracks.

D1f   Mass oscillating vertically on spring

expendable spring
retort stand base, rod, boss, and clamp
G-clamp, large
hanger with masses totalling 400 g

D1g   Undamped light beam galvanometer

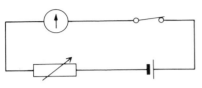

light beam galvanometer
cell holder with one cell
switch
resistance substitution box
leads

Figure D37

Set up the circuit with the galvanometer on its least sensitive scale; then
increase the sensitivity until, with a resistance of over 500 kΩ, the spot
reaches almost a full-scale deflection with the switch closed. Then, with
the galvanometer on its 'direct' setting, open the switch: the spot will
oscillate about its central zero position.

D1h   Bar magnet suspended over another magnet

cylindrical magnet
horseshoe magnet
retort stand base, rod, boss, and clamp
nylon fishing line

Hang the bar magnet on nylon or cotton so that it is horizontal and lies
just over the poles of the horseshoe magnet resting on its back. You
might use a small piece of mirror attached to the suspension to observe
the oscillations by optical means.

D1i   Large-amplitude pendulum

turntable clamped vertically (or large gyroscope)
boss (or small G-clamp)
retort stand base, rod, and boss

With the boss or G-clamp on the edge of the turntable or gyroscope, the
system can be made to execute large-amplitude oscillations (see figure
D38).

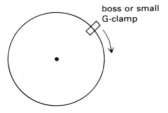

Figure D38
Large-amplitude pendulum.

D1j   Air track vehicle running between elastic barriers

air track with rubber bands at both ends, and blower
air track vehicle

D1k   U-tube containing liquid

large U-tube filled with water or potassium manganate(VII) solution

DEMONSTRATION
D2   Oscillators and circular motion

either
record-player turntable
or
fractional horsepower motor, with gearbox, turntable, and band
l.t. variable voltage supply

2 pendulum bobs
retort stand base, rod, boss, and clamp
compact light source
screen

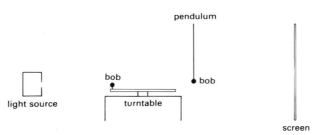

Figure D39
Oscillator and circular motion.

Observe the shadows of the two pendulum bobs as they move across the screen (figure D39). What do you see on the screen if the pendulum is exactly the right length to synchronize with the rotating bob, and the pendulum is released just as the rotating bob passes it? Why do you think this happens?

What happens if the pendulum is not released at the correct moment? What happens if the pendulum is not of exactly the right length? Use the words 'phase difference' in describing your observations.

DEMONSTRATION
D3   Basic ideas about waves

either
ripple tank kit
or
long spring
or
large Slinky spring
or
any other wave machine

Figure D40
Wave profile.

Define the terms *wavelength*, *frequency*, and *amplitude* for a wave lik
that shown in figure D40. What is meant by the *undisturbed positic*
and the *displacement* of a point such as P? What determines tl
frequency of the wave, and what unit is frequency measured in?

How are wave speed, wavelength, and frequency related? How do
P move as the wave travels along?

GROUP OF EXPERIMENTS
D4   Properties of mechanical waves

D4a   Transverse waves on a long narrow spring

D4b   Transverse waves on a Slinky spring

long spring
Slinky spring
metre rule
optional
string
large curtain ring
retort stand base, rod

It is easier to see what is happening if you make single hump-like pulse
by giving the end of the spring or Slinky a single sideways flick.

Observe the pulse as it travels, with a view to answering question
such as:
Does the speed depend on the shape of the pulse – its height or length'
Does the speed depend on how rapidly you flick the end of the spring?
Does the speed depend on the spring – how could the speed be mad
larger or smaller?
Does friction make any difference to the speed of the pulse? to its shape
What decides the shape of a pulse?
What happens when pulses, starting from opposite ends of the sprin;
meet?

What happens when the pulse reaches the far end of the spring, and that end is not free to move?

If you have time, try the following:

i Attach a large nylon curtain ring to the end of the spring and slide the ring onto a retort stand rod. When the pulse reaches this end, the end is free to move. What happens to the pulse?

ii Tie a piece of thick string or cord onto the end of the spring. What happens to the pulse as it moves from the spring to the string or vice versa? Can you explain this?

D4c   Transverse waves on a trolleys-and-springs model

12 dynamics trolleys
44 expendable steel springs
12 masses, 1 kg (or 12 more trolleys)
stopwatch or stopclock

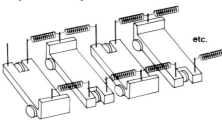

etc.

Figure D41
Transverse wave model made of trolleys and springs.

The model is made of a row of trolleys linked by springs, as shown in figure D41, with the trolleys spaced out so that the springs are in tension. It is best to set the model up on the floor, or on a surface with raised barriers along the edges to prevent the trolleys running off.

Make a transverse pulse travel along the model by moving the end trolley sharply from side to side. You should be able to answer these questions:

Does the speed of the pulse depend on its shape – its height or length?

Does the speed of the pulse depend on how quickly you move the end trolley?

Does the speed depend on the spacing of the trolleys?

What happens to the speed if the mass of each trolley is doubled? (Do this by adding loads, or stacking a second trolley on each one.)

What happens to the speed if the tension between the trolleys is doubled? (Do this by putting an extra spring in parallel with each existing spring.)

What happens to the speed if both of these changes are made at the same time?

By what factor does the speed change in each of these cases?

Explain why these changes affect the speed.

DEMONSTRATION
D5 Longitudinal waves on a Slinky spring

large Slinky spring

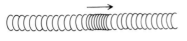

Figure D42
Longitudinal pulse on a Slinky.

The Slinky should be on a smooth surface.
Make a longitudinal pulse by moving the end of the spring sharpl
to and fro once only in the direction of the spring. (See figure D42.)
Look carefully at a pulse as it passes and see if you can answer th
questions:
What makes the pulse move along the spring?
Do compression and expansion pulses travel at the same speed?
How does the speed at which the individual coils of the spring mov
compare with the speed at which the pulse moves along the spring?
Use this model to explain how sound, which is a longitudinal wave
travels through a gas.
The pulse travels along the spring from one end to the other. How d
the individual coils of the spring move?

DEMONSTRATION/EXPERIMENT
D6 What happens when waves meet?

D6a Waves on a spring

long spring

What happens where transverse pulses meet? Does this meeting hav
any permanent effect on the pulses?
What happens when two wave trains of equal frequency and simila
amplitude meet? (You can create this situation by sending a single wav
train from one end of the spring with the other end fixed: the secon
wave train is caused by reflection of the first at the fixed end.)

D6b Ripples on water

ripple tank kit

Waves of the same frequency spread from each dipper. Observe th
effects on the resulting pattern of adjusting the frequency of the wave
and of altering the separation of the two dippers. Explain these effects.

DEMONSTRATION
D7 Path differences and phase difference

signal generator
loudspeaker
2 microphones
double beam oscilloscope
metre rule
leads

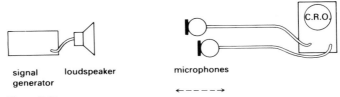

Figure D43
Path differences and phase difference.

The oscilloscope traces show the electrical oscillations produced in the two microphones by the sound waves (see figure D43). The sound from the single loudspeaker has to travel different distances to the two microphones. Because of this, the two traces will probably not be in phase. For what path differences will the oscillations be *i* in phase, *ii* in antiphase (phase difference of half a cycle)? Use this to determine the wavelength of the sound waves.

Some double beam oscilloscopes are able to display the resultant of adding the two input signals. What resultant would you expect if you add two vibrations which are *i* in phase, *ii* in antiphase? If you have such an oscilloscope, then try this.

GROUP OF EXPERIMENTS

D8   Superposition of waves and determination of wavelength

In each of these experiments you should be able to demonstrate that the radiation you are using has wave properties; you should also try to measure the wavelength. Then if you know the frequency you can calculate the wave speed.

D8a   1 GHz radio waves

15 cm dipoles and oscillator

either
microammeter
or
galvanometer (*e.g.*, internal light beam)
or
general purpose amplifier and loudspeaker

2 metal screens
leads
metre rule

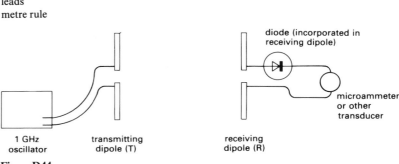

Figure D44
Transmitting and receiving 1 GHz radio waves.

Start with a simple investigation of the waves (figure D44). For example are they blocked by your arm, or by metal screens? Does their strength diminish with distance? Does the orientation of the receiving dipole (vertical, horizontal) affect the magnitude of the signal? Does the position of the receiver (different positions around the transmitter; above the level of the table) affect the signal received?

Now look for superposition. Use the metal screen to reflect radiation to the receiving dipole so that waves can follow two different paths to reach it: directly from the transmitter and indirectly via the screen.

Figures D45 and D46 show two possible arrangements.

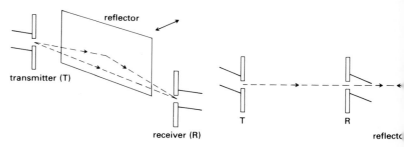

Figure D45   **Figure D46**

Look for maxima and minima. Hence measure the wavelength of the 1 GHz waves, and estimate the percentage uncertainty in your result.

From your measurement calculate the speed of 1 GHz waves.

D8b Microwaves

microwave transmitter
microwave receiver
2 metal reflectors
narrow metal plate
general purpose amplifier
loudspeaker
microammeter (may be incorporated in receiver)
diode probe receiver
metre rule
leads

transmitter (T)   2-slit arrangement   diode (D)   to microammeter *or* amplifier/speaker

Figure D47
2-slit experiment with microwaves.

Various arrangements for superposition are possible, one of which is shown in figure D47. The slits should be a few centimetres wide, and a similar distance apart.

Will the waves be in phase at the two slits?

At positions where the signal falls to a minimum the intensity is not necessarily zero. Why not?

Microwaves from the transmitter must travel in different directions to reach the two slits. If they continued in these directions they would not overlap and no interference would result. Why do they interfere? What must have happened? Can you use the equipment to check your answer?

Other arrangements to try are suggested in figures D48(a) and (b).

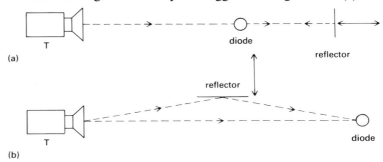

(a)

(b)

Figure D48
Simple superposition experiments with microwaves.

The apparatus you use may be labelled '3 cm wave equipment', but the actual wavelength is unlikely to be exactly 3 cm. Make the best measurement of the wavelength that you can (estimate the percentage uncertainty in your result).

D8c   v.h.f. radio waves or u.h.f. television waves

portable radio capable of receiving v.h.f.
television set
television aerial
coaxial cable
metal reflector
metre rule

You will need to find out the direction to the radio or television transmitter and then place a reflecting screen so that both direct and reflected radiation arrives at the receiver (figure D49).

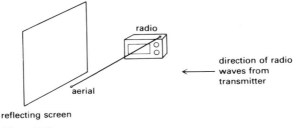

Figure D49

A similar arrangement may be used for u.h.f. television broadcasts but if the aerial is not rigidly connected to the television set then the set itself need not be moved.

For the v.h.f. radio the reflecting screen should be as large as possible: say 1 m × 1 m. For the television waves, the same screens as were used in experiments D8a and D8b will do.

With the aerial very close to the reflector, is the signal received a maximum or a minimum? Explain why.

Try to find evidence that radio/television radiations have wave properties. Measure the wavelength(s), and estimate the percentage uncertainty in the result. If you know or can find out the frequency calculate the wave speed.

D8d   Light waves

sodium lamp or sodium flame pencil
2 microscope slides
micrometer screw gauge
thin paper, *e.g.*, cigarette paper
retort stand base, rod, boss, and clamp
glass plate
hand lens

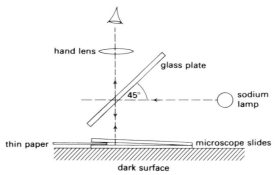

Figure D50
Superposition of light waves by reflection at a wedge.

Take care that your head, hair, or clothes do not get too close to the Bunsen flame.

The microscope slides must be clean.

The light and dark fringes are caused by the superposition of light waves and are clearer when viewed through a microscope.
Where do the two sets of waves come from?
In which direction do the fringes run? Why?
How would the appearance of the fringes differ if a different colour of light were used?
What would you need to measure, or find out, in order to determine the wavelength of the light from this arrangement? Make the necessary measurements, calculate the wavelength, and estimate the percentage uncertainty involved.

D8e   Sound waves

signal generator
2 loudspeakers
microphone
pre-amplifier
oscilloscope
metal reflector
metre rule

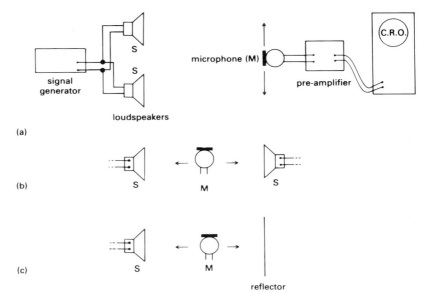

(a)

(b)

(c)

reflector

Figure D51
Superposition of sound waves.

Two sets of overlapping waves may be produced either by using two loudspeakers, figure D51(a) and (b), or by using one loudspeaker and a reflector, figure D51(c). Set up a demonstration of superposition and use it to measure the wavelength of the sound. Try with a frequency of about 3000 Hz; measure λ, calculate c, and estimate the percentage uncertainty. Then halve the frequency, and repeat the experiment. Do you arrive at the same value for c? Is this what you expect?

EXPERIMENT
D9   Factors affecting the period of an oscillator

D9a   Mass on spring

4 expendable steel springs
hanger with 8 slotted masses, 100 g
retort stand base, rod, boss, and clamp
stopwatch or stopclock
stiff wire and pliers

Try the following investigations:

i How (for a given arrangement of masses and springs) does T, the time for one oscillation, depend on the amplitude, A?

ii How does T depend on m (for a given arrangement of springs)?

iii How (for a given mass, m) does T depend on k, the spring constant of the arrangement? (What is the effective k of several springs arranged in series and in parallel as in figure D52?)

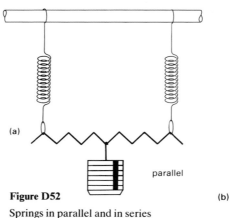

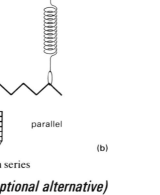

(a)

parallel

(b)

series

Figure D52

Springs in parallel and in series

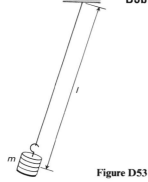

D9b **Simple pendulum** *(optional alternative)*

string, 2 m length
hanger with 8 slotted masses, 100 g
retort stand base, rod, boss, and clamp
2 metal strips (as jaws)
stopwatch or stopclock
metre rule

l

m

Figure D53

Try the following investigations:

i How (for a given mass m and length l) does T, the time for one oscillation, depend on the amplitude, A?

ii How does T depend on m (for a given length l)?

iii How does T depend on l (for a given mass m)?

EXPERIMENT

D10 **Oscillation of a tethered trolley**

dynamics trolley
2 retort stand bases and rods
2 G-clamps, large
runway for trolley
6 expendable steel springs
newton spring balance, 10 N
ticker-tape vibrator, carbon paper disc, gummed ticker-tape
transformer
leads
metre rule
masses, 100 g and 10 g
Plasticine

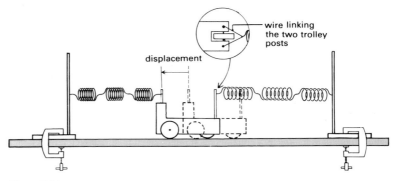

Figure D54
Oscillation of a tethered trolley.

Find the effective k of the complete system of springs by measuring the force needed to displace the trolley a suitable distance, with all the springs connected as shown in figure D54.

Find the mass, m, of the trolley.

It is desirable to adjust the mass of the trolley so that k/m has a simple value – your teacher may suggest a value for the whole class. Since the trolley is only going to travel one way, you should friction-compensate the slope. Obtain a tape for half an oscillation of the trolley. You can use this to obtain a displacement against time graph for the motion; or your teacher may have a different plan for it.

DEMONSTRATION
D11a Longitudinal wave on trolleys-and-springs model

11 dynamics trolleys
11 masses, 1 kg (or 11 extra trolleys)
20 compression springs
20 spring holders
metre rule
newton spring balance, 10 N

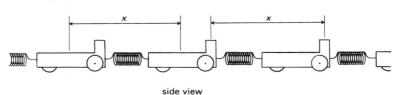

side view

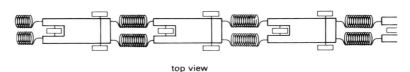

top view

Figure D55
Trolleys linked by compression springs.

Send longitudinal pulses down the line of trolleys shown in figure D55 and observe the effect on the wave speed of increasing the mass of the trolleys. Several kinds of pulse are possible; the one analysed later is created by moving the end trolley at a steady speed towards the others, generating compression in each successive spring. Do different kinds of pulse all travel at the same speed?

It is useful to measure the mass of a trolley, the spacing between trolleys, and the spring constant of the springs now, as they will be used later to calculate the speed of the wave theoretically.

The trolley spacing (x) must be measured with the trolleys in equilibrium (see figure D55).

$$\text{The spring constant } (k) = \frac{\text{force needed to extend one pair of springs}}{\text{extension produced by force}}$$

DEMONSTRATION
D11b   Measuring the speed of the wave

Apparatus as for experiment D11a with:
aluminium block
timer, resolution 10 ms
insulated copper wire
2 crocodile clips
leads

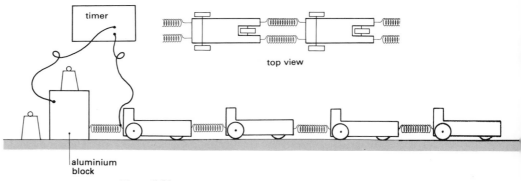

Figure D56
Timing a pulse up and down a row of trolleys.

The four trolleys are moved together towards the block. When the first spring makes contact with the block, a compression pulse sets out down the line of trolleys; it reflects off the open end as a rarefaction pulse, which travels back, finally pulling the end spring away from the block. The timer measures the total contact time.

Some points to think about:

Why is it important that all four trolleys move at the same speed towards the barrier? What time is the timer measuring?

How far does the wave travel in this time? Hence what is the wave speed?

What will be the major source of uncertainty in this experiment?

How would you expect the time recorded to change if the trolleys were more massive? If the springs were stiffer?

D12 Speed of sound in a metal rod

oscilloscope
signal generator
2 retort stand rods, 1 m long
retort stand base, rod, and boss
2 rubber bands, about 10 cm long

either
crocodile clip
or
adhesive tape

hammer, club or claw head, at least 0.5 kg
leads

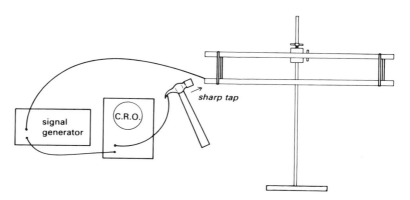

Figure D57
Speed of sound in a metal rod.

Compare the situation shown in figure D57 with the one in demonstration D11b. The steel rod replaces the row of trolleys, and the hammer acts as the aluminium block. What determines how long the rod remains in contact with the hammer? (Remember what happened when the row of trolleys hit the block.) Does it matter that the hammer moves towards the rod, rather than vice versa?

The contact time is very small, which is why the timer has been replaced by a signal generator and oscilloscope. When the hammer and rod are in contact the high frequency signal from the signal generator is fed to the oscilloscope. A frequency of about 25 kHz is used, so the time for one oscillation is 1/25 ms. So if four cycles of the signal appear on the screen, then the hammer and rod have been in contact for $4 \times 1/25$ ms. Measure the contact time, and the distance the wave has travelled in this time. Hence calculate the speed of the wave, and estimate the percentage uncertainty involved. Suggest reasons why the percentage uncertainty in this experiment is high. How could the experiment be improved?

D13 Forced vibration of a mass on a spring

2 expendable springs
slotted masses and hanger
Perspex tube or glass tube (wide
enough just to fit outside slotted
masses)
light string

either
vibrator and signal generator
or
wheel with pin offset about 1 cm
fractional horse power motor
variable voltage supply
single pulley

leads

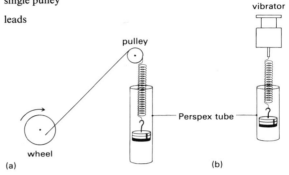

Figure D58
Forced vibration.

How does the mass on a spring behave when it is subjected to a periodi force? How does its behaviour depend on the frequency of the drivin force? Why is the tube necessary?

EXPERIMENTS

D14 Investigations of resonance

Find out what you can about forced oscillations. Devise your ow experiment(s) using one of the two sets of equipment shown belov Make observations and measurements which reveal in detail som aspect(s) of the relationship between the motion of the forced oscillatc and the driver. Investigate the effect of damping (energy loss).

D14a Resonance of a pendulum

resonance kit
metre rule
retort stand base, rod, boss, and clamp
stopclock
card

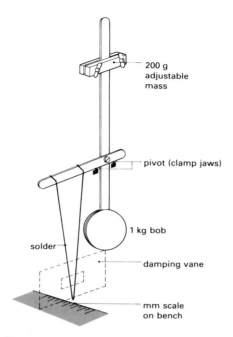

Figure D59
Resonance of a pendulum.

The solder pendulum is the driven oscillator; the horizontal strip provides the driving force. The driving frequency can be varied by moving the adjustable mass.

D14b Resonance of a mass on a spring

2 expendable springs
mass, 50 g
thread
single pulley with suitable support
large beaker or measuring cylinder
metre rule

either
signal generator
vibrator
or
wheel with pin offset about 1 cm
motor
power supply for motor

drinking straws

The mass on the spring is the driven oscillator. The thread supplies the driving force. Try the experiment with and without damping.

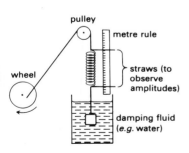

Figure D60
Resonance of a mass on a spring.

DEMONSTRATION
D15 Barton's pendulums

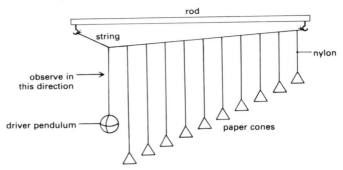

Figure D61
Barton's pendulums.

This demonstration provides the opportunity to observe and compare the motion of several forced oscillators with differing natural frequencies.

The driver swings in a direction in and out of the paper. Relative damping can be varied by adding extra mass to the paper cones, *e.g.* using plastic curtain rings. What factors control the amplitude, phase and frequency of the forced vibrations of the paper cones?

EXPERIMENT
D16 Standing waves on a rubber cord

signal generator
vibrator
xenon flasher
rubber cord (0.5 m long, 3 mm square cross-section)
2 retort stand bases, rods, bosses, and clamps
4 metal strips (as jaws)
2 G-clamps, large
leads

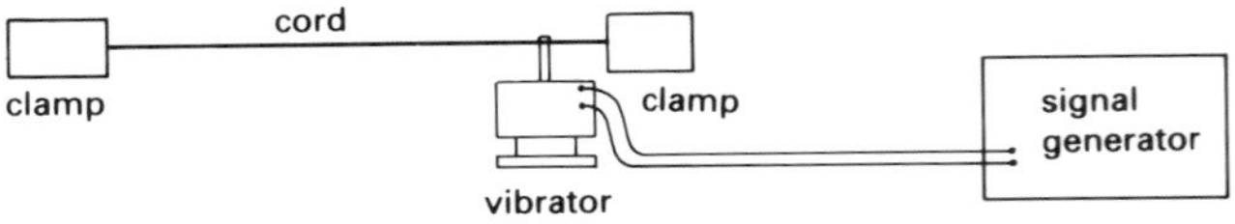

Figure D62
Standing waves on a rubber cord.

Why does the cord show a large response at certain frequencies, but not at other frequencies?
How are the resonant frequencies related to each other and to the length of the cord?
When the vibrator is first switched on a wave travels along the cord. How does this develop into a standing wave?
What factors affect the amplitude of vibration at the antinodes?
Is there an optimum position for the vibrator? Is the vibrator always at a node or an antinode?

DEMONSTRATIONS
D17   More complicated standing waves

Many types of standing wave can be demonstrated. There will generally be more than one resonant frequency, and in some cases these frequencies will have a simple relationship. At each resonant frequency (for each mode of vibration) you should try to understand how the system is vibrating. You should be able to make an estimate of wavelength, and possibly wave speed.

D17a   The Kundt dust tube

signal generator
small loudspeaker, about 60 mm diameter
measuring cylinder
cork dust
paper

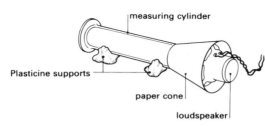

Figure D63

A thin layer of cork dust on the bottom of the glass tube shows the position of nodes and antinodes.

D17b   Longitudinal standing waves in rods

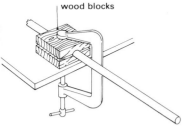

Figure D64

glass, steel, or brass rod, about 10 mm diameter and about 1.5 m long
G-clamp
wooden blocks
cloth
rosin (for metals)
alcohol (for glass)

Rub the rod with the rosined or dampened cloth. If you can find a way of estimating the frequency of the sound emitted, then you can go on to estimate the speed of sound in the rod. What is the wavelength of the standing wave being produced?

D17c   Vibrations of circular wire rings

signal generator
vibrator
xenon flasher
copper wire, 0.9 mm diameter, or thinner steel wire
leads

Figure D65

To get standing waves on the wire loop, it must be vibrated at specific frequencies. How are these frequencies related to each other? To the number of nodes on the loop?

D17d Longitudinal standing waves

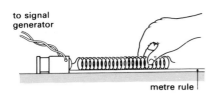

to signal generator

metre rule

Figure D66

signal generator
vibrator
xenon flasher
long spring
metre rule
leads

The spring should be stretched. What frequencies give standing waves? How are these frequencies related to each other? To the wavelengths?

D17e Vibrations in a rubber sheet

signal generator
large loudspeaker
xenon flasher
sheet of rubber
2 retort stand bases and rods
big metal ring (*e.g.*, embroidery ring)
leads

A rubber diaphragm stretched over a ring can be excited by placing it over a loudspeaker. Lines drawn on the rubber help to show up the vibration patterns. You should be able to see several modes of vibration. Is there a simple relationship between the resonant frequencies? Between the patterns obtained?

D17f   Chladni figures

signal generator
vibrator
square or round metal plate
sand
leads

The metal plate is attached centrally to the vibrator. Sand is used to reveal the vibration pattern. Many modes of vibration exist. Measure the resonant frequencies. Are they simply related?

D17g   Vibrations of a loudspeaker cone

signal generator
large loudspeaker
xenon flasher
leads

Watch the resonance of a loudspeaker cone under stroboscopic illumination.

D17h   Standing waves in a round bowl

signal generator
vibrator with dipper attached
Petri dish
large plastic bowl
wooden block

Arrange a dipper to generate circular ripples at the centre of the bowl or dish. You may be able to see stationary ring patterns on the water surface at various frequencies.

D17i   Standing waves – musical instruments

oscilloscope
microphone
assorted musical instruments

Standing waves are set up when musical instruments are either plucked, blown, struck, or stroked. Usually the standing wave pattern is a complex one, and waves of several different frequencies are present. You can see the different waveforms of the sounds produced by different instruments on the oscilloscope.

D

HOME EXPERIMENTS

DH1 Making a chronometer

Throughout the history of science and technology one of the mos
difficult problems has been the development of an accurate, robus
chronometer. The task is to make a device which can measure, accurat
to 1 second, any time interval between 0 and 3 minutes. Compare you
design with others in the class and see who can produce the mos
accurate device.

DH2 Make your own wave machine

Fix drinking straws at intervals along a piece of sewing tape, as show
in figure D67.

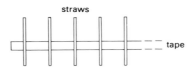

Figure D67
Simple wave machine.

You can add mass to each straw by putting small screws into th
ends. Suspend the system vertically from one end of the tape; or clam
each end of the tape so that it is horizontal. You are now ready to d
experiments with the wave machine; try, for example, sharply twistin
one of the straws around the tape at an angle of about 45°.

DH3 A mechanical oscillator

Using either a home-made spring or combinations of springs that yo
can find in the laboratory, make a mechanical oscillator that vibrate
say, at 5 Hz. You should not approach this problem on a 'Trial an
error' basis, but rather from detailed knowledge of the spring constar
of your device. You must also try to think of accurate ways to measur
the frequency of your oscillator. Maybe you might use a strobe, o
perhaps some voltage-inducing device coupled with an oscilloscope.

DH4 Standing waves in a rectangular tank

plunger

Figure D68

It is easy to excite water into a 'slopping' mode, which is why it is hard to carry pans of water. What can you say about the Q (quality factor) of this system? You may be able to excite other modes of oscillation using the plunger, for example, by moving it up and down in the middle of a large tank of water, and in other positions.

DH5 Standing waves under a running tap

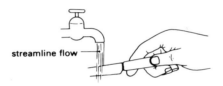

streamline flow

Figure D69

Hold a knife-blade about 2 cm below a smoothly running tap. Look for standing waves in the water flow. How does the pattern depend on the speed of water flow? On the separation of knife and tap?

DH6 Step waves under a running tap

Water flowing down from a tap onto a flat surface, like the under-surface of a baking tray, shows a strange discontinuity: as the water flows away from the impact point, there is a step-like increase in its depth, at a distance r all round the impact point. (This is nothing to do with standing waves, but is a nice piece of physics relevant to this Unit!) Find out what factors affect r and try to explain this phenomenon.

QUESTIONS

The motion of oscillators

1(I)a A car body oscillating vertically gives a time trace as shown in figure D70. Why is it almost certain that the lefthand side of the graph is earlier in time than the righthand side?

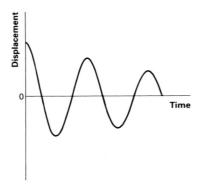

Figure D70

b Sketch time traces of these motions:
i a tennis-ball's displacement measured from the net during a rally;
ii a yo-yo moving vertically;
iii a pendulum hanging against a wall, after it is pulled away from the wall and released (it loses a fraction of its energy on each collision with the wall).

2(P) The graph in figure D71 shows the time trace of a pendulum (about 10 m long) during one complete oscillation.

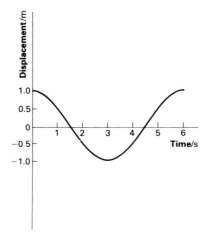

Figure D71

a What is the amplitude of this oscillation?

b Consider another identical pendulum oscillating with the same amplitude as the first one, but with a phase lag of $\pi/2$ radians.
i What is this phase difference as a fraction of one cycle?
ii What would be the displacement of this second pendulum after 3 s? And after 4.5 s?

c The second identical pendulum is now stopped, and released again in phase with the first and with an amplitude of 0.5 m.
i What is its displacement at $t = 3$ s?
ii At what time(s) will the two pendulums have the same displacement?

Time and its measurement

3(E) In 1761, the Board of Longitude, which had been set up to consider claims for a government prize of £20 000 for a clock which could keep accurate time at sea, arranged a test of a clock made by John Harrison. This clock was more accurate and reliable than any other mechanical clock that had ever been made (except pendulum clocks, which will not work well on a rolling ship). Harrison and his clock were sent on a sea voyage to Jamaica, to test the clock. How could a test possibly be made, if there were no better clocks to compare it with?

D

4(E) Suppose that an inventor brings to a standards laboratory a clock that he claims will keep very accurate time. His clock punches dots onto a roll of moving paper tape, and he claims:
1 That it does so at very regular intervals.
2 That these intervals are accurately $\frac{1}{5}$ second.

The laboratory has a standard clock of its own which produces an audible 'pip' once every half second.

a How would the laboratory set about testing the two claims made by the inventor?

b Is it possible that claim *1* can be true, but claim *2* not true? Is the reverse possible?

c The laboratory finds that claim *1* is not true; it judges that the time-dots come irregularly. The inventor replies that it is the laboratory clock that is irregular, not his. Can the conflict of opinion be resolved?

5(E) Until 1964, a time interval of one second was defined as $\dfrac{1}{24 \times 60 \times 60}$ of a 'mean solar day'. To-day one second is the time for 9 192 631 770 oscillations of a particular radiation from a caesium atom.
Discuss:

a whether one of these definitions is more 'correct' than the other;

b whether one definition is more practical and useful than the other;

c why the change was made.

Angles

6(P) A circle has a radius of 2 m. What angle is subtended at the centre of the circle by an arc of length 4 m? Give your answer in radians and also to the nearest degree.

Angle approximations

7(L)a In terms of θ (in radians) and r, what is the length (in figure D72) of:
  i the arc AQ
  ii the straight line AP?

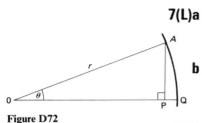

Figure D72

b If θ is very small, it can be seen from the diagram that arc length AQ and straight line AP are approximately equal. From your answer to **a**, what does this suggest about $\sin \theta$ and θ, if θ is small?

8(L) Use a calculator to make a table of θ (in radians) and $\sin \theta$, $\cos \theta$, and $\tan \theta$ for $\theta = 20°$, $10°$, $5°$, $2°$, $1°$, $0.5°$, $0.1°$.

a What is the percentage error made in assuming $\sin \theta = \theta$ (radians) for:
  i $\theta = 10°$
  ii $\theta = 1°$
  iii $\theta = 0.1°$?

b As θ becomes small, what are suitable approximations for
  i $\cos \theta$, *ii* $\tan \theta$?

The behaviour of waves

9(P) Two fishing floats on a lake are 20 m apart. Waves travel along the water surface from a point in line with the floats, so that each float bobs up and down 30 times per minute. Someone notices that when one of the floats is on a wave crest, the other is in a trough, and there is one crest between them. Calculate the speed of the wave.

10(P) A wave pulse travels along a Slinky. An instant in the process is sketched in figure D73. Q, R, and S are points on the Slinky.

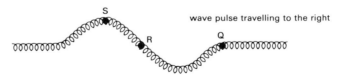

wave pulse travelling to the right

Figure D73

a Describe the states of motion (velocity and acceleration) of the points Q, R, and S at this instant.

b What two physical factors affect the speed of the wave along the Slinky? Explain qualitatively why each factor affects the speed.

11(E) A compression pulse travelling along a Slinky spring carries energy. Is it kinetic energy (because the coils of the spring are moving) or potential energy (because the spring is squashed by the pulse) or both?

What happens to the energy when a compression pulse going one way coincides with an expansion pulse going the opposite way, and the two superpose to give no net compression or expansion?

12(P) Figure D74 shows three hypothetical graphs of displacement, s, against distance for three different travelling waves at time $t = 0$. In each case, draw a graph of displacement, s, against time, t, for the point marked P, for the range $t = 0$ to $t = 6$ seconds. Show as much numerical information on the graphs as you can.

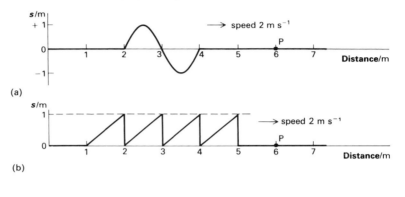

(a)

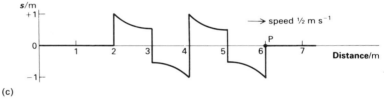

(b)

(c)

Figure D74

13(R) This question is about the behaviour of water waves.

Figure D75 shows a wave in water deep enough for the wavelength to be much less than the depth. As the wave moves to the right, the water at P acquires in succession the velocities v_1, v_2, v_3, v_4, and v_5 ($v_5 = v_1$) and it can be shown that it moves in a circle with constant speed, where the radius of the circle is equal to the wave amplitude, A.

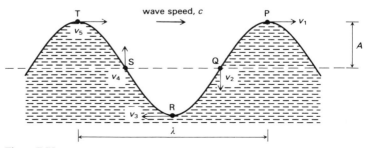

Figure D75

a *i* How would you find the time it takes for a water particle to go once round in a circle?

ii If the water particles are moving in circles, how would you find the magnitudes, and what are the directions, of their accelerations at P, and at Q?

b The speed, c, of deep water waves is given by $c^2 = \lambda g/2\pi$.

Sketch graphs of *i* speed, c, and *ii* frequency, against wavelength, λ, for deep water waves in the range $\lambda = 1$ m to $\lambda = 100$ m. Indicate on your sketch the orders of magnitude of speed and frequency for wavelengths 1 m and 100 m. What would you plot in order to obtain straight line graphs relating wavelength, λ, to *i* speed, *ii* frequency?

c *i* Suppose a storm at sea generates waves of wavelengths in the range 1 m to 100 m. What wavelengths of waves from the storm will be felt by a ship 100 km from the storm during the 24 hours following the onset of the storm?

ii A small boat at sea has to ride such waves. What will be the speed of the water around circles in waves of wavelengths 100 m and amplitude 10 m? Describe the motion of such a boat (short compared to the wavelength) which rides such waves and also travels forward in the direction of travel of the waves at a mean speed of about $2 \, \mathrm{m \, s^{-1}}$.

iii What will be the maximum vertical acceleration of the water in such waves? How will the maximum force on the yacht causing this acceleration compare with the weight of the yacht?

d If the wavelength, λ, is larger than the depth, d, the water no longer moves in circles and the wave speed for such shallow water waves is given by $c^2 = gd$.

Calculate the wave speed for waves of wavelengths 1 m and 1000 m in a sea of depth 100 m.

(Long answer paper, 1981)

Superposition of waves

14(P) In figure D76, S_1 and S_2 are two water wave sources in a ripple tank. They are vibrating at the same frequency and amplitude. There is a maximum disturbance at A, a minimum at B, another maximum at C, and so on.

a Write an expression for the wavelength of the ripples.

b How will the pattern of maxima and minima change if:

i the two sources are moved closer to each other?

ii the frequency of vibration is increased?

iii the velocity of the ripples is decreased (by reducing the depth of the water in the tank)?

iv S_1 and S_2 vibrate in antiphase?

v (Harder) the frequency of S_1 is slightly greater than that of S_2?

c Why is the amplitude unlikely to fall to zero at the minima?

Figure D76

15(P) With the arrangement of wave transmitter, T, receiver, R, and reflector shown in figure D77(a), the signal strength received at R is a maximum. When the reflector is removed to the position shown in figure D77(b), the signal reaches a minimum. Why? Suggest a value for the wavelength of the radiation. Why can you not be sure of the value? What are some other possible values?

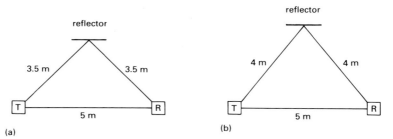

Figure D77 (a)   (b)

16(P) Figure D78 shows three hypothetical graphs of displacement, s, against distance at time $t = 0$ for wave pulses on a stretched spring. Use the principle of superposition to draw the displacement against distance graphs for each spring at time $t = 1$ s.

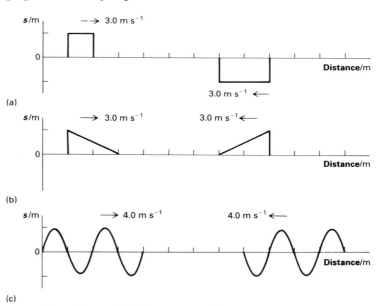

Figure D78   The markings on the distance axes are 1.0 m apart

17(E) How would a researcher who wants to measure wavelengths in *a* the visible region, *b* the microwave region, *c* the X-ray region, and *d* the v.h.f. radio region set about the task? In your answer be sure to give some idea of the order of size of the vital parts of the apparatus.

18(P) A transmitter, T, emits radiation, some of which is reflected from a partially reflecting screen, S_1, and some of which carries on to be reflected from a second screen, S_2 (figure D79). The radiation reflected

Figure D79

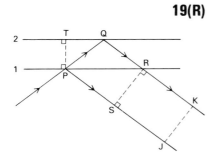

Figure D80

back from S_1 and S_2 is detected by a receiver, R, placed alongside T. At a certain separation of S_1 and S_2 the receiver records zero signal. S_2 is then moved away from S_1. As S_2 is being moved, the detector records signal minimum and S_2 is moved on until the detector again records a minimum signal, a total movement of 120 mm.

a What is the wavelength of the radiation?

b At the original separation the signal detected was very nearly zero; but after S_2 had moved 120 mm the minimum signal was quite perceptible. Why?

19(R) In figure D80, a wave reaching surface 1 is partly reflected at P, but part of it passes to a second parallel surface where it is reflected at Q. There are no phase changes on reflection or transmission. The waves from P and Q are in step at JK if

A $\lambda = 2PT$ **B** $\lambda = PQ$ **C** $\lambda = PS$
D $\lambda = PQ + QR - PS$ **E** $\lambda = PS - QR$

(Coded answer paper, 1979)

Applications of superposition

20(P) In designing a camera lens, it is desirable that as little light as possible is reflected from the front of the lens, so that as much light as possible is transmitted. This can be achieved by 'blooming', coating the outer surface of the lens with transparent material one-quarter of a wavelength thick (see figure D81) — as the following questions will illustrate. (If light strikes either boundary from *above* and is reflected back *upwards*, its phase is changed by π; if it strikes either boundary from *below* and is reflected back *downwards*, its phase is unchanged.) Possible routes for light are shown in figure D82.

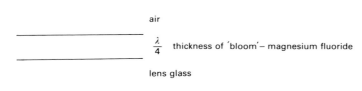

Figure D81

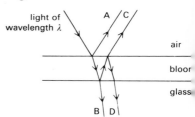

Figure D82

a Is the light following path A in phase or out of phase with the light following path C?

b Compare similarly light following paths B and D.

c Explain why the $\lambda/4$ bloom reduces the light reflected by the lens surface, but increases the light transmitted.

d The thickness of the bloom can only be exactly $\lambda/4$ for one particular value of λ. The chosen value of λ for cameras is usually in the middle o the visible band (4×10^{-7} m to 7×10^{-7} m) – that is about 5.5×10^{-7} m. Explain why the lenses of good cameras usually look purple.

21(P) In the medium wave radio band, waves may reach a receiver by two routes: the ground wave travels direct, while the sky wave is reflected off the ionosphere – see figure D83.

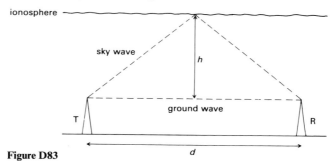

Figure D83

Thus superposition may occur. At night the sky wave is particularly strong; its amplitude is comparable with that of the ground wave. The receiver may receive a strong signal, or almost none at all, depending on the effective height, h, of the ionosphere at that moment. Since h varies over a few seconds, the signal rises and falls – an irritating phenomenon called 'fading'.

Suppose that $\lambda = 250$ m, $d = 120$ km, and h at one moment is effectively 80 km. If, at that moment, the receiver is receiving a maximum signal, then by what distance would h have to change in order for this signal to become a minimum?

22(E) Why do soap films and oily patches on roads appear brightly coloured? (Look this up if you don't know the answer.)

Qualitative motion of oscillators

23(L) Figure D84 illustrates a simple kind of oscillator, for which the restoring force is proportional to the displacement.

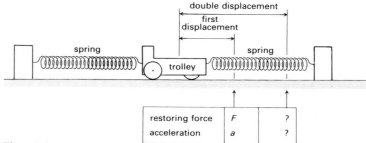

Figure D84

Suppose the trolley is given a small displacement and released. It oscillates, taking a certain time to go from the extreme position to the centre of the motion (one-quarter of an oscillation).

Now suppose it is given twice that initial displacement.

a How has the average restoring force on it been changed?

b How must its average acceleration have changed?

c In the same time, how will the speed acquired compare with the first trial?

d How long will the trolley take to cover the double distance to the centre of the motion, by comparison with the first trial?

24(L) Figure D85(a) shows a trolley tied by two springs, and its displacement–time graph. Figure D85(b) shows the same trolley, with different springs. It oscillates twice as rapidly as before.

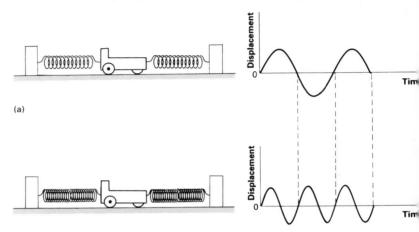

(a)

(b)

Figure D85

a Are the springs in figure D85(b) stiffer or weaker than those in figure D85(a)?

b In figure D85(b), the oscillation time is halved. For comparable positions, how does the speed of this second oscillator compare with that of the first oscillator?

c The speed of the second oscillator must be attained in half the time taken by the first oscillator. How do the accelerations of the two oscillators compare?

d $F = ma$, $F = -ks$ where F is the force, m the mass, a the acceleration, k the spring constant, and s the displacement.

  If the masses of the two trolleys are the same, how do the spring stiffnesses k compare?

e If T is the oscillation time, which relationship below agrees with the answer to **d**?

$$T \propto k^2; \quad T^2 \propto k; \quad T \propto \frac{1}{k^2}; \quad T^2 \propto \frac{1}{k}.$$

f To *decrease* the oscillation time, by changing the mass of the trolley, would one *increase* or *decrease* the mass?

g The answers to **a** to **c** above show that halving the oscillation time means quadrupling the acceleration, or that doubling the time means having one-quarter the acceleration. How would the mass have to be changed to achieve one quarter the acceleration (and so twice the oscillation time)?

h Which of the following relationships agrees with the answer to **g**?

$$T^2 \propto m; \ T \propto m^2; \ T^2 \propto \frac{1}{m}; \ T \propto \frac{1}{m^2}.$$

25(P) The graphs in figure D86 indicate how the force, F, necessary to displace a mass varied with displacement, s, from the rest position for different cases. Each mass oscillates when it is released.

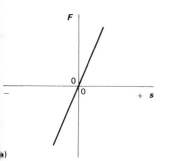

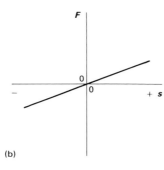
(b)

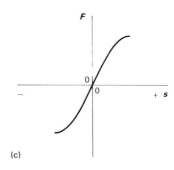
(c)

Figure D86

The second set of graphs (figure D87) shows the displacement–time traces for the above cases. Which trace corresponds with which force–displacement graph?

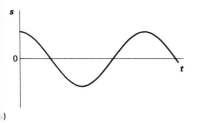

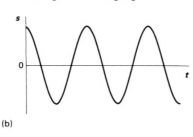
(b)

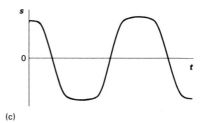
(c)

Figure D87

26(P) Figure D88 shows the displacement–time graph of an oscillator.

a Consider the speed of the oscillator at the four times labelled A, B, C, D. Arrange the times A, B, C, D in order of *decreasing speed*.

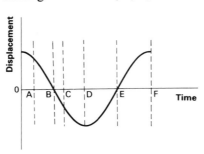

Figure D88

b How does the velocity at time B compare with that at time E?

c How does the velocity at time D compare with that at time F?

d At which of the times 0 to F is the acceleration at its largest value?

e At which of the times 0 to F is the displacement equal in size to the amplitude of the motion?

f Consider the time intervals 0B, 0D, 0F, BE, DF. If the periodic time of the oscillator is T, write down each interval in terms of T. ('0F $= 3T$' is the sort of answer expected, though this particular answer would be wrong.)

27(E) Figure D89 shows three things which would oscillate in a laboratory on Earth. Which, if any, would oscillate in a spacecraft going at steady speed a long way from the Earth and from any planet or star?

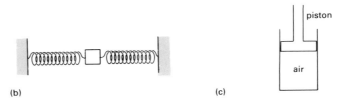

(a)   (b)   (c)

Figure D89

Analysis of a simple harmonic oscillation

28(L) Figure D90 represents a multiflash photograph of the motion of a trolley held between two springs. The distances from the mean position are given in table D1.

Figure D90

Point	Time/s	Displacement/cm
0	0	5.0
1	0.01	4.9
2	0.02	4.6
3	0.03	4.1
4	0.04	3.5
5	0.05	2.7
6	0.06	1.8
7	0.07	0.9
8	0.08	0

Table D1

a What is the value of T for the motion?

b What is the value of ω for the motion?

c Draw a semicircle of radius 5 cm. Mark the points 0 to 8 on to the diameter (8 is the centre of the circle), then project them vertically upwards onto the semicircle, as illustrated in figure D91.

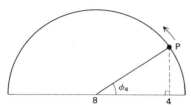

Figure D91

Measure the angles ϕ_1, ϕ_2, etc. They should be equal, showing that ω is constant. Hence the position of the trolley at any moment is the projection of P onto the diameter of the circle. P moves round the imaginary circle at constant angular rate ω. Calculate ω
i in degrees s^{-1}
ii in rad s^{-1}.

d In the experiment, the mass of the trolley was 0.1 kg, and the spring constant was 40 N m^{-1}. Show that this gives a value of ω corresponding to the results of the experiment.

29(L) This question leads to a numerical solution of the harmonic oscillator equation $a = -\dfrac{k}{m}s$. It shows how, if you know the values of k and m and also the initial position and velocity, you can calculate values of position, s, at different times. The method is quite general and can be used to find how a body controlled by a more complicated force law moves. The essence of the numerical method is to calculate what happens during each of a series of short time intervals. (The shorter the intervals the more accurate the result.) The body is assumed to move with constant velocity during each of these short time intervals.

The acceleration of the harmonic oscillator, a, depends on its position s $(a = -\dfrac{k}{m}s)$. From the acceleration we can find the change in velocity during a short time interval Δt:

$$\Delta v = a\Delta t$$

If we know the velocity at the beginning of this time interval, we can calculate the velocity at the end:

$$v_{new} = v_{old} + \Delta v$$

We can use the velocity to find how far the object moves in one time interval:

$$\Delta s = v \Delta t$$

And hence the new position from:

$$s_{new} = s_{old} + \Delta s$$

One step is now complete. We have a new s from which we can get a new a, and we can repeat the whole process for the next time interval. The accuracy of the calculation is improved if we work out the velocity at times midway between the times at which we want to calculate displacement and acceleration.

We know the displacement, s, at $t = 0$, and hence can work out the acceleration, a, at $t = 0$. We will use this value of a to work out the speed, v, in the middle of the first interval Δt, that is, at $t = \frac{1}{2}\Delta t$. Later values of v will be calculated for $t = 1\frac{1}{2}\Delta t$, $2\frac{1}{2}\Delta t$, etc.

Consider a specific case:

$k = 10\,\mathrm{N\,m^{-1}}$
$m = 1\,\mathrm{kg}$

initial values $s = 0.1\,\mathrm{m}$
$v = 0\,\mathrm{m\,s^{-1}}$

at $t = 0\,\mathrm{s}$

a What is the acceleration at $t = 0\,\mathrm{s}$?

Take time increments $\Delta t = 0.1$ s.

b What is the change in velocity in the first *half* interval, that is, between $t = 0\,\mathrm{s}$ and $t = 0.05\,\mathrm{s}$?

The oscillator is at rest at $t = 0\,\mathrm{s}$.

c What therefore is the velocity at $t = 0.05$ s?

d Use this value to work out the distance travelled, Δs, between $t = 0\,\mathrm{s}$ and $t = 0.1$ s.

The displacement at $t = 0\,\mathrm{s}$ was 0.1 m.

e What is the displacement at $t = 0.1\,\mathrm{s}$?

f What is the acceleration at $t = 0.1\,\mathrm{s}$?

g From the acceleration at $t = 0.1$ s, work out the change in velocity between $t = 0.05$ and $t = 0.15\,\mathrm{s}$.

h What therefore is the velocity at $t = 0.15\,\mathrm{s}$?

The calculations now become routine, repeating steps **c** to **h** for each successive time step.

It is convenient to record your values in a table. The starting values and the velocity at 0.05 s have been entered. Other values you need to calculate are shown ▒▒▒▒

t/s	s/m	a/m s^{-2}	v/m s^{-1}	Δs/m
0	0.1	-1	0	
0.05			-0.05	-0.005
0.1	▒▒	▒▒		
0.15			▒▒	▒▒
0.20	▒▒	▒▒		
0.25			▒▒	▒▒
0.30	▒▒	▒▒		

Continue in this way at least until s becomes zero, preferably until it reaches its greatest negative value.

i Use your results to sketch graphs of s against t, v against t, and a against t.

j Estimate the period of oscillation from your results, and compare this estimate with the value calculated from

$$T = 2\pi \sqrt{\frac{m}{k}}$$

D

k Suppose k were twice as great, $20\,\mathrm{N\,m^{-1}}$. How would the answers to **a**, **b**, **c**, and **d** change? Would the period of this oscillator be greater or smaller?

l If m were 2 kg (but k still $10\,\mathrm{N\,m^{-1}}$) how would the answers to **a**, **b**, **c**, and **d** change? What would be the effect on the period of the oscillator?

30(E) The routine for solving the problem of the harmonic oscillator in question **29** can easily be made into a computer program.
     Write a program that will print out values of v at $t = \Delta t/2,\ 3\Delta t/2,\ 5\Delta t/2 \ldots$, and values of s and a at $t = 0,\ \Delta t,\ 2\Delta t \ldots$.
     Some investigations you can make with the program:

a Does it show that the oscillator is isochronous, that is that the period does not depend on the initial displacement?

b Explore the effect on period T of changing k and m.

c How sensitive is the program to the value of Δt?

d Add a line to your program so that it represents a damped oscillator. Explore the effect of a constant frictional force, friction proportional to velocity, or friction proportional to (velocity)^2.

31(R) 'Figure D92 is a stroboscopic photograph of the motion of a harmonic oscillator, released from rest at the extreme left of the photograph. It is, however, detached from its restoring force when it reaches its central position, and as a result continues on at constant velocity.'

Check, by drawing a graph of velocity against time, whether the above description is plausible.

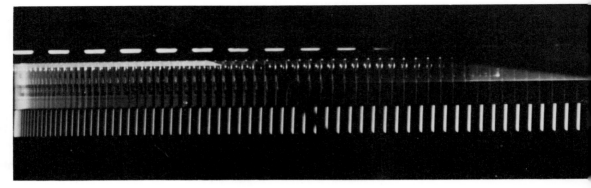

Figure D92

Further examples of oscillations and S.H.M.

32(R) Figure D93 shows a liquid in a U-tube of uniform cross-section. In figure D93(b) the liquid is displaced as shown.

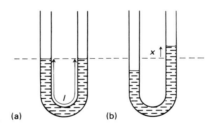

liquid density, ρ
tube cross-section, A
total length of liquid, l
gravitational field strength, g

Figure D93

a In figure D93(b), what is the value of the force tending to return the liquid to its equilibrium position?

b Explain how this situation fulfils the necessary condition for simple harmonic motion.

c Find expressions, in terms of the symbols given, for
i the acceleration, a, of the liquid;
ii the angular velocity, ω, of the S.H.M.;
iii the periodic time, T, of the S.H.M.

33(R) Figure D94 shows a cylindrical buoy floating in water. In equilibrium, figure D94(a), the buoy floats so that a length L, of its total height, h, i below the water surface. The buoy is displaced a further depth, x, then released, figure D94(b). Find an expression for the periodic time of the oscillations, in terms of the symbols given, and show that it has the dimensions of time.

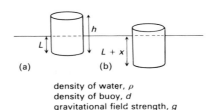

density of water, ρ
density of buoy, d
gravitational field strength, g

Figure D94

34(P) An object oscillates on the end of a spring. If the mass of the object is multiplied by four, what change must be made in the amplitude of the oscillations for the maximum speed of the object to be unaltered?

35(P) In a small harbour, the depth of water at high tide is 10 m, and at low tide is exactly zero. The depth of water follows approximate S.H.M., with amplitude 5 m and periodic time 12 hours. (Really about $12\frac{1}{2}$ hours, but use 12 hours for this question.) Fishing boats can only leave the harbour when the depth of water is 9 m or more. On one day, high tide is at noon.

   a What is the value of ω (in radians per hour)?

   b If d is the depth of water *above* the mean depth (5 m), write an equation of the form $d = A \cos \omega t$, substituting the appropriate values for A and ω.

   c Calculate the latest time in the afternoon at which boats can leave the harbour.

36(L) At room temperature the atoms of a particular solid vibrate with S.H.M. of frequency 10^{13} Hz and amplitude 12×10^{-12} m. (This is a much simplified model of a solid, but it is a useful starting point.) The mass of each atom is 10^{-25} kg.

   a About what fraction of a typical atomic separation is this amplitude?

   b Calculate the approximate value of the force constant, k, between two atoms.

   c What is the total energy of vibration of one atom
     i in joules,
     ii in eV?

37(P) A baby in a 'baby-bouncer' is a real-life example of a mass-on-spring oscillator. The baby sits in a sling suspended from a stout rubber cord, and can bounce himself up and down if his feet are just in contact with the ground (an example of resonance). If suspended out of contact with the ground, he oscillates if displaced and released. Suppose a baby of mass 5 kg is suspended from a cord with spring constant $500 \, \text{N m}^{-1}$.

   a What is the initial (equilibrium) extension of the cord?

   b The baby is pulled down a further distance, 0.1 m and released. What is the value of ω?

c How long after his release does he pass through his equilibrium position?

d With what speed does he pass through his equilibrium position?

38,39(R) Figure D95 shows the tip of a stereo record-player stylus resting in the right-angled groove of a record. It rests on the lefthand and righthand walls of the groove, labelled L and R.

   As the record turns, the stylus is moved by either wall moving at right angles to itself, as indicated by the broken lines. Suppose that both walls move with simple harmonic motion. Here are five possible combinations of wall motions, of the same frequency, and except where specified, the same amplitude.
A L and R move up and down in phase.
B L and R move up and down out of phase (180° difference).
C L oscillates, but R has zero amplitude of motion.
D The upward displacement of L is a maximum when that of R passes through zero going downwards.
E The upward displacement of R is a maximum when that of L passes through zero going downwards.

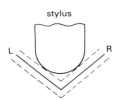

Figure D95

38 In which one of the above does the stylus move only in the vertical direction?

39 In which one of the above does the stylus move only in the horizontal direction?

(Coded answer paper, 1979)

The speed of compression waves

40(R) Figure D96(a) shows a snapshot of a row of identical trolleys joined to each other by identical springs. The trolleys here are at rest and their positions are shown by the scale above the trolleys. Figure D96(b) shows the same trolleys at a later instant. Trolley A has been pushed in such a way that, at this instant, each of the trolleys A, B, and C is moving at a steady speed to the right. Figure D97(a) shows the displacement of each trolley from its original position, at this instant.

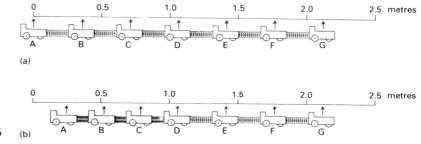

Figure D96

Figure D97(b) will show the speed (in arbitary units) of each trolley at this same instant. The speed of trolley A has been plotted as a cross (X) on this graph.

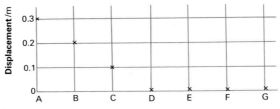

(a)

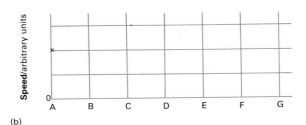

(b)

Figure D97

a Mark crosses on a copy of figure D97(b) to indicate approximately the speeds of all the other trolleys at this instant.

b State what factors affect the speed with which the initial displacement of the first trolley passes along the line of trolleys.

It is suggested that a row of spring-connected trolleys might be a model which would help us to understand the way in which a compression pulse travels along a metal bar.

c Give one reason why you consider that such a row of trolleys would be a good model for this purpose.

d Suggest a way in which you think the model is unlike a metal bar.

(Short answer paper, 1974)

41(L) This question shows how the wave speed for a row of trolleys, discussed in question **40**, can be applied to discussing the speed of sound in a solid.

The speed, c in a compression wave travelling along a row of trolleys linked by springs is given by:

$$c = x\sqrt{k/m}$$

where x is the distance between the centres of successive undisturbed trolleys, m is the mass of a trolley, and k the constant relating the force causing the compression of a spring and the amount of compression. Trolleys linked by springs might be compared to the atoms in a solid linked by interatomic forces. Table D2 gives some comparative values of x and m.

k is roughly $50\,\mathrm{N\,m^{-1}}$ both for the spring linking a pair of trolleys and for the bond between a pair of atoms in steel.

	Trolleys and springs	Atoms in steel
x	0.35 m	2.5×10^{-10} m
m	0.95 kg	9.3×10^{-26} kg
c	2.5 m s^{-1}	?

Table D2

a Suppose that, without changing anything else, the mass of a trolley were reduced by a factor of 10^{-25} (to about the mass of an atom in steel). What would be the speed of a compression wave in such an arrangement?

b Suppose, with this new arrangement, that the distance between trolley is reduced by a factor of 7×10^{-10} (or 10^{-9} if you don't mind a rougher answer). What would be the speed of a compression pulse in such an arrangement?

c As the value of k is much the same for both steel and spring, the answer to **b** should be the speed of a compression wave in steel if it is permissible to scale down from the trolleys to atoms, and if the scaling has been done correctly. The measured speed of sound in steel is about $5100 \, \mathrm{m \, s^{-1}}$. How does your estimate compare?

42(L) This question shows how the speed of sound in a metal can be written down in terms of the Young modulus, E, and the density, ρ.

The speed, c, of a compression pulse travelling along a row of trolleys linked by springs is given by:

$$c = x\sqrt{k/m}$$

where x is the distance between the centres of successive undisturbed trolleys, m is the mass of a trolley, and k is the spring constant in the equation

force $= k \times$ change in length

Unit A, 'Materials and mechanics', question **30** showed, for a specially simplified case, that if E is the Young modulus and x the spacing between atom centres, the atomic bond spring constant, k, is given by $k = Ex$ if the atoms are in a simple cubic array. Consult that question again to see why.

a Suppose the density of the material is ρ, and that each atom occupies a volume x^3. What is the mass, m, of each atom?

b Find a new expression for the speed, c, of compression waves (or sound waves) in a solid in terms of E and ρ, by substituting for k and for m.

c What is the speed of sound in an aluminium rod?

Young modulus $= 7.0 \times 10^{10} \, \mathrm{N \, m^{-2}}$

density $= 2700 \, \mathrm{kg \, m^{-3}}$

Note: The atoms in aluminium are not arranged in a simple cubic array, but the answer to **b** does give a correct expression for the speed of sound in an aluminium rod. For arrays more complicated than cubical arrays, the relations between k and E and between m and ρ are both more complicated than above, but the spacing, x, still cancels out as it did in **b**, together with all the extra geometrical factors which allow for the more complex atomic arrangement.

d The speed of sound in aluminium is very nearly the same as its speed in steel, but steel is several times more dense than aluminium. What must be true of their Young moduli?

Resonance

These questions have to do with various practical problems involving vibration and resonance. You will need to use $T = 2\pi\sqrt{\dfrac{m}{k}}$ or $2\pi f = \sqrt{k/m}$, and to know that the total energy of a harmonic oscillator is equal to $\frac{1}{2}kA^2$.

43(E) Estimate the spring constant of the suspension of a car. Imagine a man sitting over one wheel: how much might the suspension deflect? What frequency of oscillation might a wheel have, considered as a mass on the end of this spring? What sort of repeated ruts on a road would give trouble at, say, 50 kilometres an hour? Why does the car body move with smaller amplitude than the wheels?

44(R) This question is about explaining to non-scientists some applications of scientific ideas which they find confusing. Give an explanation, suitable for a non-scientist, which explains the ideas involved and clears up any errors or confusions in the statement below. Assume that your explanation will include experimental demonstrations and describe these in detail.

D

'People take it for granted that a car needs to be "well sprung" so that the ride is comfortable. It is not so obvious how to get the suspension right: it can be too hard or stiff, so that the car jolts on every bump; it can be too soft so that the suspension sags when the car is loaded and the car sways about over the wheels. It must certainly not make the car body oscillate up and down after the wheels hit a bump, and there could be a danger that the suspension will resonate if the car travels over a regular series of bumps.'

What does need to be taken into account in choosing a suspension, and how are these difficulties avoided?

(Long answer paper, 1981, part question)

45(P) Diatomic molecules such as HF or HCl can vibrate by extension and compression of the bond between the atoms. They behave like a pair of masses held by a spring. For these two molecules, the H atom is much less massive than the F atom or the Cl atom to which it is bonded, and so, roughly, it will be good enough to imagine the H atom vibrating at the end of its bond. (Compare a small mass linked with a spring to a large mass – will the large mass move much?)

Now HF absorbs infra-red radiation very strongly at a wavelength of 2.4×10^{-6} m. The corresponding wavelength for HCl is 3.3×10^{-6} m.

Why can you say straight away that the HF bond is probably rather stiffer (larger force for the same extension) than the HCl bond? Show that the bond stiffnesses are roughly:

$1000\,\text{N m}^{-1}$ for HF $500\,\text{N m}^{-1}$ for HCl

The mass of a hydrogen atom is nearly $1.7 \times 10^{-27}\,\text{kg}$. The velocity of light is $3 \times 10^8\,\text{m s}^{-1}$.

46(L) The model of sodium chloride illustrated in figure D98 shows just one line of ions in a crystal.

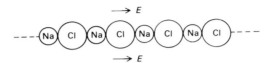

Figure D98

a If an electric field, E, is switched on in the direction shown, what would be its effect on
i the Na^+ ions,
ii the Cl^- ions?

b Calculate an approximate value for the mass of each Na ion. (Atomic mass of $Na = 23$ u; the Avogadro constant $= 6 \times 10^{23}\,\text{mol}^{-1}$.)

c The 'spring constant' between *each* pair of ions is roughly $100\,\text{N m}^{-1}$. Calculate a rough value for the frequency (in Hz) at which a Na^+ ion would oscillate. (Treat each ion as a mass tethered by two springs t the negative ions which, because of their greater mass, move only slightly. Comment on the validity of this assumption.)

d If the electric field oscillates at this frequency, resonance can occur. This can be achieved by directing electromagnetic radiation of the right frequency at the crystal.
i What would be the wavelength of the radiation corresponding to the required frequency?
ii To what part of the electromagnetic spectrum do these waves belong?

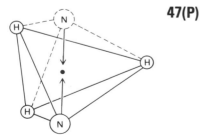

Figure D99

47(P) The ammonia molecule, NH_3, can vibrate with the nitrogen atom passing to and fro 'between' the three hydrogen atoms (figure D99). The frequency happens to be 23 870 MHz. This vibration has been used as the basis of the 'atomic clock'.
    Would the rate of vibration be affected by using molecules of ND_3, with deuterium (heavy hydrogen, ^2H) atoms in place of the hydrogen atoms? (A deuterium atom has a nucleus with one neutron and one proton, whereas the hydrogen nucleus is just one proton.)

48(E) Stand on one foot and allow your other leg to swing freely and easily like a pendulum. What connection do you think there is between the time of these swings and the speed at which you usually walk?

49(E) A car accelerates away from traffic lights, and the driver notices that as the car accelerates past a certain speed, his view in the driving mirror goes 'blurred' and then becomes sharp again. Suggest a reason for this, and a practical way of reducing or eliminating the effect.

50(E) Ultrasonic vibrations may be used to kill bacteria in liquids. How big are bacteria? Discuss the choice of a suitable frequency.

51(E) A record-player pick-up has a stylus or needle that runs in the record groove and is oscillated sideways by the wavy walls of the groove (see figure D100). The performance of a pick-up is often specified by giving the effective tip mass of the stylus and the compliance or flexibility of the stylus when it is pushed sideways. The compliance is the reciprocal of the stiffness, k, which is measured in newtons per metre. The compliance is thus measured in metres per newton (deflection for a given force).

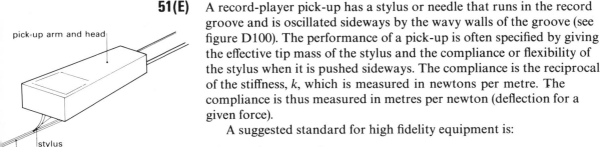

pick-up arm and head

stylus

groove

Figure D100

A suggested standard for high fidelity equipment is:

effective tip mass $< 2\,\mathrm{mg}$
compliance $> 4 \times 10^{-3}\,\mathrm{m\,N^{-1}}$

If the mass and compliance have these values, at what frequency will the stylus resonate? Would you regard this as satisfactory? You might go on to consider whether it would be desirable to have low or high values of tip mass and compliance.

52(E)a Some high fidelity sound systems have the loudspeaker mounted in a 'bass reflex cabinet'. The cabinet in figure D101 is sealed except for a port, and the air in the port (shaded) behaves like a mass acted on by the springiness of the air in the cabinet. Suppose you made such a cabinet and found that it 'boomed' whenever notes of about 300 Hz were reproduced. How might it be improved? Can you suggest any reason why the speaker is not just used on its own, without a cabinet?

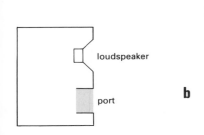

loudspeaker

port

Figure D101

b A violin, or a guitar, has its strings held taut over a hollow wooden box which contains air and has holes in it (or a hole). Air in such a box can vibrate, the air near the hole acting as a mass driven in or out by the springiness of the air in the box.

What will be the effect on the sound produced when a string is sounded at a frequency near to the resonant frequency of the air in the body of the instrument? What other parts of a violin or guitar can resonate to notes from the strings?

53(R) This question is about measuring the acceleration, and so the velocity and displacement, of a moving vehicle, by making observations on masses carried within the vehicle.

Figure D102 shows the principle of one sort of accelerometer (a device for measuring acceleration). A mass, m, is free to move horizontally within a case, but is restrained by springs fixed to the case. A pointer on the mass can move over a scale fixed to the case. When the case and mass are at rest, the pointer is opposite the zero mark on

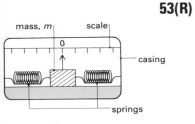

mass, m scale

casing

springs

Figure D102

the scale. When the pointer shows a displacement, x, from zero, the net force exerted by the springs is kx.

a Explain why, when the mass and case are moving at constant velocity in the horizontal direction, the pointer still reads zero. Assume that the velocity has been constant for a long time.

b If the casing is in a state of steady acceleration a to the *left*, explain carefully in words why the pointer now has a fixed displacement, saying whether the displacement is to the left or to the right and explaining why. Assume that the acceleration has been constant for a long time.

c Give an expression for the magnitude of the displacement in **b**.

d If the casing were to be suddenly displaced from rest by a sharp blow from a hammer, for example, and then held at rest, describe the subsequent motion of the mass if there is a small amount of friction between it and the casing.

e It is suggested that, in use to measure varying accelerations, it would be good to have zero friction between the mass and the casing. Argue briefly for or against this idea.

f Suppose that, in use, appreciable changes of acceleration are expected to occur over times not exceeding time t. Give an argument to help decide whether the period, T, of natural oscillation of the mass and springs should be large, or should be small, compared with t.

g In designing an accelerometer for use in a car, a period, T, of $\pi/5$ seconds was chosen and it was assumed that accelerations up to $2\,\mathrm{m\,s^{-2}}$ should be measured. What would be the displacement at an acceleration of $2\,\mathrm{m\,s^{-2}}$? (The values of m and k are not needed.)

h Suppose that it is decided that an accelerometer for use in a car accelerating at up to $2\,\mathrm{m\,s^{-2}}$ should have a period of 2π seconds. What problems would arise in designing this accelerometer?

(Long answer paper, 1979)

Quality factor, Q

54(L) When an excited atom emits its excess energy, this is equivalent to an oscillating electron losing energy by damping. The energy is emitted as light or other electromagnetic wave radiation, each cycle of the wave corresponding to one oscillation of the electron. Suppose that one particular atom returns to its unexcited (ground) state, emitting a burst of radiation with $\lambda = 5 \times 10^{-7}$ m; and that Q for the oscillating electron is 10^7.

a Roughly how many complete cycles of the radiation occur?

b The speed of the electromagnetic wave is $3 \times 10^8\,\mathrm{m\,s^{-1}}$. What is its frequency?

c How long does it take for the atom to lose its excess energy?

d What length in space does the burst of radiation occupy?

Notice that this burst of radiation takes a finite time to be emitted (part **c**). During this time the atom itself (if gaseous) will have moved a significant distance. This causes a change in λ for the radiation, due to the Doppler effect. This is one reason why spectral lines, even using the best possible optical equipment, are not perfectly sharp.

55(E)   Each of the following objects oscillates, possibly in resonance, during its normal operation. Discuss whether each should be designed with a value of Q which is high (>100), intermediate, or low (<2).

a   The platform of a suspension bridge.

b   The mountings for a machine like a lathe.

c   A car-body on its suspension.

d   The balance wheel of a mechanical watch.

e   The tuning circuit of a radio.

f   The moving arm of a ticker-timer.

Standing waves

56(P)   An elastic string is clamped at both ends as shown in figure D103. Near one end a vibrator is loosely attached to it. The vibrator oscillates with fixed amplitude and variable frequency. A graph of maximum oscillatory amplitude along the string against frequency of the vibrator looks like figure D104.

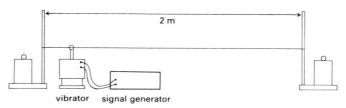

vibrator   signal generator

Figure D103

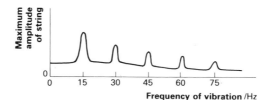

Figure D104

a   Sketch the instantaneous appearance of the string at
i   15 Hz,
ii   30 Hz, and
iii   45 Hz

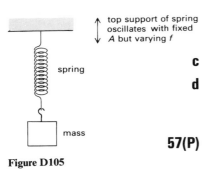

Figure D105

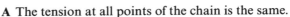

N   A

Figure D106

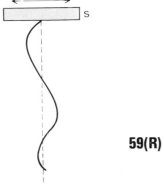

Figure D107

b What is the wavelength of the waves on the string at
i 15 Hz,
ii 30 Hz, and
iii 45 Hz?

c What is the speed of travel of the waves along the string?

d How is the above situation
i similar to,
ii different from the case in figure D105.

57(P) An organ pipe (or any wind instrument) closed at one end can allow standing waves which have a node (N) at that end and an antinode (A) at the other (neglecting a small 'end-correction'); see figure D106. One such pipe has a fundamental note of 64 Hz. Consider what other wavelengths are possible for standing waves in this pipe, and calculate the next two higher frequencies at which the pipe vibrates.

58(R) Figure D107 represents a hanging chain of constant mass per unit length. The support at S is gently oscillated from side to side sending transverse waves down the chain. These are reflected at the free bottom end and a stationary mode of oscillation (a standing wave) is set up.
    Which one of the following statements is correct?

A The tension at all points of the chain is the same.
B The speed of a wave at all points along the chain is constant.
C The frequency of oscillation of each part of the chain is constant.
D The wavelength of a wave travelling along the chain is constant.
E Every point on the chain oscillates with the same amplitude.

(Coded answer paper, 1974)

59(R) In this question you are required to estimate the tension in a violin string which vibrates at a natural frequency of 650 Hz. The string is made of steel. The speed of transverse waves along a stretched wire is given by $c = \sqrt{T/\mu}$, where T is the tension in the wire, and μ is its mass per unit length.

a Starting from the formulae $c = f\lambda$ and $c = \sqrt{T/\mu}$, obtain an expression for the frequency (f) in terms of the tension (T), the length (L), the cross-sectional area (A), and the density (ρ) of the string.

b Estimate values for the quantities you will need to know in order to calculate the tension using the above expression.

c Combine your estimates in order to obtain a value for the tension in the string.

d Say in a few words how you decide on the appropriate number of significant figures to give in the answer.

(Short answer paper, 1981)

Unit E
FIELD AND POTENTIAL

Trevor Sandford
Henbury School, Bristol

SUMMARY OF THE UNIT

E

SUMMARY OF THE UNIT

INTRODUCTION

This Unit is about electric and gravitational fields and potential. You have probably used the word 'field' before to describe effects due to magnetism, electricity, or gravity, but here we develop further what is meant by the term and explore the relationships between fields, forces, energy, and potential for both charges and masses. Electricity and gravity have much in common – their inverse-square force laws, for example, and the mathematics that describe them. They also have enormous differences: gravity only becomes noticeable when huge masses are involved but its effect is felt over astronomical distances. By contrast, electrical forces dominate the behaviour of atoms and molecules on the microscopic scale.

Although this area of physics can be treated in a rather abstract mathematical way, the emphasis in this course is on understanding the physics and being able to apply fundamental ideas to a variety of situations. These range from space travel and satellites to the structure of atoms, molecules, and crystals; from printing, painting, and photo copying to sparks, 'static', and thunderstorms.

The Unit uses ideas about capacitors, charge, and potential difference from Unit B, 'Currents, circuits, and charge'; vectors and dimensions, introduced in Unit A, 'Materials and mechanics'; and Newton's Laws of Motion. Ideas from this Unit are used later in Unit F 'Radioactivity and the nuclear atom', Unit J, 'Electromagnetic waves' and Unit L, 'Waves, particles, and atoms'.

Section E1   THE UNIFORM ELECTRIC FIELD

Fundamental ideas

New ideas:
$E, V, \Delta V/\Delta x$

Parallel plates:
$V, d, A, Q, \varepsilon_0, \varepsilon_r$

The fundamental ideas developed here are the concept of field strength, electric potential, equipotentials, and potential gradient. There is a detailed study of the field between charged parallel plates and how it relates to the p.d. between the plates, their separation and their area, the charge stored, and the medium which separates them. Applications of these ideas cover a range of situations from industrial processes to domestic problems.

Observing a uniform electric field

DEMONSTRATION E1
The 'shuttling ball'

DEMONSTRATION E2
Charged foil strip as a field detector

Experiments on a charged ball between charged parallel plates show that the force, F, on the ball depends on the p.d. between the plates, and on their separation, d (figure E1). A charged foil 'detector' shows that the force is uniform over most of the region between the plates, and that its size and direction depend on the size and sign of the charge, Q on the foil as well as on V and d (figure E2).

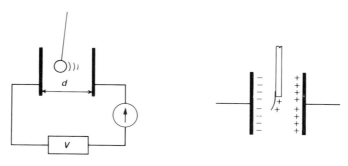

Figure E1

Figure E2

Definition of electric field strength, E

QUESTIONS 1 and 2

The strength of a gravitational field is the force per unit mass; by analogy electric field strength is defined as the force per unit charge,

$$E = \frac{F}{Q}.$$

units of E are NC^{-1}

If, then, we could measure the charge on the foil and the electric force exerted on it, we could calculate the field strength. But both of these measurements prove impracticable: a small charge has too small a force on it, but a larger charge disturbs the field and gives a misleading result. This problem is solved by considering the links between F, Q, V, and d.

QUESTION 3

A charged drop is moved across the space between two charged plates. If a constant force, F, is exerted over distance d, then the energy transformed $= Fd$. But charge Q is then moved through a p.d. V, so the electrical energy transformed is QV.

Millikan's experiment, Unit B

energy = force × distance

Hence

$$Fd = QV.$$

and

$$\frac{F}{Q} = \frac{V}{d}$$

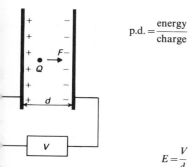

Figure E3

p.d. $= \dfrac{\text{energy}}{\text{charge}}$

But $\dfrac{F}{Q}$ is equal to the electric field strength E.

$E = \dfrac{V}{d}$
for a *uniform* field *only*

Note that this is only the case when the force, F, is uniform over all the region – in other words, for a *uniform field*. However, it is extremely useful since the field strength can be measured simply using a voltmeter and a ruler, and expressed in $V\,m^{-1}$.

Alternative units of $E = V\,m^{-1}$
(equivalent to $N\,C^{-1}$)

$E = \dfrac{V}{d}$ is true *only* for a uniform field, but $E = \dfrac{F}{Q}$ is true in every

QUESTIONS 4 and 6

situation: it is quite general, though in practice less useful.

E

Patterns of electric field

DEMONSTRATION E3
Electric field patterns

QUESTION 5

Semolina particles floating on an organic liquid orientate themselves in the presence of an electric field to reveal its shape, rather as iron filings do in a magnetic field. Parallel, point, and circular electrodes may be used to show different shapes of field (figures E4 and E5).

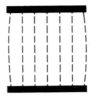
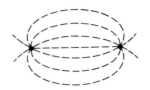

Figure E4

Using a flame probe

DEMONSTRATION E4
Measuring potentials in a uniform field
using a flame probe

The field strength between charged parallel plates, then, is uniform over the region between them and at all points is equal to the potential difference per metre in that region. The p.d. between the *plates* can be measured with a voltmeter or oscilloscope; the p.d. between one of the plates and a point *in space* between them can be measured with the flame probe.

QUESTION 7
The flame prevents any charge building up on the probe and disturbing the field. If one of the plates (usually the earthed one)

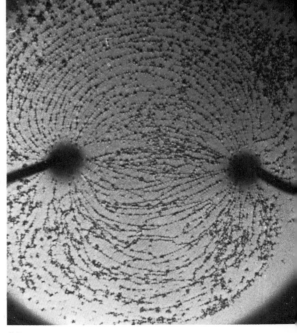

(a)   (b)

Figure E5 (part)

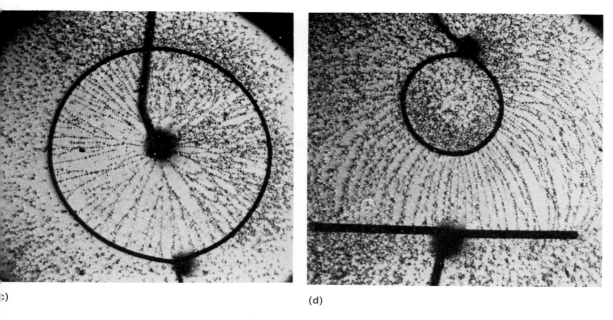

(d)

Figure E5 (part)
Electric field patterns. (Note that there is no field inside the ring in (d).)
Colin Price

referred to as our zero of potential, then we can define the *potential* at any point in the space as being equal to the potential difference between that point and the reference plate (figure E6).

Field and potential gradient

Moving the probe steadily from one plate to the other reveals a steady change in potential. A graph of potential, *V*, against distance, *x*, from the reference plate is a straight line; its gradient $\dfrac{\Delta V}{\Delta x}$ is equal in magnitude to $\dfrac{V}{d}$, the field strength. So the *potential gradient* at any point gives the strength of the field: the uniform field has uniform potential gradient (figure E7). Since *V* increases towards the positive plate, and *E* is

Figure E6
Flame probe.

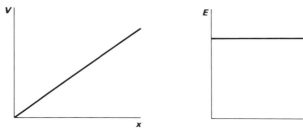

Figure E7

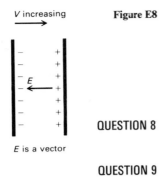

V increasing

Figure E8

E is a vector

QUESTION 8

QUESTION 9

EXPERIMENT E5
Plotting equipotentials

QUESTIONS 10 and 11

Figure E9

EXPERIMENTS E6, E7
Charge on parallel plates
(coulombmeter, reed switch)

Figure E10

directed toward the negative plate (the direction of the force on
positive charge, see figure E8), we must insert a negative sign to show th
direction in which *E*, a vector, acts:

$$E = -\frac{\Delta V}{\Delta x}$$

Now since we can consider any bit of any field to be uniform over a
infinitesimally small distance Δx, this relationship can be applied to an
shape of field, not just the uniform field, provided we speak of th
potential gradient at a *point* only. This is very useful later in Sections
and 3 of this Unit when we deal with fields which vary with distance.

Equipotentials

Moving the flame probe along surfaces parallel to the plates reveals n
change in potential. Such surfaces are called equipotentials. Equipoten
tials appear as lines when the field is two dimensional, betwee
electrodes drawn or placed on conducting paper. They can be plotted a
given intervals to reveal the variation in potential between th
electrodes and are analogous to contours drawn on a map. Th
direction of the electric field is always at right angles to the equipoten
tials; the field has no component along the equipotential (figure E9).

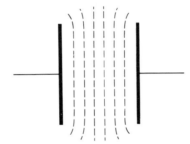

Factors affecting the strength of the field between parallel plates

Charged parallel plates form a capacitor, storing a certain charge
when a p.d. *V* is applied. For a capacitor, $Q \propto V$. Other factors whic
affect the charge stored are the area of the plates, *A*, their separation,

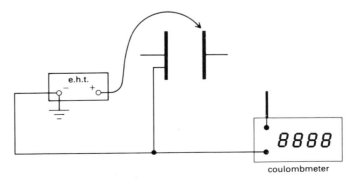

coulombmeter

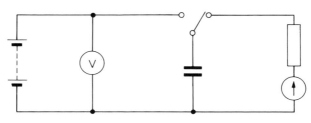

Figure E11

and the medium which separates them. These can be varied in turn and the charge Q measured using a coulombmeter or reed switch circuit (figures E10, E11).

QUESTION 13

Results show that Q is proportional to V, to A, and to $\dfrac{1}{d}$:

$$Q \propto \frac{VA}{d}$$

or $Q = \varepsilon_0 \dfrac{VA}{d}$

where ε_0 is a constant.
    This relationship can be arranged in two useful ways:

i $\dfrac{Q}{A} = \varepsilon_0 \dfrac{V}{d}$

$\dfrac{Q}{A}$ is the charge per unit area, or surface charge density usually denoted

QUESTION 12

by σ. The quantity $\dfrac{V}{d}$ is of course the field strength E.

$\varepsilon_0 \approx 8.85 \times 10^{-12} \text{ F m}^{-1}$

So $\sigma = \varepsilon_0 E$

ε_0 is called the 'permittivity of free space' and is a fundamental constant linking field strength and charge density.

ii $\dfrac{Q}{V} = \varepsilon_0 \dfrac{A}{d}$

$\dfrac{Q}{V}$ is the capacitance C so

$$C = \varepsilon_0 \frac{A}{d}$$

The presence of a medium other than air (or a vacuum) increases the charge Q by some factor ε_r called the 'relative permittivity', which is

QUESTIONS 14 to 18

different for each material but, being purely a factor, has no units. For air $\varepsilon_r \approx 1$, for paper ε_r is 2 to 3, for water $\varepsilon_r \approx 80$. The modified equations are then

$\sigma = \varepsilon_r \varepsilon_0 E$ and $C = \varepsilon_r \varepsilon_0 \dfrac{A}{d}$

The second equation is used in the design of capacitors: one can be made from household materials.

Applications

Many industrial processes make use of electric fields, while 'static' can be a big problem in industry and in the home.

Section E2 GRAVITATIONAL FIELD AND POTENTIAL

This Section is concerned with gravitational fields – in particular that of the Earth. The fundamental ideas and the way they are linked are very similar to those in the last Section – except that, whereas electric fields act on charges, gravity acts on masses.

A uniform gravitational field

$$g = \frac{F}{m}$$
$$g \approx 9.81 \, \text{N kg}^{-1}$$
on Earth's surface

The gravitational field strength, g, is defined as the force on unit mass. Measurements show this to be virtually uniform near the surface of the Earth, and certainly so in the laboratory.

Gravitational potential difference

$$\Delta(\text{g.p.e.}) = mg\Delta h$$

$$\Delta V_g = \frac{\Delta(\text{g.p.e.})}{\text{mass}} = g\Delta h$$
in uniform field

$$\Delta(\text{g.p.e.}) = \Delta V_g \times \text{mass}$$

QUESTIONS 19 to 21

In the uniform field, changes in gravitational potential energy (g.p.e.) are given by $mg\Delta h$, where Δh indicates a change in height. It is useful to know the change in potential energy per kilogram; this is called *gravitational potential difference* (ΔV_g). In this context it is simply $g\Delta h$.

Contours on a map join points at the same height above sea level. The energy involved in moving from one contour to another can be calculated from the gravitational potential difference (ΔV_g) between the two contours. Movement along a contour involves no change in g.p.e. contours can be called gravitational *equipotentials*.

The inverse-square law for gravitational fields

Newton first deduced that the gravitational force obeys an inverse square law. The force, F, between two masses m_1 and m_2 separated by distance r is

$$G \approx 6.67 \times 10^{-11} \, \text{N m}^2 \text{kg}^{-2} \qquad F = -G\frac{m_1 m_2}{r^2}$$

G is a universal constant, which applies to all masses everywhere. The '−' sign is used to indicate that the force, F, is inwards, that is attractive.

QUESTIONS 22 to 24

Because gravitational forces between laboratory sized objects are so small, very sensitive equipment is needed to measure G.

Testing the inverse-square law

Since field strength is defined as force per unit mass, the field strength, g, due to a mass M is given by

$$g = -\frac{GM}{r^2}$$

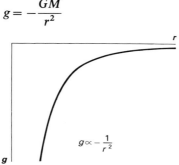

$$g \propto -\frac{1}{r^2}$$

Figure E12

The inverse-square law can be tested over large distances by analysing spaceflight data. Indeed it is used to plan the trajectories of spacecraft with great accuracy.

Calculating energy changes from force–distance graphs

Energy changes can be calculated by computing areas under a force–distance graph as shown in figure E14(a). Since field strength is force per unit mass, the area under a graph of field strength against distance indicates the energy change per unit mass, or the *gravitational potential difference*, ΔV_g, shown in figure E14(b).

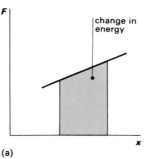

change in energy

(a)

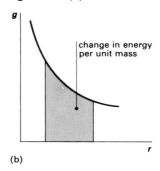

change in energy per unit mass

(b)

Figure E14

Figure E13
Sir Isaac Newton (1642–1727).
The Mansell Collection

E

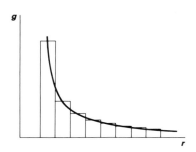

Figure E15
Approximate method of finding area.

QUESTION 25

Computer program 'GFIELD'

Calculating areas is a tedious process: approximate results can b obtained by adding up strips under the graph (figure E15). A compute can do the job quickly and accurately.

Calculating ΔV_g from field–distance graphs

QUESTION 26

QUESTION 27

The amount of energy required to move unit mass from the Earth' surface to various distances r_2 from the centre of the Earth increase with r_2, but approaches a finite limit, even if r_2 is very large (figure E16 Conversely, the energy required to move unit mass from a poin distance r_1 from the Earth's centre, to some larger distance r_2 decrease as r_1 is increased (figure E17). Measurements from graphs show tha this value of ΔV_g is given by

$$\Delta V_g \approx 4 \times 10^{14}\left(\frac{1}{r_1} - \frac{1}{r_2}\right)$$

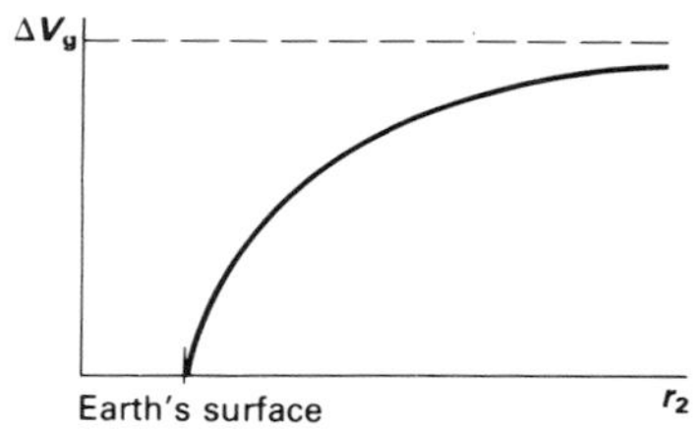

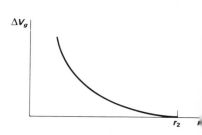

Figure E16
Energy needed to reach different distances r_2, starting from Earth's surface.

Figure E17
Energy needed to reach distance r_2 depends on starting position r_1.

Gravitational potential V_g

QUESTION 28

Potential is negative for all attractive fields

If we make r_2 '*infinity*' or 'as far away as we like' we can calculate the energy required to pull one kilogram completely free of the Earth' gravity. It seems sensible to agree on some reference point to use as ou zero of gravitational potential energy. This 'point' is agreed to be 'a infinity', or 'as far away from any mass as you like'. There the gravitational potential energy of any object is zero. But since it lose potential energy and gains kinetic energy by 'falling' towards the Eart (or any other planet), then we must consider the potential energy a

QUESTIONS 29 and 30

points nearer the Earth to be negative. (This is so for all attractive fields.) The formula for gravitational potential at distance r from mass M is

$$V_g = -\frac{GM}{r}$$

Equipotentials, lines joining points at equal potential, can be drawn around the Earth (figure E18). From these the energy required to move a mass, m, between various points can be calculated (ΔV_g between the points × mass). For example, the energy for complete escape from the Earth's surface is $\dfrac{GMm}{r_E}$, where r_E = radius of the Earth. Such energy changes do not depend on the route taken between the points. This means that any energy 'stored up' on an outward journey can be regained on the return.

QUESTION 31

g field is 'conservative'

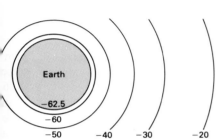

Figure E18
Equipotentials around the Earth
(at intervals of $10 \times 10^6\,\mathrm{J\,kg^{-1}}$).

Figure E19
$\dfrac{1}{r}$ profile potential well or rubber sheet model.

E

Field and potential gradient

Potential can be found from the area under a field strength–distance graph. Field strength can be found from a potential–distance graph by measuring the gradient. As in the electrical case 'field strength = − potential gradient'. It is easy to show by measuring gradients that if V_g varies as $\dfrac{1}{r}$, then g varies as $\dfrac{1}{r^2}$. This connection will be used in the next Section, for electric fields.

for gravity: $\quad g = -\dfrac{\mathrm{d}V_g}{\mathrm{d}r}$

for electricity: $\quad E = -\dfrac{\mathrm{d}V}{\mathrm{d}x}$

for both: $\quad \dfrac{1}{r}$ potential $\Leftrightarrow \dfrac{1}{r^2}$ field

QUESTION 32

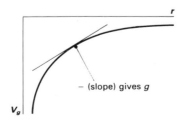

− (slope) gives g

Figure E20

Formulae and relationships for gravity

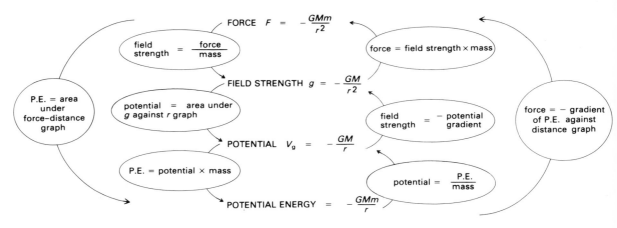

Figure E21
Formulae and relationships for gravity.

You do not need to *remember* the formulae, but you do need to *understand* the connections between them.

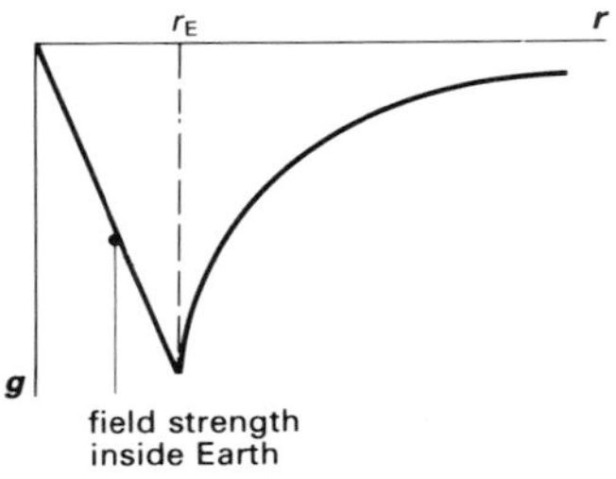

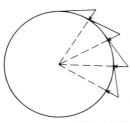

Figure E22
Field strength outside and inside a uniform sphere.

QUESTIONS 33 to 35

Field outside and inside a sphere

The inverse-square law, defined for 'point' masses, works outside a solid sphere of uniform density (which the Earth is, very approximately), or a hollow sphere, as if all the mass were concentrated at the centre. This is an invaluable simplification, enabling field and potential to be calculated even quite close to the surface of a large sphere. It will be of even greater use later as it applies also to electric fields around a charged sphere.

The value of g falls to zero at the centre of a sphere of uniform density (figure E22) with some interesting, if theoretical, consequences.

Inside a *hollow* sphere there is no field.

Circular motion

QUESTIONS 36 and 37

An object moving in a circle at constant speed changes direction and therefore *velocity*. Its acceleration is towards the centre or 'centripetal' (figure E23), and must be caused by an overall *centripetal force* if circular motion is to be maintained.

For an object of mass m to move at speed v in a circle of radius r, a centripetal force of size $\dfrac{mv^2}{r}$ must be provided towards the centre.

For planets and satellites this force is provided by gravity. We can make quite accurate calculations of the positions and energies of satellites, assuming circular orbits, though we know that satellite and planetary orbits are in fact elliptical, with varying degrees of eccentricity.

Figure E23
Centripetal force acts inwards.

QUESTIONS 38 to 40

Section E3   THE ELECTRICAL INVERSE-SQUARE LAW

Unit F, 'Radioactivity and the nuclear atom'

Unit L, 'Waves, particles, and atoms'

Many applications of electrostatics can be understood using the uniform field alone. However, if we want to delve into the structure of atoms, which are in essence made up of charged particles, we must know how spherical and point charges behave.

Potential near a charged sphere

The field of a spherical mass varies as $\dfrac{1}{r^2}$ and the associated potential as $\dfrac{1}{r}$. The electric potential around a charged sphere is found to vary as $\dfrac{1}{r}$ (figures E24 and E25). By analogy with gravity we deduce that the electric field varies as $\dfrac{1}{r^2}$.

DEMONSTRATION E8a
Investigating the variation of potential around a charged sphere

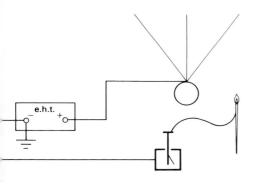

Figure E24

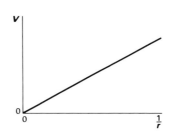

Figure E25

DEMONSTRATION E8b

Measuring the value of k in $V = \dfrac{kQ}{r}$

The potential, V, depends also on the charge, Q, on the sphere (figure E26).

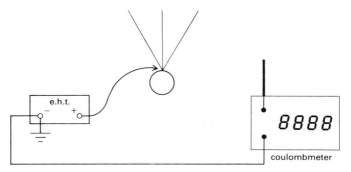

Figure E26

If Q is positive, the potential is positive: a positive charge nearby ha▮ positive potential energy and will be repelled. Conversely a negativel▮ charged sphere has negative potential and would attract a nearb▮ positive charge (figure E27).

By contrast all gravitational potentials are negative because a▮ gravitational forces are attractive.

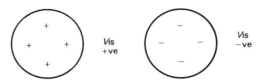

Figure E27

Analogy between electricity and gravity

	Force	Field	Potential	Potential Energy
Gravity	$-G\dfrac{m_1 m_2}{r^2}$	$-\dfrac{GM}{r^2}$	$-\dfrac{GM}{r}$	$-G\dfrac{m_1 m_2}{r}$
Electricity	$k\dfrac{Q_1 Q_2}{r^2}$	$\dfrac{kQ}{r^2}$	$\dfrac{kQ}{r}$	$k\dfrac{Q_1 Q_2}{r}$

Comparing the expressions for potential which have been es▮ tablished we can deduce the other expressions above for the $\dfrac{1}{r^2}$ electri▮

EXPERIMENT E9
Experiments to test the inverse-square law for electric forces

Compare Newton's Law for Gravity

QUESTIONS 41 to 44

field. $E = -\dfrac{dV}{dr}$ of course applies here and we can deduce the fiel▮ strength from the gradient of a graph of potential against distance. Th▮ force expression may be tested experimentally with some care (figur▮ E28). Coulomb first established it as a fundamental law of nature i▮ 1785.

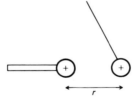

Figure E28

The radial field and the uniform field

Single charges have a $\dfrac{1}{r^2}$ field: yet many charges together on a large fla▮ plate give a uniform field (Section E1). This is because of the way i▮ which contributions from each bit of charge add up to give an overal▮ effect (figure E29). The mathematics of this adding up process show tha▮ the constant, k, is in fact $\dfrac{1}{4\pi\varepsilon_0}$.

QUESTION 45

QUESTIONS 46 to 51

Computer program
'EFIELD'

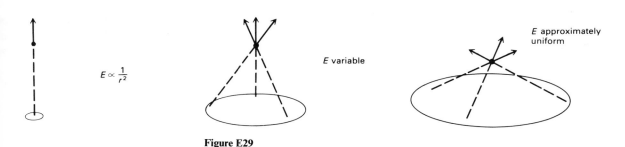

$E \propto \dfrac{1}{r^2}$

E variable

E approximately uniform

Figure E29

Sizes of electrical and gravitational forces

$G \approx 7 \times 10^{-11} \, \text{N m}^2 \, \text{kg}^{-2}$

$\dfrac{1}{4\pi\varepsilon_0} \approx 9 \times 10^9 \, \text{N m}^2 \, \text{C}^{-2}$

The electrical and gravitational force constants are very different in size. Comparison of the two forces between an electron and a proton within an atom shows that the electrical forces are bigger by a factor of nearly 10^{40}. For the charged spheres of the Coulomb's Law experiment the ratio is less, but still about 10^{10}. This means that very few of the atoms of the ball carry charge.

QUESTIONS 52 and 53

Although very large forces exist between neighbouring ions within a crystal, say, the almost perfect balance of $+$ and $-$ charges ensures that the whole crystal is neutral and exerts no electrical force on other crystals. More massive objects experience gravitational forces which are much greater than the electrical forces between them.

The four known interactions

QUESTIONS 54 and 55

'Forces and particles' in the Reader
Particles, imaging, and nuclei

Although they both obey inverse-square laws, electric and gravitational forces are quite distinct, acting as they do on charges and masses respectively. All other 'everyday' forces, (friction, surface tension, 'contact' forces between objects and even magnetism) are due to the interactions of charged particles surrounding atoms at rest or in motion. However, two other apparently distinct kinds of force do exist – the 'strong' and the 'weak' nuclear forces. Their range is limited to nuclear dimensions ($\lesssim 10^{-15}$ m) within which they are capable of holding together nuclei with enormous energy.

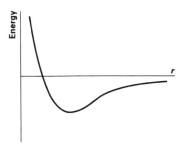

Figure E30
Intermolecular energy.

Unit F, 'Radioactivity and the nuclear atom'

Unit L, 'Waves, particles, and atoms'

The structure and behaviour of materials resulting from inter molecular electric forces has been studied in Unit A, 'Materials and mechanics' (figure E30). Knowledge of the fields and potentials around charges is used in constructing models of the atoms themselves.

READING

APPLICATIONS OF ELECTROSTATICS

In part adapted from an article by Dr Jean Cross in *Physics in technology* **12,** 1981, pages 54–59.

Introduction

Electrostatic forces are generally weak compared with gravitational effects and have little influence on macroscopic bodies. However, for small particles (1–100 µm), electrostatic forces can exceed gravitational and aerodynamic effects and this has led to the development of a wide range of industrial processes which rely upon electrostatic phenomena.

Charging small particles

In the macroscopic world electrostatic forces have little impact. Sparks may be observed which demonstrate the high potential which can be achieved when insulating materials are rubbed, but the energy stored is low and although the sparks may be a hazard in a flammable atmosphere, their energy cannot be usefully harnessed to control large bodies. However, the electrostatic charge stored (and hence the force experienced in an electric field) depends on surface area. The gravitational force, on the other hand, depends on mass or volume, so, as size decreases, electrostatic forces become gradually more important and can dominate the motion of particles less than a few hundred micrometres in diameter.

This principle has been used in a wide range of applications in which the motion of small charged particles is controlled with considerable accuracy by an electric field. Particles and droplets are nearly always naturally charged but for most applications charge is added artificially both to maximize the magnitude and to achieve uniformity. Particles can be given a charge by a number of different techniques of which friction is the simplest and best known.

This phenomenon was first observed more than 2500 years ago by Thales, who noticed that small particles were attracted to amber which had been rubbed with silk. It is now known that charge is transferred when any two materials are brought into contact as electrons move to more favourable energy levels. If both materials are conducting and earthed, charges flow away instantaneously when the surfaces are separated, but, if an insulating material is involved, the charge cannot disperse and rapidly accumulates until the electric field above the surface reaches the level where the air ionizes and charge can build up no further. Powdered materials easily acquire charge by friction in passing through a tube and even most metals are sufficiently insulating in powdered form for charge to accumulate if the particles move rapidly.

Frictional charging is not easy to control and may vary considerably from day to day, therefore many industrial applications of

E

electrostatics rely on corona charging. In this process a high voltage is applied to a sharp point, causing the air to ionize in the high field region around the point. Ions of the opposite polarity are attracted by the point but those of the same polarity are repelled, forming a stream of unipolar ions. These ions will be attracted to any powder or dust particles present because the difference in relative permittivity between the powder and air causes a distortion in the electric field (figure E31) and thus ions are pulled in to the particle. Charging stops when the repulsion due to the charges on the particle balances the attraction due to the field distortion.

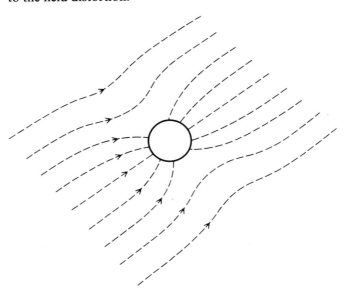

Figure E31
Distortion of field by a dust particle.

A third method of applying charge – induction – can be used only for conducting particles and is most often used in practice to charge liquid droplets. The principle is illustrated in figure E32. A particle on a positive electrode in an electric field will acquire a positive charge simply because the charge is attracted towards the negative electrode. The force of attraction may be sufficient to detach the particles from the electrode, creating a free charged particle. A liquid surface is distorted in an electric field and is pulled up into a cone. The induced charge on this cone can cause droplets to break off – the process of electrostatic atomization. The resistivity of the liquid is critical for the formation of uniformly charged droplets. If it is too conducting, a corona discharge may form or the cone break up erratically. If it is too insulating, charge cannot flow to help form the cone.

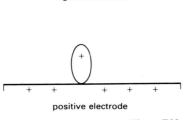

Figure E32
Induction charging.

Electrostatic generators

One of the earliest applications was the design of electrostatic generators which use mechanical energy to build up a high voltage. Most are high voltage, low current devices. The Van de Graaff machine is probably the best known and is still used today for the production of voltages exceeding 1 MV for the acceleration of elementary particles in nuclear research.

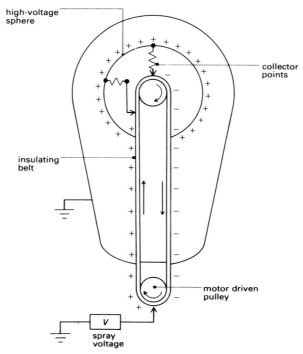

Figure E33
Van de Graaff generator.

A Van de Graaff generator is shown schematically in figure E33. Charge is sprayed onto the bottom of a conveyor belt by corona from a sharp point. (This charge may also be produced by friction.) As the belt moves round, charge is transferred to the large sphere by ionization from collector points. Resistors maintain a voltage between the sphere and the upper pulley so as to allow a negative discharge from the collector points, leaving a positive charge on the sphere. The larger the sphere, the higher the voltage which can build up without discharge. The leakage from the sphere limits the voltage which is produced. Mechanical belt generators have been produced for high voltage laboratory supplies of 50–200 kV, although at these levels conventional multipliers are more commonly used.

Electrogasdynamic generators also produce voltages of this level and have been used in electrostatic paint and powder spray guns. In this device ions are created, usually by a corona discharge between a point and an attractor, and then forced down a tube by a high velocity air stream. Usually a vapour or a powder is introduced to collect ions and reduce their mobility so they are not collected at the attractor but are swept on to the collector. It can be seen from figure E34 that this results

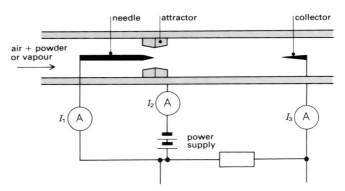

Figure E34
Electrogasdynamic generator: I_1 needle current; I_2 attractor current; I_3 collector curre

in a current flow through the load which is higher than the curre
drawn from the attractor power supply. The voltage on the collector
several times higher than that applied to the attractor and the ext
energy is supplied by the gas stream motion.

Electrostatic coating techniques

Commercially the most useful applications of electrostatics involve tl
control of fine particles. The earliest device to be put into industrial u
was an electrostatic precipitator built in 1890 and powered by
Wimshurst machine. In its simplest form an electrostatic precipitator
an earthed cylinder with a fine wire along its axis which gives a coro
discharge. As dusty air is passed through the cylinder, the particles a
charged by the corona ions and repelled by the wire to be deposited c
the cylinder walls. Commercial precipitators are now generally tw
stage devices with separate charging and collecting zones. The di
charge electrodes are wires suspended between plates (figure E35) a
the planar collecting region can be cleaned without disturbing tl
electrodes. Efficiencies of over 99 per cent can be achieved down
particle sizes of a few micrometres. Precipitators were first develope
for large scale industrial processes such as steel works, but small devic

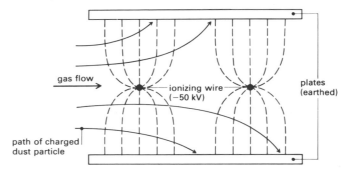

Figure E35
Electrostatic precipitation.

have also been developed for collecting cigarette smoke from public rooms.

In the electrostatic powder coating process shown diagrammatically in figure E36, a thermoplastic or thermosetting resin powder is blown out of a spray gun past a high voltage needle at which a corona discharge forms. The powder cloud is charged by the ions from the corona and carried by the electric field to the earthed workpiece. Since the powder is highly insulating it retains its charge and adheres even when the electric field is removed. The workpiece can then be transported to an oven where the resin is fused to give continuous paint film. The most important advantage of this technique over conventional wet paint spraying is that the wasteful use of flammable and toxic solvents is eliminated. Since there is no need to evaporate a solvent, better cross-linking resin formulations can be used and high scratch and chip resistance is obtained. Unlike wet paints, dry powders can be collected and re-used, giving a high overall efficiency which often compensates for the higher cost of materials.

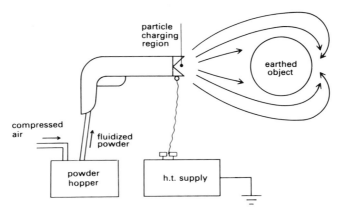

Figure E36
Electrostatic powder coating.

Electrostatic deposition may also be used with wet paints. The paint is atomized by an air jet, by a rapidly spinning disc or bell, or occasionally simply by electrostatic forces which tend to break up large drops until the electrostatic repulsion is counteracted by the surface tension of the drop. As with any electrostatic process, the paint travels along the electric field lines to areas not directly within the field of view of the gun – a phenomenon known as wrap round. Recently the techniques developed in the 1960s for electrostatic paint spraying have also been applied to crop spraying. Efficiencies have been increased considerably and there is a great improvement in deposition on the underside of leaves, which is particularly important for fungicides.

The ink-jet printer – printing without pressing

One of the problems with modern computers and data processing is that whilst a computer can happily churn out 10^6 numbers per second, a printer which can produce only 100–200 characters per second obviously slowing down production somewhat. The development of printer capable of dealing with over 1000 characters per second with resolution of over thirty points per centimetre is clearly an 'order magnitude' improvement.

There are several types of ink-jet printers being developed or in us In the 'deflect-to-print' type, which is shown in figure E37, ink from reservoir is pushed through a narrow jet (about 35μm in diamete which is modulated ultrasonically at a frequency of about 500 kHz a breaks up into a fine stream of droplets. A charging cylinder throug which they pass induces on each drop a charge which varies accordi to the p.d. applied to the cylinder. This p.d. is determined directly by t computer so that each drop can be given a unique charge. The dro now pass between deflector plates, rather as in an oscilloscope, acro which there is a steady p.d. so that the deflection depends on the char of the drop. As the jet moves over the paper (or vice versa), characte can be built up at extremely high speed and with great precisio Uncharged droplets (representing spaces) are collected in a gutter ar may be returned to the reservoir. Since no contact is made, the jet m be used to print on very delicate surfaces – even butterfly wings!

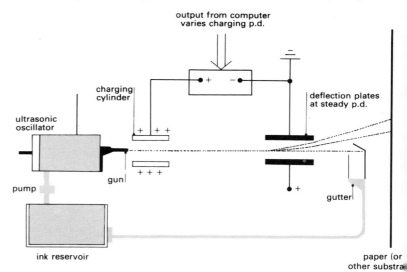

Figure E37
Ink-jet printer. Deflect-to-print type.

This 'continuous flow' method has some advantages over t simpler and cheaper 'drop-on-demand' printer where the ink jet is he back by surface tension and switched on at low pressure only whe required. The drop-on-demand printer shown in figure E38 has bee designed for use with personal computers. Considerable research an development has gone into producing an ink which does not dry in t jet, blocking it up, but which *does* dry quickly on paper. A third metho charges the drops *not* required for printing and deflects them away an earthed plate. With the use of three coloured inks a large full-colo page can be printed to a high degree of resolution in a few minutes. Th

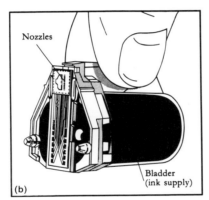

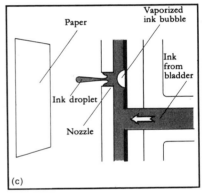

Figure E38
'ThinkJet', a drop-on-demand type ink-jet printer. The printer is small and portable and is used here with an office microcomputer. It also has the advantage of operating below 50 dB. Diagrams (b) and (c) show how the printer works.
Hewlett-Packard Ltd.

opens up potential applications in cartography and the production of hard copies of aerial and satellite images.

Xerography

This is the process used in almost all copying machines. In the U.K. one leading firm's machines alone produce 12 billion copies a year; typical office machines handle up to 120 copies per minute.

The name derives from two Greek words meaning 'dry writing', and it was indeed a great step forward when copies of documents could be made almost at once on ordinary dry paper without the messiness of earlier methods of printing. The basic process has about six main stages: charging, exposure, development, transfer, fusion and cleaning.

Charging The plate or rotating drum on which the image of the

Figure E39
The photoreceptor plate.

document is to be formed is made of a thin (60 μm) layer of selenium c
an aluminium substrate separated by a thin layer of oxide (figure E39

Selenium is photoconductive, that is, it is effectively an insulator i
darkness but conducts well when exposed to light. The plate or dru
surface is first coated with positive charge by being traversed by
corotron, or fine wire at high positive potential (typically 850 V
surrounded by an earthed metal shield. Since the plate is also earthe
positive ions in the air formed by corona discharge are forced onto i
surface (figure E40).

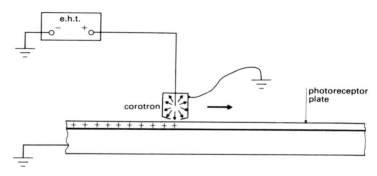

Figure E40
Charging the photoreceptor plate.

Exposure   Now the document to be copied is illuminated. A lens, an
usually also mirrors, to reduce the physical size of the device, focus
real image (laterally inverted) onto the photoreceptor plate. Where ligh
falls on the selenium it conducts and the charge drains away throug
the earthed aluminium substrate (the oxide layer is thin enough to allo
this). However, dark areas (for example writing) on the document caus

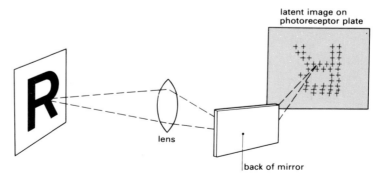

Figure E41
Exposure.

dark areas on the image so the charge remains at these points. So a 'latent image' is formed on the plate (figure E41).

The selenium is most sensitive to blue light, whose photon energy is greatest (see Unit L, 'Waves, particles, and atoms') and least to red, so that red areas on a photograph will behave like black, but light blue areas, like white, will be difficult to reproduce.

Development   A dry black thermoplastic powder called toner is used to develop this latent image. The toner powder is mixed with carrier beads (metal shot or glass); the carrier and toner obtain opposite charge by frictional contact and the now negative powder adheres lightly to the positive carrier. When this mixture is sprinkled over the latent image, most of the toner powder sticks to the positive image and the carrier falls off to be re-used (figure E42).

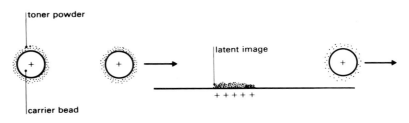

Figure E42
Development using toner.

Transfer   A sheet of paper is given a positive charge by a separate corona wire and placed carefully in contact with the photoreceptor plate. A good deal of the toner powder is attracted to the paper forming on it an image which is now like the original, that is, not inverted.

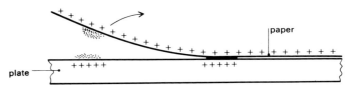

Figure E43
Transfer onto paper.

Fusion   The paper is transported underneath a radiant heater or between hot rollers to melt the plastic toner (at about 140 °C) giving a firm, permanent coating to the paper (figure E44). Alternatively, the fusing may be done using a chemical vapour which breaks down the plastic in a similar way to melting.

Cleaning   Some toner remains sticking to the plate or drum. This 'residual image' must be cleaned off before the next copy is taken. First the plate or drum is neutralized, generally by a corona discharge from an alternating current source, to loosen the powder. Next the powder is

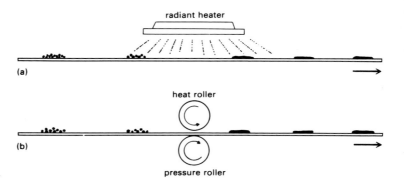

(a)

(b)

Figure E44
Fusing to make a permanent copy.

heat roller

pressure roller

scraped, wiped, or brushed off. Since this process may charge up the plate again, it is finally exposed to light to render the selenium surface free of charge and ready to be used again (figure E45).

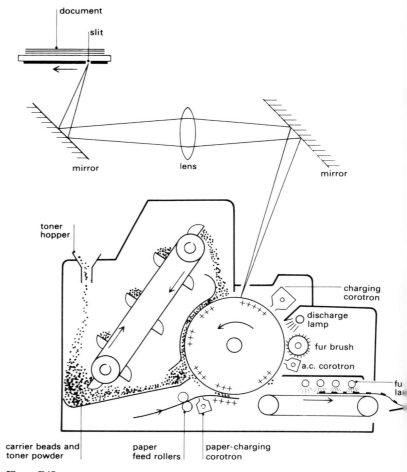

Figure E45
The complete process for a rotating drum copier. In this case a slit traverses the document forming an image on the drum as it rotates.
(*Information courtesy of Rank Xerox (U.K.) Limited, who may supply further details on request.*)

Questions

a *i* The article states that 'charge stored depends on surface area'. This implies a limit on the charge density on a surface. Why is this limited? (Think about the field strength at the surface and support your argument with relevant formulae.)

ii Why do electrostatic forces have more effect on small particles?

b Describe the three most common methods by which particles are given an electric charge, explaining the limitations of each method.

c Discuss the effect on the potential attained by a Van de Graaff generator of each of the following (separately):

i increasing the radius of the sphere;

ii increasing the resistance between the sphere and the upper pulley;

iii increasing the speed of the belt;

iv increasing the spray voltage.

d In an electrogasdynamic generator, why is the current flowing through the load greater than that flowing between the needle and the attractor?

How will the gas stream be affected by giving energy to the generator?

e What are the advantages of electrostatic dry powder paint spraying over conventional liquid paint spraying? Are there any disadvantages? Explain how the technique could be applied to crop spraying.

f The article states that 'paint travels along the electric field lines'. Sketch some equipotentials and lines of electric field between a paint nozzle and a spherical earthed object.

g In the diagram of the deflect-to-print ink-jet printer (figure E37) the drops do not appear to be travelling *along* field lines between the deflecting plates but at right angles to them. Why is this and what is the shape of the path of a drop whilst it is between the plates?

h Some photocopiers are poor at reproducing large black areas.

i Sketch the 'latent image' of a large black circle.

ii Considering the mutual repulsion of the charge on this image, suggest why the central area may not come out fully black.

i Describe what happens during one complete rotation of the drum in figure E45.

j 'Static' can be a hindrance, indeed a hazard, as well as being of use in commerce and industry. Find out about some problems of static, for example, on plastic-hulled ships, in pumping liquids in oil tankers, in the manufacture of semiconductor devices.

E

LABORATORY NOTES

DEMONSTRATION

E1 Forces on a charged ball between charged plates – the 'shuttling ball'

e.h.t. power supply
2 metal plates with insulating handles
table tennis ball coated with colloidal graphite
nylon sewing thread
galvanometer (*e.g.*, internal light beam)
polythene strip
2 retort stand bases, rods, bosses, and clamps
leads

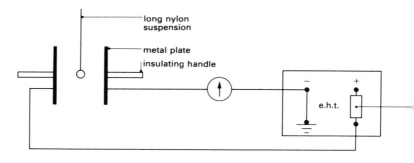

Figure E46
The 'shuttling' ball.

What causes the ball to move from one plate to the other?
Why does it keep changing direction?
What sign of charge is being 'ferried' across the gap by the ball? Why does the galvanometer not show pulses of current in *both* directions?
What factors affect the force on the charged ball? How do they affect it?

DEMONSTRATION

E2 Using a charged foil strip as a field detector

pair of capacitor plates
2 slotted bases
2 square polythene tiles
polythene strip
foil
adhesive tape
razor blade or scissors
e.h.t. power supply
leads
means of projection (optional)

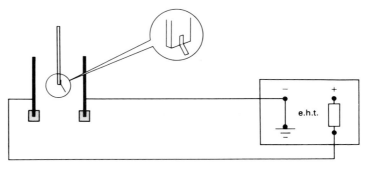

Figure E47
Charged strip field detector.

How can you judge the size of the electric force acting on the foil? How
does this force vary from place to place
a between the plates,
b outside the plates?
How are the direction and size of the force on the foil affected by
a the p.d. between the plates,
b the distance between them, and
c the charge on the foil itself?
How might you set about measuring both the charge and the force on
the foil? Explain whether or not these would be easy tasks.

DEMONSTRATION

E3 Patterns of electric field for different geometries (using semolina)

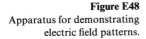

Figure E48
Apparatus for demonstrating
electric field patterns.

either
e.h.t. power supply
or
Van de Graaff generator

electric field apparatus
1,1,1-trichloroethane
castor oil
semolina
bare copper wire, about 2 mm diameter, for electrodes
leads

You should be familiar with the pattern produced between two parallel
electrodes, two point electrodes, one point and one straight electrode,
and a point in the centre of a circular electrode.
What might happen if the liquid conducted electricity? Would you still
obtain a field pattern? Explain your answer.

DEMONSTRATION

E4 Measuring potentials in a uniform field using a flame probe

The electroscope, calibrated as a voltmeter, measures the potential
difference between the tip of the probe, connected to the cap, and the

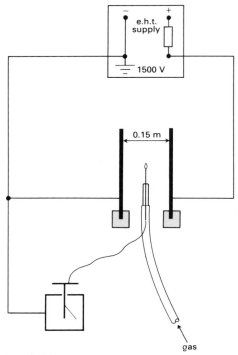

Figure E49
Using a flame probe to measure the potential between parallel plates.

case, which is earthed. The flame ionizes the air around the tip, neutralizing any charge on it, so that there is no p.d. between it and its surroundings. In effect, then, the electroscope measures the potential difference between a point in the space between the plates and earth. We call this the 'potential' at that point. The potential at different points is found by moving the probe around.

Does the potential change when the probe is moved parallel to the plates?

How does it change along a line perpendicular to the plates?

Draw a graph of potential, V, against distance, x, from one plate.

EXPERIMENT

E5 Plotting equipotentials in two dimensions for various shapes of field

4 cells in holder or l.t. variable voltage supply and smoothing unit
oscilloscope or voltmeter
pencil or ball point pen adapted as probe
copper or aluminium sheet, 0.1–0.5 mm thick
conducting paper (*e.g.* 'teledeltos' type) cut to A5 or A6 sheets
conducting putty
stapler and staples
carbon paper and white paper
drawing board or hardboard
bulldog clips
silver conducting paint (optional)

Use the probe connected to the oscilloscope or voltmeter to find a series of points at the same potential, and use the carbon paper to mark these points on the plain paper underneath the conducting paper. (Do not mark the conducting paper itself.) Try various shapes of electrodes.

For 'point' electrodes, there may be difficulty in obtaining reliable equipotentials close to the point. In practice a slightly larger electrode such as the *head* of a drawing pin works better than just a point. You might like to think why this is so: why are equipotentials closer together near such points – what is this telling you about the strength of the field there? In all your patterns, look for areas where the equipotentials are equally spaced: how does the field strength vary in these regions?

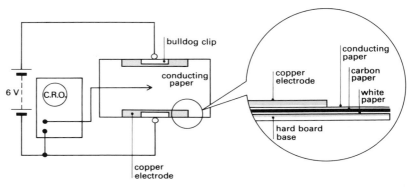

Figure E50
Plotting equipotentials on conducting paper.

Consider also the effect of the shape of the electrodes. How do sharp curves and corners affect the spacing of the equipotentials nearby?

Further investigations could include modelling some 'real' situations such as plotting equipotentials between different shapes of electrode in a cathode ray tube, say, or between the base of a thundercloud and the roofs of buildings beneath it. If silver paint is available such shapes can be easily created, though they can also be achieved using copper strip.

An alternative method for plotting equipotentials uses copper electrodes in copper sulphate solution. Teachers will be able to provide details.

EXPERIMENT

E6 Investigating factors affecting the charge on parallel plates using a coulombmeter

coulombmeter with probe rod
2 metal plates with insulating handles
e.h.t. power supply
polythene strip for use as insulating handle
2 retort stand bases, rods, and bosses
metre rule
leads

The unearthed plate is charged by a flying lead from the positive of th
e.h.t. supply (with limiting resistor). The lead is held by an insulatin;
handle. Other earthed conductors (the bench, hands, etc.) should b
kept well away.

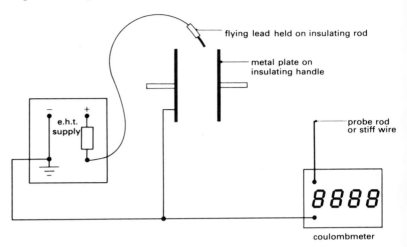

flying lead held on insulating rod

metal plate on
insulating handle

probe rod
or stiff wire

e.h.t.
supply

coulombmeter

Figure E51
Using a coulombmeter to measure Q.

Caution: The flying lead *must be removed* from the plate before th
charge is measured. The coulombmeter is then brought up so that th
probe rod touches the plate; the charge can be deduced from th
resulting reading.

Experiments to test whether $Q \propto \dfrac{1}{d}$ and $Q \propto V$ are possible. $Q \propto$
would require a set of plates of various sizes. The effect of a plastic shee
or paper between the plates can also be explored.
Does the coulombmeter reading return reliably to zero when shortec
or does it suffer from zero drift?
How accurately repeatable is any given reading?
Is any charge left on the plate? How could you check this?

Plot a graph for Q against $\dfrac{1}{d}$. What significance has the Q intercept?

EXPERIMENT

E7 Investigating factors affecting the charge on parallel plates using a reed switch

reed switch
signal generator
1 pair capacitor plates with 16 polythene spacers, $10 \times 10 \times 1.5\,\text{mm}$
polythene sheet, 1.5 mm thickness, and paper sheets

either
l.t. variable voltage supply and smoothing unit
or
3 cell holders with four cells

voltmeter, 100 V and 10 V
galvanometer (*e.g.* internal light beam)
resistance substitution box
class oscilloscope
metre rule
mass, 1 kg
leads

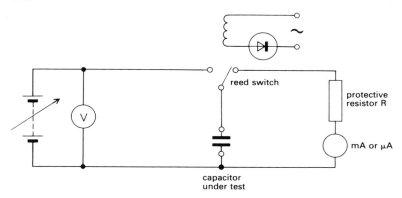

Figure E52
Using a reed switch to measure Q.

Caution: The signal generator output p.d. should be raised until the switch is heard to vibrate cleanly, and not increased further. Nor should the switch's stated operating p.d. be exceeded. The capacitor plates should never touch whilst the switch is working; turn off the signal generator if you are not taking readings, and disconnect the power supply before adjusting the plates. Keep hands and earthed objects well away from the plates when taking readings.

Use an oscilloscope across the protective resistor R to check that discharge of the capacitor is complete for each cycle. Use $R = 100\,\mathrm{k\Omega}$ and $f = 400$ Hz; try raising R or f, observing what happens.

$Q \propto V$ can be easily tested for p.d.s up to 25 V.

$Q \propto \dfrac{1}{d}$ can be tested by spacing the plates at different separations using spacers.

$Q \propto A$ can be tested by allowing different amounts of overlap between the plates.

Estimate the percentage uncertainties in your readings and draw error bars in your graphs. Are graphs of Q against A and Q against $\dfrac{1}{d}$ straight lines, within the limits of uncertainty? What significance has the Q intercept of each?

DEMONSTRATION

E8a Investigating the variation of potential around a charged sphere using a flame probe

The flame probe is constructed and the electroscope calibrated as in demonstration E4. A triple suspension of nylon thread keeps the ball

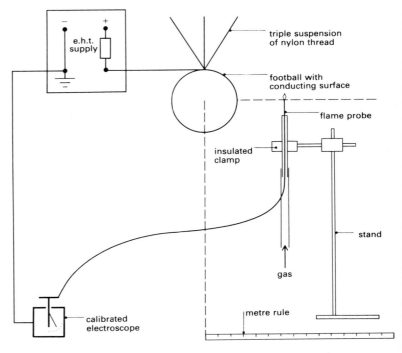

Figure E53
Using a flame probe to explore potential near a charged sphere.

stable. The ball should be as far as possible from walls, benches, and other conducting surfaces.

What changes of potential are indicated as the probe is kept a constant distance r from the centre of the sphere (*i.e.*, moved around a spherical surface concentric with the ball)? Explain this result.

How does the potential vary as the distance r is varied? Explain. In particular, note what happens when r is doubled (or halved). Do this for two values of r, one of which is equal to the radius of the sphere.

Measure the potential V at different distances r and plot a graph which you think will yield a straight line. Deduce what you can about the relationship between V and r.

DEMONSTRATION

E8b   Measuring the value of k in $V = k\dfrac{Q}{r}$

A potential of rather less than 1000 V is suitable, say 500 V. The sphere is charged by touching with a flying lead from the e.h.t. positive terminal held, as shown, on an insulating handle. This flying lead is then removed. To measure the charge it is transferred to the coulombmeter by touching the probe rod to the sphere.

If an earthed hand were nearby whilst the sphere was being charged, how might this affect the amount of charge stored on it?

Why should the flying lead always be removed before the probe rod touches the sphere?

Measure the charge, Q, stored for one or more values of V and deduce a value for the constant k.

This experiment depends on all (or very nearly all) of the charge on the sphere being transferred to the coulombmeter. Should the coulombmeter's capacitance be large or small compared with that of the sphere?

Estimate what percentage of the original charge stays on the sphere if the coulombmeter has a capacitance of $10^{-2}\,\mu\mathrm{F}$. (Use $C = Q/V$ to deduce the capacitance of the sphere.)

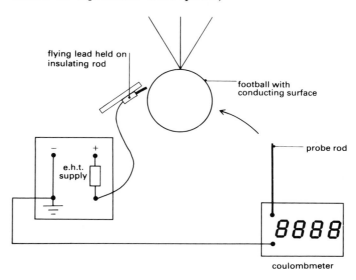

flying lead held on insulating rod

football with conducting surface

probe rod

e.h.t. supply

coulombmeter

Figure E54
Measuring the value of k.

EXPERIMENT

E9 Experiments to test the inverse-square law for electric forces

small proof plane, or ball point pen barrel (to act as insulating handle)
2 metallized polystyrene balls
nylon thread for suspension
lamp, holder, and stand
transformer

either
e.h.t. power supply
or
Van de Graaff generator
or
electrophorus plate, 'rubber', and polythene tile

retort stand base, rod, boss, and clamp
graph paper
glue (Durafix or Evo-Stik 863)
adhesive tape
leads
hair dryer (if air is humid)
balance (optional)

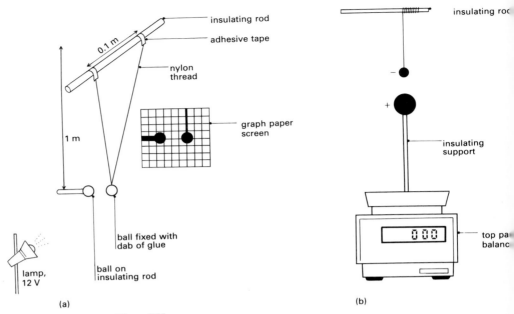

Figure E55
Testing Coulomb's Law.

One ball should be fixed to the nylon suspension using a dab of quick drying glue, the other attached to the insulating handle. Both the nylon and the insulating rods should be clean and free of finger grease and well dried before, and if necessary during, the experiment using the hair dryer. Shadows of the balls cast onto a screen allow the magnified movements to be measured without touching the balls.

A pendulum performs simple harmonic motion when displaced. How does the sideways (*i.e.*, restoring) force vary with displacement of the ball?

Is $F \propto \dfrac{1}{r^2}$? Use the sideways displacement, d, of the suspended ball to investigate the force exerted at different separations, r.

Is $F \propto Q_1, Q_2$? Alter the charge on one ball (how?) to investigate this relationship. Knowing the angle at which the ball hangs and its weight how could you find the actual force exerted on it? How could you measure the charge on each ball?

It is difficult, but not impossible, to make these measurements and thus deduce a value for k in $F = k\dfrac{Q_1 Q_2}{r^2}$.

HOME EXPERIMENTS

EH1 The capacitor

Using only household materials, make a capacitor with the biggest capacitance to volume ratio that you can contrive. For safety reasons your capacitor volume must not be larger than that of a *small* match box.

In order to help you in your design, consider the capacitance equation

$$C = \frac{\varepsilon_r \varepsilon_0 A}{d}$$

Consider each variable in turn and decide what must be done to it to make the capacitance big whilst the volume of the device is kept small. You may recall that ε_r represents the relative permittivity of the material that separates the capacitor plates. Use data books to help you to determine the best material to use.

Having made your capacitor you will need to find its actual capacitance. You might consider using a 'time constant' method – either charging or discharging the capacitor through a known resistance, plotting charge against time, and using the equation

time constant $= CR$

Compare your results with others in the class – perhaps on a competition basis.

EH2 The potential hill/well

In dealing with certain physical phenomena or events it is very often helpful to try to visualize these things in concrete terms that we can readily appreciate. The Rutherford model of the atom, for example, provides us with an 'image' of the atom which is extremely helpful though often incomplete or limiting.

In this task a three-dimensional model of a $\frac{1}{r^2}$ field may be created and you may find it very helpful.

On a large piece of graph paper draw a $\frac{1}{r}$ curve – maybe the graph of gravitational potential against r.

Put this graph on eleven other sheets of paper and cut out the graph – cutting along both axes and the line of the graph so that you have twelve graph 'envelopes'.

Stick the vertical axes of the graphs together, and arrange the graphs to radiate outwards. This is the skeleton of your potential hill or well. Put the skeleton on its wide base and clad the shape in strips of paper to make a detachable funnel shape.

When it is placed on its wide base, you see a potential or 'coulomb' hill. Holding it with its wide base up you see a potential well.

QUESTIONS

Uniform electric fields

1(L) A table tennis ball covered in conducting paint is suspended between two metal plates, A and B, which are connected via a galvanometer to an e.h.t. power supply. The ball is initially given a positive charge.

a Which way does it move? Why?

b How is the charge on the ball affected when it touches plate B? How does it move as a result?

c As the ball continues to 'shuttle' back and forth, what sign of charge does plate B lose?

d What sign of charge does plate A gain?

e In which direction do electrons flow in the circuit (clockwise or anticlockwise)?

f In which direction does conventional current flow?

g Why does the surface of the ball have to be conducting?

h What two factors affect the frequency at which the ball shuttles backwards and forwards? How is the frequency altered by changing each factor?

i (Harder) Describe and explain what happens if the ball is initially *uncharged* but very near to one of the plates.

2(L) A small charged strip of foil on an insulating handle is held between two large charged plates which are connected to a 5000 V supply. The plates are 80 mm apart; the angle at which the strip hangs is noted.

The plates are now moved until they are only 40 mm apart, leaving them connected to the supply at the same p.d. Will the strip hang at a larger angle to the vertical (showing more force on its test charge) or a smaller angle (showing less force)?

How strong are the electric fields in the two cases?

To what value must the p.d. be changed to get the same force as at first on the charged test strip at the new, smaller spacing?

3(L) This question introduces the volt per metre as an alternative unit for E

A ball carrying charge Q is moved a distance d by an electrical force, F. The force arises from a field created by a p.d. V between the plates; the force is the same everywhere between the plates as the field is uniform.

a Write down an expression for the energy transformed by the force F in moving the ball a distance d.

b Write down the energy transformed when a charge Q crosses a p.d. V.

c Use your answers to **a** and **b** to obtain an expression for the force per unit charge on the ball.

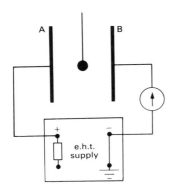

A B

e.h.t. supply

Figure E56

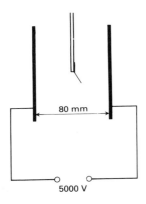

80 mm

5000 V

Figure E57

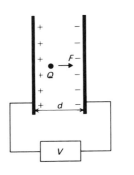

+ −
+ −
+ − F−
+ Q −
+ −
+ −
+ d −

V

Figure E58

d Instead of $N\,C^{-1}$, what other units could be used for field strength?

e *i* Express volts in terms of joules and coulombs.
ii Express joules in terms of newtons and metres.
iii Hence show that $V\,m^{-1}$ can be written $N\,C^{-1}$.

f *i* What is the strength of the electric field in experiment E2, where 1500 V is applied across plates 0.15 m apart?
ii If the force on the foil is about equal to the weight of a 1 mg fly, what is its charge?
iii What would be the force on 1 coulomb?

4(P)a In the sparking plug of a motor car engine there are two electrodes separated by a spacing of about 0.67 mm (figure E59). If air begins to ionize when the electric field is about 3×10^6 volts per metre (at atmospheric pressure), roughly what p.d. must be applied across the electrodes to cause a spark in the air?

b The p.d. needed to cause a spark will depend on the gas pressure. Explain what effect you think a change of pressure will have on this p.d. Is the gas pressure in a motor car engine greater or less than atmospheric pressure when the spark is needed?

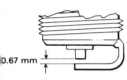

0.67 mm

Figure E59

Some examples of non-uniform fields

5(P) Copy the diagrams of different electrodes and on them sketch lines to illustrate the shape of the electric fields between them. Draw arrows to indicate the direction of the field. Some you may already know; others you will have to guess.

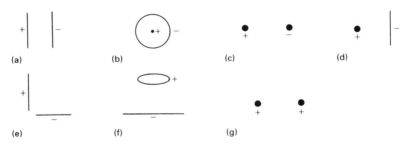

Figure E60

6(P) A small, charged ball weighing 10^{-3} N is attracted to a charged plate as shown in figure E61, and hangs at 45° to the vertical.

a Draw a diagram showing all the forces acting on the ball.

b Deduce the size of the electric force on the ball.

c If the electric field strength near the plate is $10^5\,V\,m^{-1}$, calculate the size and sign of the charge on the ball.

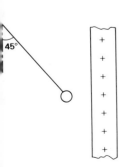

Figure E61

Electric potential

7(E) A 'flame probe' is used to measure potentials between parallel plates connected to an e.h.t. supply. The potential is indicated by an electroscope, whose case is connected to the earthed negative terminal of the e.h.t. supply.

a Why is it not practicable to measure an electric field strength by placing a charged object between the plates and measuring the force on it?

b Why is the flame necessary? What does the electroscope indicate if there is no flame?

c No gas supply is available. What could be used in place of the flame?

d What sign of charge does the electroscope acquire during the experiment?

e In which direction does current flow in the wire leading from the probe to the electroscope cap?

f Is this current steady?

8(L) A small object carrying charge, $+Q$, is placed at X between parallel plates as shown, a distance x from the lefthand plate (figure E62). In this question, vectors to the right will be considered positive, those to the left, negative.

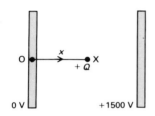

0 V   +1500 V

Figure E62

a What sign has the vector, x, indicating the displacement of the charge from O?

b In what direction is the force, F, on the charge, and the electric field strength, E, at X?

c What sign is therefore given to the field E?

d As one moves in the direction of increasing x, does the potential increase or decrease?

e Sketch a rough graph of V against x.

f What sign has the potential gradient $\dfrac{\Delta V}{\Delta x}$?

g Refer to your answers to **c** and **f**. Write a correct expression relating the field strength, E, to the potential gradient, $\dfrac{\Delta V}{\Delta x}$.

9(P) Figure E63 shows a scale drawing of the electrodes inside a gas filled tube. G_1 and G_2 are of thin wire gauze and are externally connected.

a Draw a graph showing how the potential varies along the line joining C and A.

b What is the strength of the electric field in the three regions CG_1, G_1G_2, and G_2A?

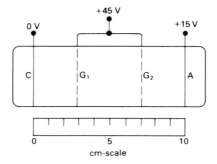

Figure E63

c Describe the motion of an electron through these three regions, stating its energy (in eV) at C, G_1, G_2, and A. Assume it starts at rest at C and does not hit either G_1 or G_2.

d (Harder) Suppose another electron ionizes a gas atom in the G_1G_2 region and loses 35 eV of energy in so doing. Describe and explain its subsequent behaviour.

10(P) The figure shows a full-scale section of equipotentials at 1 V intervals between two conductors.

On a tracing of figure E64, starting from A on the upper conductor, construct a field line until it terminates on the lower conductor. By taking measurements, plot a graph of the variation of electric potential with distance from A along this line. How can the electric field strength at a point be deduced from this graph?

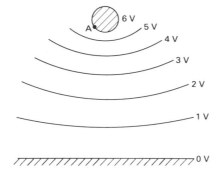

Figure E64

11(E) Figure E65 shows equipotentials drawn on a two-dimensional model representing a thundercloud over a churchyard. Assuming the charge is evenly spread over the base of the cloud, sketch lines of electric field between the cloud and earth on a copy of this diagram.

Comment on the safety of the following locations during a thunderstorm:

a On top of the church spire.

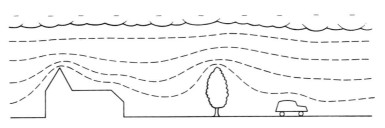

Figure E65

b Inside the church.

c Leaning against the tree.

d In a car.

e On a bicycle.

f On horseback.

Parallel plates and capacitance

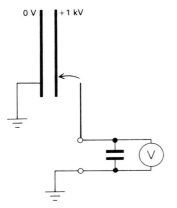

Figure E66

12(L) Two circular metal plates of radius 0.15 m are placed parallel to one another, 50 mm apart. A p.d. of 1 kV is established between them by momentary connection to an e.h.t. supply. A high resistance voltmeter is to be used to measure the charge on one of the plates, the other being earthed.

This is done by transferring its charge, via a thick wire, to a capacitor connected across the voltmeter, and measuring the resulting p.d.

a Estimate the capacitance of the parallel plates.

The experimenter has several capacitors available in the range 10^{-11} F to 10^{-8} F. He first chooses the 10^{-8} F capacitor.

b What is the effective capacitance of the parallel plates and the 10^{-8} F capacitor (think whether they are in series or parallel)?

c What percentage of the original charge on the plate is retained, not transferred?

d Answer **b** and **c** if the 10^{-11} F capacitor is used and comment on any problems or hazards which may arise as a result.

13(E)a Check the accuracy of the following statement. 'If a reed switch is used to discharge a 1 μF capacitor through an ammeter 50 times a second, a 1 mA meter can be used safely if the capacitor is charged to 10 V.'

b What order of magnitude of capacitance can be used in a similar experiment, again using 10 V, if the reed switch discharges it 100 times a second through a meter which gives a measurable deflection for a steady current of 1 μA?

14(P) A pair of horizontal parallel plates, each measuring 0.1 m × 0.1 m, 0.01 m apart, is connected to a 10 V battery as shown in figure E67.

Figure E67

a What is the electric field strength between the plates?

b Estimate the charge density on the plates.

c How many excess electrons are there on the negative plate?

d Estimate the upward force on a water drop bearing one excess electron positioned midway between the plates.

e Treating the drop as a cube, estimate its 'diameter' if it is held stationary.

f Approximately how many atoms would it contain?

15(E) A 1 µF capacitor is to be made as follows. Long, 50 mm wide strips of thin metal foil, B and D, and of insulating paper 0.1 mm thick, A and C, are arranged in a sandwich as shown (figure E68), and then rolled up to make a cylinder. The relative permittivity of the paper is 2.

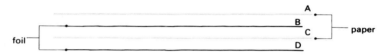

Figure E68

a At what point in the manufacture does the need for the top sheet of paper, A, become obvious?

b About how long a sandwich would be needed to get a final capacitance of 1 µF?

Figure E69 shows a small area of the rolled up cylinder in cross-section. Now as well as capacitance between 'plates' B_1 and D_1 of the same layer, there is also capacitance between 'plates' D_1 and B_2 of adjacent layers, allowing extra charge to be stored on the foils for the same p.d.

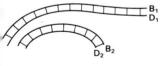

Figure E69

c By roughly what factor will the charge stored increase when the cylinder is completely rolled up?

d How does this affect your answer to **b** – the total length needed?

e Estimate roughly the diameter of the rolled up cylinder. You might then compare its size with that of a commercial 'paper' capacitor of about the same value.

16(P) Two conducting plates a distance d apart are connected to a battery of p.d. V. In figure E70(b) the separation is increased to $2d$ while the battery is still connected; in figure E70(c) the battery is disconnected before separation.

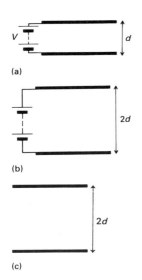

(a)

(b)

(c)

Figure E70

a In (b) what changes occur in:
i the capacitance, *ii* the p.d., *iii* the charge stored, *iv* the energy stored, *v* the electric field in the space.

b In (c) what changes occur in:
i the capacitance, *ii* the charge stored, *iii* the p.d., *iv* the energy stored, *v* the electric field.

c Account for the changes in energy in each case.

17(R) A plate of area $2 \times 10^{-3}\,\text{m}^2$ is held by a very well insulated support 50 mm above the bench, which conducts well (figure E71). It is charged to 4 kV, but its potential drops to 3 kV over 10 minutes. Estimate the conductivity of air, assuming that a uniform conduction across the plate area is the cause of the leakage. (You may be able to show that the question has given you two pieces of data that you need not use.)

Figure E71

18(R) Figure E72 shows a section through a capacitor microphone; figure E73 shows a circuit with which the microphone is used.

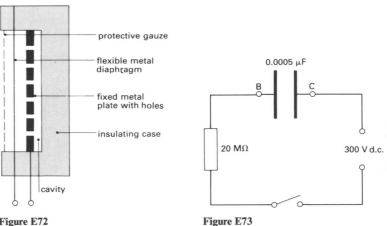

Figure E72   **Figure E73**

a The switch is initially closed. If it is opened, and the diaphragm pushed slightly inwards, explain what would happen to:
i the capacitance of the microphone,
ii the p.d. between B and C.

b Explain what would happen if, with the diaphragm still pushed in, the switch were closed.

c Why is the instrument constructed so that the diaphragm is as close to the first plate as possible?

d What is the time constant of the circuit in figure E73?

e Assuming that the switch is closed, state the changes of p.d. between B and C that you would expect to occur if a compression wave moved the diaphragm inwards in a time which was:
i short compared with the time constant of the circuit – say about 10^{-5} s.
ii long compared with the time constant of the circuit – say about 1 s.

f Two sources of sound, one of frequency 10 kHz, the other 50 Hz, are each found to produce the same amplitude of mechanical vibration in the diaphragm.
i Why is the amplitude of the resulting variations of p.d. across BC smaller for the 50 Hz vibrations than for the 10 kHz?
ii Explain what change you could make in the circuit to bring the amplitude of the electrical output from the microphone, when responding to the 50 Hz note, nearer to that produced by the 10 kHz note.

(Long answer paper, 1970)

Uniform gravitational field

19(I)a A child of mass 40 kg on a swing loses 0.8 m in height in swinging from A to B as shown. What is the child's speed at B? (Take $g = 10 \, \text{N kg}^{-1}$ in this question.)

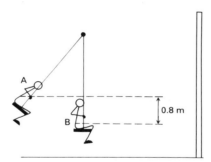

Figure E74

b Repeat **a** for an adult, in place of the child.
You may have found **b** more difficult than **a** if you tried to first work out the potential energy of the adult since you were not given the adult's mass. However, since in the equation $mgh = \frac{1}{2}mv^2$, the m cancels out, it was not required, even for **a**.

c *i* What is the extra P.E. of each kilogram of the child at A compared with B?
ii What is the extra P.E. of each kilogram of the adult at A?
iii What is the extra P.E. of the adult if his or her mass is 80 kg?

d Suppose you had to paint on the wall of the playground marks to show the P.E. gained by 1 kg at different heights. If you put a mark each time the P.E. increased by 10 J how far apart vertically would they be? How far apart for 1000 J intervals of energy? (You might need a rather high wall!)

e Suppose you continued painting lines at these vertical intervals all over the surface of the countryside. What would the overall pattern look like and what would it represent?

20(I) Figure E75 shows several cars parked in a multi-storey car park. The floors rise in 5 m steps. (Take $g = 10 \, \text{N kg}^{-1}$.)

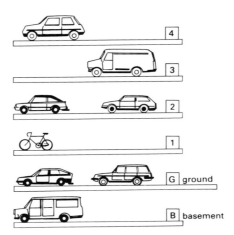

Figure E75

a A 'Sierra' has a mass of 1 tonne (1000 kg). How much potential energy does it gain in moving from the ground floor to
i the first floor, *ii* the fourth floor, *iii* the basement?

b What is the gain in energy of each kilogram of the car in *i*, *ii*, and *iii* above?

c What is the gravitational potential difference between the ground floor and *i* the first floor, *ii* the fourth floor, *iii* the basement?

d Write down the gravitational p.d. between *i* the first and the fourth floors, *ii* the first floor and the basement.

e How much potential energy does a 'Fiesta' of mass 800 kg lose in coming down from the fourth to the first floor?

21(P) Figure E76 shows part of the gravitational field near the surface of the Earth where $g = 10 \, \text{N kg}^{-1}$, and is uniform over this small region.

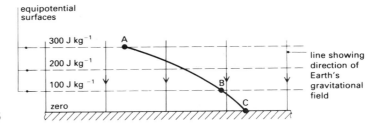

Figure E76

a How far apart are successive equipotential surfaces drawn in the figure?

b What is the gravitational force on a mass of 5 kg
   i at A,   *ii* at B?

c What is the gravitational P.E. of a mass of 5 kg
   i at A,   *ii* at B?

d How much energy must be supplied to move the mass from B to A?

e A cannonball of mass 5 kg follows the path ABC shown in figure E76. At A its speed is 20 m s^{-1}. What is its speed
   i at B,   *ii* at C?

The gravitational inverse-square law

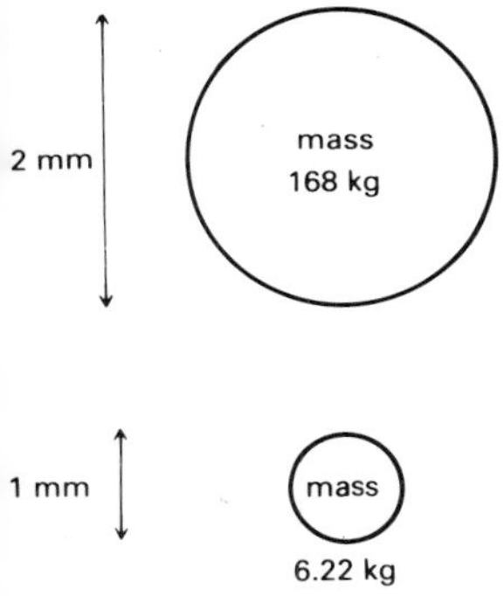

22(P) Figure E77 shows the two lead spheres used by Cavendish in 1798 to measure *G*.

a How far apart would the centres of the spheres be to give the maximum force of attraction between them?

b Calculate a value for *G* if this force was measured as 6.76×10^{-6} N.

The next two questions use data from the spaceflight of Apollo 11. This made the first manned landing on the Moon in July 1969. You can use the data to test whether the Earth's gravitational field strength obeys the inverse-square law at quite large distances.

E

23(L) On a section of its outward flight, Apollo 11 was coasting on a path almost directly away from the Earth and well outside its atmosphere. Over a period of 6 hours and 13 minutes its distance from the Earth's centre changed from 209×10^6 m to 241×10^6 m, and its velocity fell from 1527 m s^{-1} to 1356 m s^{-1}. The thrust motors were not used during this time.

a Why did the velocity decrease, even though the rocket motors were not used for forward or reverse thrust?

b What was the average acceleration of the spacecraft over the 22 380 s period?

c Without performing any further calculation, write down the average value of *g* (the Earth's gravitational field strength) over this period.

d Newton's Law of Gravitation gives the value of *g* as $-\dfrac{GM}{r^2}$, where $GM \approx 4.0 \times 10^{14}$ N m^2 kg^{-1} for the Earth. What value does *g* have if *r* is taken as the average distance of the spacecraft from Earth (225×10^6 m) during the period being considered?

e The Moon also pulls on the spacecraft. *GM* for the Moon is 4.9×10^{12} N m^2 kg^{-1}. Calculate its contribution to the field strength $\left(-\dfrac{GM}{r^2}\right)$, at the same point (about 150×10^6 m from the Moon). Does it have a significant effect?

mass
168 kg

2 mm

1 mm

mass

6.22 kg

Figure E77

24(L) Table E1 gives four other similar pairs of points between which the Earth's gravitational pull was the only significant force acting.

Time from launch h :min: s	Distance from Earth's centre $r/10^6$ m	Velocity $v/\mathrm{m\,s}^{-1}$	Mean distance $r/10^6$ m	Mean acceleration $g/\mathrm{m\,s}^{-2}$
03 : 58 : 00	26.3	5374		
04 : 08 : 00	29.0	5102		
05 : 58 : 00	54.4	3633		
06 : 08 : 00	56.4	3560		
09 : 58 : 00	95.7	2619		
10 : 08 : 00	97.2	2594		
19 : 58 : 00	169.9	1796		
20 : 08 : 00	170.9	1788		

Table E1

During the whole period covered by the table the spacecraft was travelling almost directly away from the Earth with its motors shut down.

a Calculate the average acceleration between each pair of points.

b Without further calculation deduce an estimate for the gravitational field strength at the mid-point of each interval.

c Plot a graph which enables you to check whether g does vary as $\dfrac{1}{r^2}$.

(Think which graph will most easily reveal this.)

Gravitational potential difference, ΔV_g, and potential, V_g

Questions 25 to 29 are based on a computer program which calculates values of gravitational potential difference from a graph of field strength against distance by adding up the areas of strips under the graph. It is not absolutely necessary to have seen the computer program, for all the data you will need are given with each question.

25(L) Table E2 shows various calculations of the energy required to transport one kilogram (i.e., the gravitational potential difference, ΔV_g between the surface of the Earth and a distance of 50×10^6 m from its centre. These have been computed using the number of steps shown to calculate the area under a field–distance graph by adding the rectangular strips 'under' the graph. The time taken for each calculation to be performed by the computer is also shown.

The computer program assumes the field strength g to be constant over the full distance given by the step size.

a For which number of steps is this least true? Explain why this leads to a poor estimate of ΔV_g.

b For which number of steps is this most nearly true? Which number of steps yields the most accurate estimate of ΔV_g?

c What could be done to obtain an even better estimate of ΔV_g?

Number of steps	Step size $/10^6$ m	gpd(ΔV_g)MJ kg^{-1}	Time taken/s
1	43.6	218	<1
2	21.8	120	<1
5	8.73	70.3	≈1
10	4.36	59.2	≈2
20	2.18	55.8	≈3
50	0.873	54.8	5
100	0.436	54.7	14
200	0.218	54.7	26
500	0.087	54.6	74
1000	0.044	54.6	144

Table E2

d What would be the disadvantage in doing this?

e A precise calculation (or use of a *very* small step size) leads to a value for ΔV_g of 54.52 M J kg^{-1}. What is the percentage error in the estimate of ΔV_g if just 20 steps are used?

f Which part of the graph – nearest the Earth or furthest away – makes the greatest contribution to the error? Why?

26(L) Rocket physicists need to know the amount of energy required to raise one kilogram from the Earth's surface to various heights. They might use the program to calculate values of ΔV_g between R (the Earth's surface) and various distances r_2 (from the centre of the Earth). The results are shown in table E3. ($R = 6.37 \times 10^6$ m; 200 steps are used in each calculation.)

$r_2/10^6$ m	ΔV_g/MJ kg^{-1}
6.371	0.01
7	5.64
10	22.7
14	34.1
18	40.5
22	44.5
30	49.3
40	52.7
50	54.7

Table E3

a Plot a graph of ΔV_g against r_2.

b Use the graph to estimate the amount of energy required to lift
i a 1000 kg probe from the Earth's surface to a distance of 20×10^6 m from the Earth's centre;
ii a 200 kg satellite from the Earth's surface to a height of 36×10^6 m above the Earth's surface;
iii a 1 kg mass *from* 40×10^6 m *to* 50×10^6 m from the centre of the Earth.

c The graph seems to be approaching a limit – it does not continue to rise indefinitely. The extra energy required to transport 1 kg from 50×10^6 m to a very great distance can be worked out. It comes to about 7.8 MJ. What then is the total energy needed to lift 1 kg from the Earth's surface to as great a distance as one would wish?

d Describe what will happen to a 1 kg space probe launched with an energy of

i 50 MJ

ii 70 MJ

iii an amount equal to your answer to **c**.

27(L) An interplanetary space convention is being held on a satellite 50×10^6 m from the centre of the Earth. Delegates from Earth are at present situated at the distances given in table E4 (r_1 values). The amount of energy needed to transport 1 kg from these positions to the destination, at 50×10^6 m, has been calculated by computer. These values represent the gravitational potential difference, ΔV_g, between the two points. (200 steps are used in each calculation.)

$r_1/10^6$ m	$\Delta V_g/\text{MJ kg}^{-1}$
6.37	54.7
7	49.0
10	31.9
14	20.5
18	14.2
22	10.2
30	5.3
40	2.0

Table E4

a *i* Plot a graph of ΔV_g against r_1 (or use a computer to do this for you).

ii Why does ΔV_g decrease as r_1 increases?

iii Can you suggest a mathematical form for the way ΔV_g varies with r_1? (You may be able to test your suggestion by having the computer plot a suitable graph.)

b *i* Now plot ΔV_g (*y*-axis) against $\dfrac{1}{r_1}$ (*x*-axis).

ii Calculate the gradient of this graph (it should be a straight line).

iii Measure the intercept on the horizontal axis and compare its value with $\dfrac{1}{r_2}$ where $r_2 = 50 \times 10^6$ m.

iv Suggest an expression for ΔV_g in terms of r_1 and r_2.

c (Optional) *i* Finally plot ΔV_g against $\left(\dfrac{1}{r_1} - \dfrac{1}{r_2}\right)$ where $r_2 = 50 \times 10^6$ m.

ii Confirm that this graph has the same gradient as that in **b**.

d Suppose r_2 had been much larger than 50×10^6 m. What effect would this have had on the graph in **b**?

28(L) This question discusses a zero for potential energy.

Suppose the venue of the interplanetary conference was moved to very much larger distance from the Earth. From the original ΔV_g against r_2 graph (question **26a**), you can see that not much extra energy would be needed to go *beyond* 50×10^6 m to any distance one would wish. Suppose that, in question **27**, instead of making $r_2 = 50 \times 10^6$ m we made r_2 'as far away as one could wish'. The shorthand for this is ∞, infinity.

a What would be the value of $\dfrac{1}{r_2}$?

b Would there be any difference between the graphs of ΔV_g against $1/r_1$ and ΔV_g against $\left(\dfrac{1}{r_1} - \dfrac{1}{r_2} \right)$?

c Express in words what is now represented by ΔV_g.

Delegates from other planets would, of course, have different graphs as they would have come different distances through different strengths of field. Therefore they would have required different amounts of energy per kilogram to reach the same point from their own planets. Yet it seems sensible to regard each of them as now having the same amount of potential energy per kilogram, or in other words *potential*. Indeed they agree to accept a convention on this: that as far away from all their planets as possible any object will be considered to have zero potential. Exchanging interplanetary pleasantries, they return to their respective homes. Ideally, each need only set off in the right direction: the gravitational pull of the home planet will do the rest. No fuel need be used; indeed, kinetic energy is gained on the homeward journey.

d If kinetic energy is gained then what form of energy is lost?

e If the original potential energy of 1 kg at 'infinity' is agreed to be zero, what then is the sign of the potential energy per kilogram (or 'potential') nearer a planet?

f If ΔV_g between the surface of the Earth and 'infinity' is 62.5×10^6 J kg^{-1}, what is the potential at the Earth's surface?

g Look at your graph for question **27a** and *sketch* a graph of gravitational potential, V_g (with reference to a zero at infinity), against r_1.

29(L) This question relates the graph of V_g against r_1 to a mathematical expression for V_g which can be found by integrating the expression for gravitational field strength:

$$V_g = -\int_{\infty}^{r_1} \left(-\frac{GM}{r^2} \right) dr$$

Previous work with graphs suggests that ΔV_g varies with distance from the Earth according to:

$$\Delta V_g \approx 4 \times 10^{14} \left(\frac{1}{r_1} - \frac{1}{r_2} \right)$$

a If r_2 is infinity, rewrite the expression for ΔV_g.

b From question **28**, the gravitational potential, V_g, at a point is equal to ΔV_g but negative. Write down an expression for V_g in terms of r_1.

E

c Mathematical integration yields the expression $-\dfrac{GM}{r_1}$ for V_g. Calculat
the value of GM for the Earth and compare the two expressions for V_g

d Calculate the minimum energy which must be supplied per kilogram of a spacecraft on the Earth's surface to enable it just to escape from the Earth's influence.

Potential and energy changes near the Earth

30(L) (Optional) This question uses data from the Apollo 11 spaceflight to compute changes in energy and see how such changes vary with positi
The rocket motors were not used over the period for which data are given.

Distance from centre of Earth $r/10^6$ m	$\dfrac{1}{r}/10^{-8}$ m^{-1}	Velocity $v/\text{m s}^{-1}$	Kinetic energy per kilogram $\frac{1}{2}v^2/10^6$ J kg^{-1}
11.0	9.09	8406	35.33
26.3	3.80	5374	14.44
54.4	1.84	3653	6.60
95.7	1.04	2619	3.43
169.9	0.59	1796	1.61
209.2	0.48	1532	1.17
240.6	0.42	1356	0.92

Table E5

a We cannot easily measure the change in potential energy per kilogram of the spacecraft, so we measure the change in kinetic energy per kilogram. How are the two related and what assumption is made in thus relating them?

b What is the change in kinetic energy per kilogram between the first and last points in table E5?

c What is the corresponding change in potential energy per kilogram? (That is, the gravitational potential difference ΔV_g between the two points.)

d On this 'outward' flight the spacecraft is clearly slowing down. It may stop at a distance r_0 from the Earth's centre. How much kinetic energy
would it have then?

e What would be the change in potential energy per kilogram (ΔV_g) between the first point in the table and r_0?

f Deduce values of ΔV_g between all the points in the table and r_0, and plot a graph of ΔV_g against $\dfrac{1}{r}$.

g Find the gradient of the graph and compare its value with GM_E (M_E = mass of Earth).

h Find the intercept on the $\dfrac{1}{r}$ axis and from it deduce the distance r_0.

i If r_0 had been much larger, say 'as far away as you like', what would the intercept have been?

j Suggest an equation which describes the way ΔV_g would vary with r in that case.

31(P) This question uses values of the gravitational potential at different distances from the centre of the Earth to calculate energies required to escape from the Earth. A few values of V_g and r are already given.

$r/10^6$ m		$-\dfrac{GM}{r}/10^6 \,\text{J kg}^{-1}$
6.37	(Earth's surface)	-62.5
6.38		-62.4
10		-40
400	(distance of Moon)	-1
∞	(as far as you like)	0

Table E6

a How much energy is needed to raise 1 kg to a height of 10 km? (Use the above data.)

b What force acting on 1 kg would transform this amount of energy if it were uniform over the 10 km? Comment.

c How much energy is needed to transport 1 kg from the Earth to a distance equal to that of the Moon?

d Why is the energy not equal to the product of 10 N and this distance?

e How much energy is needed for a mass of one kilogram to escape completely from the Earth's influence?

f What velocity would the one kilogram object have to have at its launch to achieve this?

g Why would a mass of any size need this same launch velocity?

h Use the $\dfrac{1}{r}$ variation of potential to deduce values of V_g at distances of 20×10^6 m, 40×10^6 m, and 80×10^6 m. Hence plot a graph of V_g against r from $r = 10 \times 10^6$ m to 80×10^6 m.

i Find the gradients of tangents at $r = 20 \times 10^6$ m and 40×10^6 m.

j What is represented by these gradients?

k Find the ratio of the gradients and comment on the result.

32(P) This question asks you to calculate the *total* potential at various points near the Earth and Moon by summing the potentials due to each body. Take the mass of the Earth as 6.0×10^{24} kg and its radius as 6.40×10^6 m. Other data can be found on page 493.

a Calculate first the separate potentials at the surfaces of the Earth and the Moon.

b What is the gravitational potential difference between the two?

c Deduce the difference in gravitational potential energy of a 10 tonne (1 tonne $= 10^3$ kg) spacecraft on the Moon and on the Earth.

d Calculate the total potential due to the Earth–Moon system at the following distances from the centre of the Earth on a line towards the Moon:
i The Earth's surface (6.40×10^6 m)
ii 1.00×10^8 m
iii 3.00×10^8 m
iv 3.40×10^8 m
v 3.60×10^8 m
vi The Moon's surface (3.78×10^8 m).

e Using these draw a graph of total potential against distance from Earth from 1.0×10^8 m outwards.

f On the same diagram sketch a dotted curve which would represent the potential if the Moon were not there.

g Mark a point X where the overall field strength of the Earth and Moon is zero.

h From the graph deduce the minimum energy needed to send the 10 tonne spacecraft from the Earth to X.

i This is greater than your answer to **c**. Nevertheless a 10 tonne spacecraft must be provided with still more energy than this if it is to land on the Moon's surface (never mind returning!). Suggest reasons for this.

j Use the potential at the Moon's surface to calculate the 'escape velocity' for the 10 tonne spacecraft. Why need it be launched with a rather smaller velocity in order to return to and make a crash landing on the Earth?

k The temperature on the Moon's surface reaches up to 400 K, at which temperature oxygen molecules have an average speed of $560 \, \text{m s}^{-1}$. Use your answer to part **j** of this question to explain why the Moon has no atmosphere.

The gravitational field *inside* the earth

33(L) This question shows why there cannot be a field inside a hollow sphere. Consider a hollow shell which has a mass per unit area of surface (or surface density) of σ. The field at P can be deduced by adding up contributions from a whole series of pairs of cones (in three dimensions) like those shown in figure E78. Let us consider one such pair of cones subtending a solid angle 2θ at P, whose bases are circles on the surface of the shell (figure E79).

a What is the radius of circle A_1, in terms of r_1 and θ?

b Deduce an expression for the area of A_1.

c The mass per unit area is σ. What is the mass of the disc A_1?

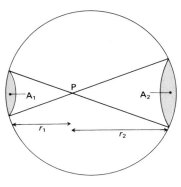

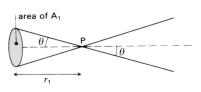

Figure E78

Figure E79

d What is the magnitude of the gravitational field strength at P due to this mass?

e Verify that this simplifies to $G\sigma\pi\theta^2$.

f Repeat **a** to **e** for the disc A_2 to find its contribution to the field strength at P.

g What is the total contribution of the two discs A_1 and A_2 to the field strength at P? Explain.

h What is the total contribution of all such pairs of discs to the field at P?

i Would the same be true if the field were not inverse-square? (Look at parts **d** and **e** again.)

34(P) This question explores the gravitational field *inside* the Earth, which is assumed to be uniformly dense.

Take a point, P, at distance r from the centre O. The shaded area represents parts of the Earth further from the centre than P. See figure E80.

This can be thought of as being made up of many thin shells centred on O.

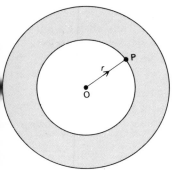

Figure E80

a What is the total field strength at P due to all these shells outside P?

b One consequence of the inverse-square law is that the field *outside* a solid or hollow sphere is the same as that produced if all the mass of the sphere were concentrated at its centre.

Write an expression for the contribution of the unshaded part of the sphere which has mass M, to the field strength g at P.

c Express M in terms of the density ρ (assumed to be uniform) and radius r of the unshaded region.

d Combine your answers to **b** and **c** to obtain an expression for g in terms of r.

e Sketch a graph showing the variation of g with distance both inside and outside the Earth (assuming, as above, that the Earth is of uniform

E

density). Sketch how the graph would change to more truly represent g for the Earth where the inner core, mainly of molten iron, has greater density than the outer mantle, the crust being the least dense.

35(R) This rather impracticable question ignores the fact that it is rather hot deep inside the Earth but imagines a tunnel drilled right through the Earth from Britain to Australia (or thereabouts).

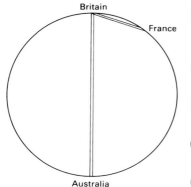

Figure E81

a Using the answer to part **d** of question **34**, write an expression for the acceleration of an object in the tunnel at different distances r from its centre (assuming spherical symmetry and uniform density).

b Deduce that an object dropped down the tunnel will perform simple harmonic motion about the centre.

c If the tunnel were pumped free of air, how much energy would be required to post a letter to Australia?

d Would more or less energy be required to send a rocket to the Moon from the centre of the Earth than from its surface?

e How does the gravitational potential at the centre compare with that on the surface? How could it be calculated (you need not do the calculation)?

f (Harder) Imagine a second tunnel cut as a chord rather than a diameter from Britain to, say, France. By considering components of gravity acting on a train in this tunnel, try to show that it should also perform simple harmonic motion.

g How much energy would be needed, in the absence of friction, for such a train to go from Britain to France?

h Imagine this (straight) track viewed from the apparently flat surface of the Earth. What shape would it appear to be? (Consider the variation of depth with distance round the curved surface and sketch what the shape of the tunnel would look like. You are not asked to derive a precise mathematical expression for the curve.) Comment.

Circular motion

36(I) Figure E82 shows a ball on a string being rotated in a horizontal circle.

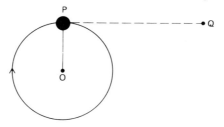

Figure E82

a If the string breaks when the ball is at P, use Newton's First Law of Motion to describe the subsequent behaviour of the ball.

b If the string does not break, use Newton's Second Law of Motion to explain the behaviour of the ball.

c Write a paragraph explaining the operation of a spin-drier without using the notion of 'centrifugal' force or without reference to the erroneous idea of water being 'flung outwards'. Diagrams will be a great help to your explanation.

37(L) This question considers an object moving in a circle of radius r at velocity v. The magnitude of its acceleration is calculated by considering the (vector) change in v after various decreasing time intervals leading to an expression for the acceleration of the object at any moment.

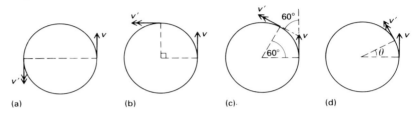

(a)　　　　(b)　　　　(c).　　　　(d)

Figure E83

a First we consider half a circle, figure E83(a), where the velocity vector has simply reversed direction, as shown in figure E84.
i What is Δt, the time interval, in terms of the period T?
ii What is the size of Δv, the change in velocity in terms of v?
iii Since acceleration $a = \dfrac{\Delta v}{\Delta t}$, derive an expression for a in terms of velocity, v, and time, T.
iv Since speed is distance travelled divided by time taken, write an expression for T in terms of r and v.
v Substitute for T in your answer to *iii* and deduce that the acceleration is $\dfrac{4}{2\pi}\dfrac{v^2}{r}$　or　$0.64\dfrac{v^2}{r}$

b Now consider a quarter of a circle, figure E83(b), and the vector diagram in figure E85.
i What is Δt now, in terms of T?
ii From the vector diagram, use Pythagoras' Theorem to deduce Δv in terms of v.
iii Hence deduce an expression for a and, by substituting for T as above express it in terms of $\dfrac{v^2}{r}$.

c Repeat **b** for figure E83(c) (a 60° or $\pi/3$ revolution). The vector diagram, figure E86, forms an equilateral triangle.

d Now consider movement through a small angle θ, figure E83(d).
i Express Δt in terms of T and θ.

Figure E84

Figure E85

Figure E86

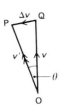

Figure E87

ii On the vector diagram (figure E87) express the *arc* length PQ in terms of v and θ. For a small angle θ this will be nearly equal to the *straight* length Δv.

iii Deduce an expression for the acceleration and simplify it as before.

iv In what direction is the acceleration, relative to the object's velocity?

e Look at the expressions for a as the time interval has decreased. They should approach a limiting value. Write down an expression for the centripetal acceleration of the object at any moment in its motion.

f If it has mass m, write down the centripetal force, F, acting on it.

g Considering the force as a vector, what addition to this formula expresses the fact that the force is inwards?

Satellites and orbits

38(R) Two physicists want to launch a satellite which orbits the Earth with a period of exactly one day. (Both realize that the most useful orbit will lie above the Equator.) One says it will have to be placed at a particular height, the other believes it can go to any height as long as its orbital speed is adjusted. They proceed to make calculations ... .

a Considering the orbit to be circular, express the speed v in terms of the radius r and period T.

b What is the centripetal acceleration required to keep an object orbiting at speed v and radius r?

c Express this in terms of r and T.

d Now this acceleration will have to be provided by the Earth's gravitational field at whatever height the orbit is. Write an expression for the acceleration, g, at distance r from the centre of the Earth.

e Using your answers to **c** and **d** deduce an expression for r. Is any value of r possible, given the required period?

f Calculate the height to which the satellite must be raised and the amount of energy per kilogram required to get it up there.

g If you wanted the satellite to remain vertically over the same point on the Earth's surface, why would this point have to be on the Equator?

h Would a lower orbit mean a longer or shorter period? (Look at your answer to **e**.)

i Rearrange the equation $-\dfrac{v^2}{r} = -\dfrac{GM}{r^2}$ to express v in terms of r.

Would a lower orbit mean faster or slower motion?

j (Harder) Explain why the action of air resistance or the impact of a shower of meteorites will speed up rather than slow down the satellite.

k Discuss the pros and cons of equatorial and non-equatorial orbits, thinking of some of the uses to which the satellite might be put.

39(P) This question shows that a satellite in orbit always has half the total energy it would need to escape completely from its orbit.

a Write down an expression for the potential energy of a satellite of mass m at a distance r from the centre of the Earth, whose mass is M.

b Write down the centripetal force required to keep the satellite in orbit at this radius with velocity v.

c Write down an expression for the Earth's gravitational pull which provides this force.

d Using your answers to **b** and **c** obtain an expression for the kinetic energy ($\frac{1}{2}mv^2$) of the satellite in terms of G, M, m, and r.

e Compare your answers to **a** and **d** and deduce the total energy of the satellite. What is the significance of the fact that this energy is negative?

f By what factor would the speed have to be increased to allow the craft to escape completely from the Earth? Explain.

40(E) Some historians credit Robert Hooke, who was Secretary of the Royal Society, with the discovery of the inverse-square law of gravitation. Hooke certainly had suggested to Newton in a letter in 1679 that the centripetal force attracting a planet to the Sun varied as $\frac{1}{r^2}$, although he could not explain why. Hooke then assumed a circular orbit: Newton subsequently proved that the orbit of an object under a $\frac{1}{r^2}$ force would in fact be elliptical, although he 'sat on' the proof for twenty years until Halley (of comet fame) questioned him on the matter.

This question shows how anyone equipped with a rudimentary knowledge of algebra and who knew Kepler's Third Law, already published by that time, could have deduced the inverse-square law for a *circular* orbit.

a Kepler's Third Law states that $\frac{r^3}{T^2}$ is a constant. What do r and T represent?

b What is the size of the centripetal acceleration of a planet moving in a circle at radius r from the Sun with velocity v?

c What is the planet's velocity in terms of r and T?

d By substitution deduce that its centripetal acceleration can be expressed as $\frac{4\pi^2 r}{T^2}$.

e This can be expressed as $4\pi^2\left(\frac{r^3}{T^2}\right)\frac{1}{r^2}$.

Use Kepler's Law to show that the acceleration is proportional to $\frac{1}{r^2}$.

f If acceleration is inverse-square, how does the force on a planet vary with distance?

g Try to find out about the events leading up to Newton's discovery of the laws of universal gravitation as outlined in his *Principia Mathematica*.

The inverse-square law

41(L) This question seeks to throw some light on the inverse-square law.

Imagine a candle illuminating room A (figure E88). Light falling on the window leaves the candle as a pyramid (a square-based 'cone') of rays. Room B (figure E89) is identical to A, except that everything apart from the candle has been scaled up in linear dimensions by a factor of 2.

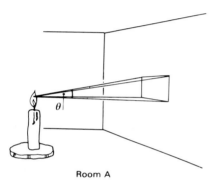

Room A

Figure E88

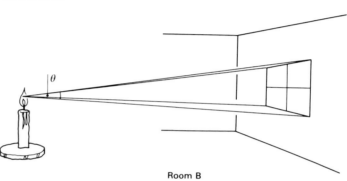

Room B

Figure E89

a How does the distance, r, of the window from the candle in room B compare with the candle-to-window distance in room A?

b How does the height of the window compare in B?

c How does the angle θ of the cone compare?

d How does the amount of light passing through this cone compare?

e How does the area of the window in B compare with the area of the window in A?

f What is the intensity of light from the candle at the window in B compared with A?

g If room C is 3 times the size of A, how will the intensity of the light hitting its window compare with that in A?

h What can you deduce about the product (intensity of light) × (area through which it passes)?

i Would the same be true if the light intensity did not obey an inverse-square law?

The electrical inverse-square law

42(L) Photographs in figure E92 show the position in which a small charged polystyrene ball, suspended like a pendulum as in figure E90, hangs when a second charged ball on an insulating rod, seen in figure E92(a)–(g), is pushed up close to it. The fine nylon thread used to suspend the first ball is not visible.

The pictures were taken with the apparatus in figure E90.

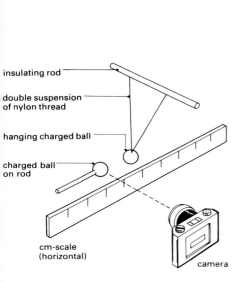

Figure E90
Taking photographs of charged balls.

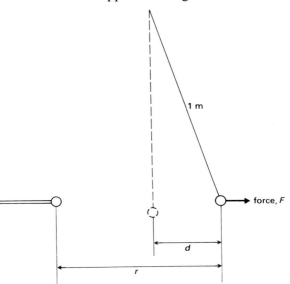

Figure E91
Sideways deflection of the ball.

When the charged ball on the rod is a long way away, the suspended ball hangs vertically, in the position shown in figure E92(a). As the second ball is brought closer, the suspended ball is pushed to the right until there is a big enough sideways force on it to balance the repulsive force due to the second ball (figure E91).

a If the sideways force on the suspended ball is doubled, how will its deflection, d, change (approximately)? How are d and the force F related?

b Measure the deflection, d, and the separation, r, between the centres of the balls, for photographs (b) to (g) in figure E92. Plot a graph to test whether the sideways force, F, on the suspended balls varies as $1/r^2$.

c The balls were given equal charges, each about 5×10^{-9} coulomb, from a high-voltage source. The charging was repeated for each picture and the charge was measured by sharing the charge on a ball with a $0.01\ \mu F$ capacitor, which was then found, using a high resistance voltmeter, to have a potential difference of 0.5 volt across it. Check that the charge stated above is correct. The measurements of charge indicated that the charge varies by a few per cent on different occasions. Would such fluctuations explain any feature of your graph?

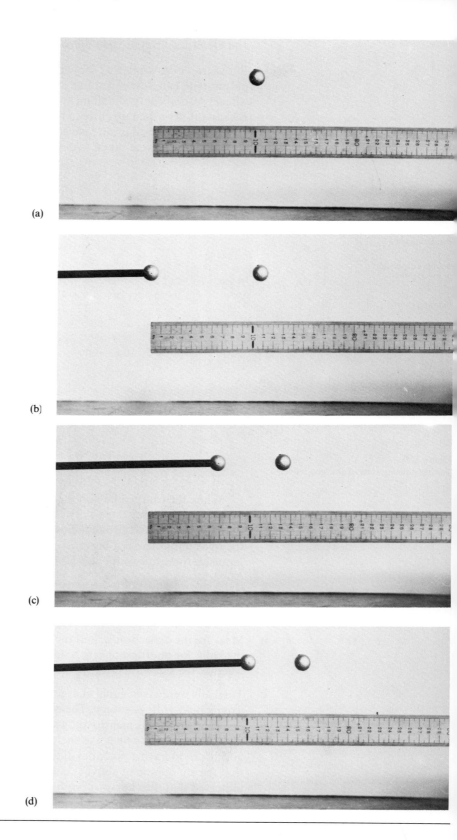

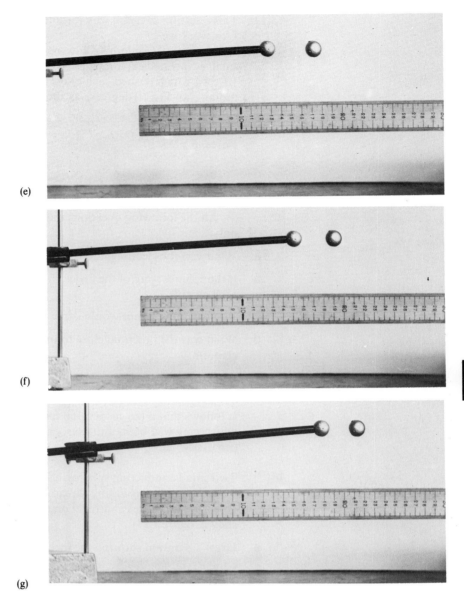

(e)

(f)

(g)

Figure E92

d Estimate the order of magnitude of the force constant in the equation:

force between balls

$$= \frac{(\text{force constant})(\text{charge on one ball})(\text{charge on other ball})}{(\text{distance between balls})^2}$$

The suspended ball weighed 1.1×10^{-3} newton.

e The charge on each ball almost certainly lies between 4 and 6×10^{-9} coulomb. Between what limits does this suggest that the force constant lies? The usual value is $9 \times 10^9 \, \text{N m}^2 \, \text{C}^{-2}$.

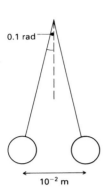

0.1 rad

10^{-2} m

Figure E93

43(L) This question suggests why it is often hard to make tests of Coulomb's Law work well.

In a test of Coulomb's Law, equal charges were placed on two small suspended expanded polystyrene spheres each weighing about 10^{-3} N, which pushed each other aside at an angle of the order of 0.1 radian (say 5 to 10 degrees), as shown in figure E93.

Coulomb's Law says that:

$$\text{Force, } F = 9 \times 10^9 \frac{Q_1 Q_2}{r^2}$$

Also, potential, $V = 9 \times 10^9 \frac{Q}{r}$

All the following questions require order of magnitude answers only.

a What was the electrical force on each sphere?

b If the spheres were of the order 10^{-2} m apart, what charge did they each carry?

c What steady current would carry this charge away in 10^2 seconds?

d What was the potential close to one sphere (say 10^{-2} m from the centre)?

e Use your answers to **c** and **d** to deduce the resistance which would allow the charge to leak away in about 100 seconds.

It follows that if the suspension has a resistance of less than 10^{14} ohms, which may well be so if the suspension is at all damp, the experiment can easily fail.

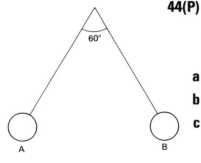

60°

A   B

Figure E94

44(P) Two small conducting spheres, each of mass 10 mg, are suspended from a point by insulating threads 10 cm in length. The spheres are given equal charges and repel one another, settling in the position shown (figure E94).

a Draw a diagram showing the forces acting on sphere A.

b Deduce the size of the electrical repulsion of A by B.

c Deduce the charge on each sphere.

Radial and uniform fields

45(L) This question helps to explain why a collection of charges each having a $\frac{1}{r^2}$ field gives a uniform field when made into a flat sheet. Imagine a carpet of charge made up of small tiles, each carrying charge Q. An observer at P, at height h above the sheet, is concerned with the field contributed by one tile at the base of a cone a distance r away, figure E95(a). The total field strength at P can be obtained by adding the contributions of all such cones.

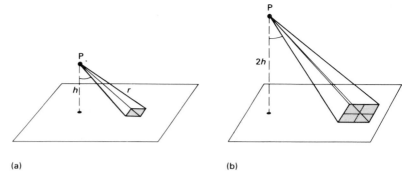

(a)

(b)

Figure E95
Field effect from a carpet of charge.

a Write an expression for the field strength at P due to the charge on the tile at distance r.

P now rises to twice the height, figure E95(b), keeping the cone pointing at the same angle.

b How far is P now from the patch of tiles at the base of the cone?

c What charge is now carried by this patch?

d What is the field strength at P (height $2h$) due to this patch?

The contribution is the same, whatever the height. Therefore the total field at P will be the same, as it is just the sum of the effects of all the cones. This assumes that the carpet is big enough for some of the cones not to 'miss' it where the height is increased (as *does* happen in figure E96).

e In figure E97(a) P is close to a large sheet so that only a few cones, at wide angles, will 'miss' the sheet if the height is doubled.
i What is the direction of the overall field strength vector at P (figure E98)?
ii Would the contributions from the missing cones have been large or small?
iii How, approximately, would the field vary close to the sheet?

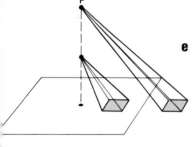

Figure E96

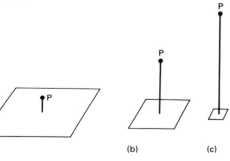

(b)

(c)

Figure E97
Variation of field with distance.

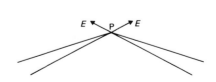

Figure E98
Contribution to the field from two cones at P.

f   In figure E97(c), P is a long way from a small sheet so that many cones 'miss' the sheet, and the field strength at P is considerably less than it is closer in. Guess the way the field strength varies with distance in this case.

46(P)   The sphere of a Van de Graaff generator 15 cm in radius is maintained at a high potential by a moving belt which carries charge to the sphere at a rate of 0.5 μA. The Perspex column which supports the sphere has a resistance of $3 \times 10^{11}\,\Omega$.

a   What is the potential of the sphere?

b   What is the charge on it?

c   What is the electric field close to its surface?

d   Estimate how close an earthed hand could be brought to the sphere before a spark crosses the gap. (Air conducts in a field strength of about $3 \times 10^6\,\mathrm{V\,m^{-1}}$.)

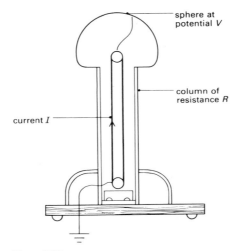

Figure E99

Electrical potential and energy in inverse-square fields

47(R)   A simplified model of a uranium nucleus is a sphere containing 92 protons and rather more neutrons, and having a radius of about 2×10^{-14} m. If the nucleus releases an α-particle (of charge $+2e$) at its 'surface', estimate:

a   the strength of the electric field experienced by the α-particle;

b   the size of the repulsive force on it;

c   the electrical potential at the surface;

d   the electrical potential energy of the α-particle;

e   the kinetic energy of the α-particle when it is a long way from the nucleus (express this in MeV where 1 MeV = 1.6×10^{-13} J). What

assumption must you make to deduce this?
(Charge on proton $= 1.6 \times 10^{-19}$ C.)

48(R) This question makes estimates concerning the structure of the hydrogen atom. Let us suppose that an electron is separated from a proton by a distance $r_0 = 0.5 \times 10^{-10}$ m. Calculate

a the electric field strength of the proton at this distance;

b the attractive force on the electron;

c the potential due to the proton at this distance;

d the electrical potential energy of the system in electon volts (eV).
(1 eV $= 1.6 \times 10^{-19}$ J.)

Figure E100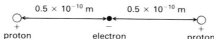

Now suppose another proton is brought up an equal distance from the electron, on the opposite side of it as shown in figure E100. Deduce:

e the electric field where the electron is;

f the force on it;

g the potential at this point due to both protons;

h the electrical potential energy of the electron.

i What would the electrical potential energy of the system be if the electron were removed?

j What is the total potential energy of this ion (2 protons plus 1 electron)?

k Comment on the role played by the electron in binding the ion together.

49(L) Figure E101 shows equipotentials drawn at 1 V intervals around two point charges. The object of this question is to deduce the overall equipotential pattern by adding together the potentials due to each charge.

a On a copy of Figure E101, first identify the contours by lightly labelling them 10, 9, 8, ... etc. (The smallest circle represents 10 V in each case.) Find the place where the two 4 V equipotentials meet and at this place mark a small figure 8 (representing 8 V, the total potential at this point). Mark a figure 8 also at all the positions where 5 V crosses 3 V, 6 V crosses 2 V, etc. Now, by using symmetry and the fact that the equipotentials will be smooth and continuous, try to draw in a complete curve joining all the points at which the potential is 8 V. (Use felt tip pen, or soft pencil.) In the same way, plot equipotentials at 7 V, 6 V, etc., as far as you can go, and also for 9 V and 10 V to reveal the overall equipotential pattern. What you have drawn is very similar to the potential experienced by an electron in the plane of the two protons in an H_2^+ ion, with the values scaled down.

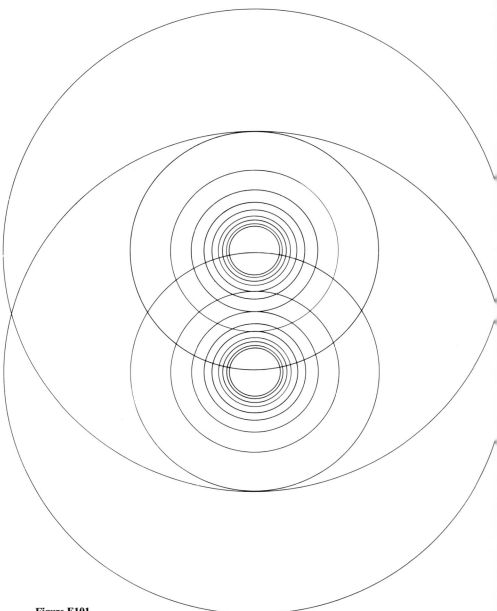

Figure E101

b Now regard one charge as positive and the other negative. On a fresh copy relabel the equipotentials $+10$, $+9$, etc. from one charge and -10, -9, etc., from the other. Using the same method as before, deduce the overall equipotential pattern. This would represent, for example, the potential between two parallel conductors carrying currents in opposite directions.

c Remembering that field is always perpendicular to the equipotentials deduce the direction of the electric field in different places and sketch in some lines to show its overall shape. (Recall experiments E3 and E:

50(R) A metal sphere, A, is connected by a long fine wire to a source of potential of 900 V. An insulated stand supports the sphere sufficiently far above the bench for the effect of the latter to be negligible.

a The potential 300 mm from the centre of the sphere is 450 V. Find the following, showing in each case the steps in your calculation:
i the potential 500 mm from the centre of the sphere;
ii the radius of the sphere;
iii the charge on the sphere.

b Another identical sphere, B, connected to the same source, is similarly supported with its centre 600 mm from that of the first sphere, A, as shown in figure E102.

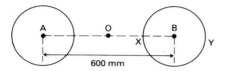

Figure E102

Would you expect the potential at O, a point midway between A and B, to be equal to, greater than, or less than 900 V? Explain why.

(Short answer paper, 1981)

51(R) (Hard) This question begins to explore the differences between conducting and insulating surfaces and the usefulness of adding up potentials to determine whether there is a field or not.

Suppose the metal spheres in question **50** were replaced by insulating spheres which separately had been given a uniform surface charge so that the potential on the surface of each was 900 V. These spheres are no longer connected to any source of charge.

a What would now be the potential at O in figure E102?

b What would be the potential at X and Y on the nearer and far surface of sphere B? How do you obtain these values?

c Would any electric field exist in the sphere B? Why?

d What would happen if B's surface were suddenly coated with conducting paint?

e Would there now be any differences in potential across the sphere B?

f What can you say about the potential at all points on the surface of a conductor? What if the conductor carries some charge?

Electrical and gravitational forces

52(E) In question **48b**, you calculated the electric force between a proton and an electron. It was assumed that this was responsible for holding the hydrogen atom together. Could not gravity be a factor here too?

a Estimate the gravitational force at separation, $r_0 = 0.5 \times 10^{-10}$ m, between a proton (mass $\approx 10^{-27}$ kg) and an electron (mass $\approx 10^{-30}$ kg)

b What is the ratio of the electric force to the gravitational force at this distance?

c Why is the ratio of these two forces the same even if the proton and electron were one light-year apart?

The enormous size of this ratio has seemed to be of fundamental significance to physicists although no one has yet been able to exploit it. Clearly gravity does not hold the atom together. However, could it have effect in the nucleus where electric forces between protons are repulsive?

d What would be the ratio of electric force to gravitational force between two *protons*?

e A helium nucleus has two protons and two neutrons. Assuming the neutrons are about as massive as the protons, would gravitational attraction between the nucleons overcome the electric repulsion?

f The force which *does* hold the nucleus together must be over 10^{36} times stronger than gravity at separations of about 10^{-15} m. Since no such force seems to dominate affairs at an astronomical, or even an atomic level, what can we say about the way it varies with distance?

53(P) If, as question **52** suggests, electrical forces are so much greater than gravitational ones, why do we not experience them much in everyday life? This question starts out to answer this problem.

   Let us consider a dipole, that is two equal and opposite charges A and B, separated by a short distance. (The distribution of charges in many molecules leads effectively to this arrangement. An ion pair, for example Na^+ and Cl^-, will also show similar properties.) Assuming charges of $+e$ and $-e$, 2×10^{-10} m apart, we shall consider how the electric field strength varies with distance from the dipole, and compare this with the corresponding variation for a single charge.

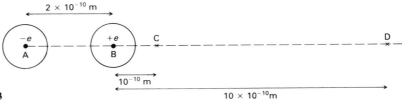

Figure E103

a First consider B alone, A being removed. Calculate the electric field strength at C due to B and call this 1 'unit'.

b D is ten times further from B than C is. Using the inverse-square law, write down the field strength in units at D, due to B alone.

c Similarly, work out the contribution to the field strength, in units, made by A (remembering the sign of its charge) at C and D.

d Work out the total field strength, in units, of the dipole at C and D.

e   Calculate the ratio now of the field strength at D to that at C.

f   Does the field of the dipole fall off following an inverse-square law, or more rapidly? Suggest a reason for this.

g   Now imagine many dipoles forming an array of positive and negative charges in equal numbers (as for instance in an ionic crystal). Explain why the very large electric forces within the array are not experienced by charges at some distance from it.

54(R)   This question is about gravity and the similarities between gravity and electric and magnetic effects.

       The passage below sets out three sets of ideas about gravity. For each of the sections **a** to **c** you are asked to write a more complete explanation of the ideas: your explanations may include
i quantitative calculations to illustrate the ideas,
ii fuller explanations of the theoretical ideas,
iii discussion of possible experiments.
       You should pay particular attention to the words and phrases that are in italics in each section.

a   There is something peculiar about gravity: it is such a *small force* that if we didn't live on a big lump of matter called the Earth we might not notice that it affected human-size objects at all. In fact the simplest calculations can show that it is very *hard to demonstrate* that the effect exists between all pieces of matter.

b   There is a close *analogy* between the theoretical ideas involved in electricity and in gravity, and this can be of great value in discussing such abstract ideas as *field* and *potential*. Thus problems such as the scattering of alpha-particles by a nucleus and the path of a spaceship round the Moon have many *similarities* though there are also *important differences*.

c   However, electrical and magnetic effects are so *much bigger*, for human-size experiments, that they swamp all effects of gravity. The fact that when we come to matter on an astronomical scale, gravity is *by far the most important force* is then hard to explain – it must be due to *electrical neutrality of big objects*.

(Long answer paper, 1979)

55(R)

Gravity		**Electricity**
$F_g = -G\dfrac{m_1 m_2}{r^2}$	Force	$F_e = \dfrac{1}{4\pi\varepsilon_0}\dfrac{Q_1 Q_2}{r^2}$
$g = -G\dfrac{M}{r^2}$	Field strength	$E = \dfrac{1}{4\pi\varepsilon_0}\dfrac{Q}{r^2}$
$V_g = -G\dfrac{M}{r}$	Potential	$V_e = \dfrac{1}{4\pi\varepsilon_0}\dfrac{Q}{r}$
$E_p = -G\dfrac{m_1 m_2}{r}$	Potential energy	$E_p = \dfrac{1}{4\pi\varepsilon_0}\dfrac{Q_1 Q_2}{r}$

This question revises the links between the estimated quantities above and some of the differences between electricity and gravity.

a Using the words *mass* or *charge* where relevant write down two sentences defining electrical and gravitational field strength.

b How can one obtain the gravitational or electrical field strength from a graph of the appropriate potential against distance?

c How is the potential difference between two points obtained from a graph of field strength against distance?

d How is the potential energy of a mass or charge at a point in a field obtained from the potential at the point?

e What would the gradient of a graph of potential energy against distance indicate?

f What does it mean to say that all gravitational forces are negative?

g Why is there no negative sign in the corresponding electrical expression, though electric force vectors may sometimes be negative?

h Why is gravitational potential (and potential energy) always negative?

i In a place where the electrical potential is positive (for example, near a proton), how could one object have a positive potential energy and yet another have a negative potential energy?

Unit F
RADIOACTIVITY AND THE NUCLEAR ATOM

Paul Jordan and **Peter Harvey**
Highfields School, Wolverhampton

SUMMARY OF THE UNIT

F

SUMMARY OF THE UNIT

THE RUTHERFORD MODEL OF THE ATOM

The nature and properties of alpha, beta, and gamma radiation

The development of our present picture of the atom began in 1896 whe[n] Becquerel discovered that uranium compounds could affect photo[o] graphic plates and also ionize a gas.

There are three types of radiation emitted and they can be identifie[d] by their penetrating power, ionizing ability, and behaviour in [a] magnetic field.

Alpha radiation has a range of a few centimetres in air and can b[e] stopped by a sheet of paper. It can cause intense ionization in gas and [is] deflected by a very strong magnetic field. The deflection by a magneti[c] field suggests that it consists of relatively heavy, positively charge[d] particles. Alpha particles are in fact helium ions with a double positiv[e] charge. The energy range of alpha particles from different sources varie[s] from about 4 to 10 MeV, and this corresponds to kinetic energies givin[g] them speeds of about 1.5 to $2 \times 10^7 \, \mathrm{m \, s^{-1}}$. The alpha particles emitte[d] from a particular source all have the same energy.

$1 \, \mathrm{MeV} = 1.6 \times 10^{-13} \, \mathrm{J}$

Beta radiation is more penetrating than alpha radiation, having [a] range up to about one metre in air, and it is able to penetrate a fe[w] millimetres of aluminium. It causes less ionization and is more easil[y] deflected by a magnetic field than is alpha radiation. The direction an[d] amount of deflection indicates that beta radiation consists of negativel[y] charged particles with a relatively small mass. In fact, beta particles ar[e] fast-moving electrons. The energy range of beta particles (from abou[t] 0.025 to 3.2 MeV) corresponds to kinetic energies giving them speed[s] very close to the speed of light ($3 \times 10^8 \, \mathrm{m \, s^{-1}}$). Beta particles are emitte[d] with an almost continuous energy spectrum.

Gamma rays have the ability to penetrate several centimetres [o]f lead: they cause only weak ionization. They are not deflected by [a] magnetic field, which indicates that this radiation does not consist [o]f charged particles. Gamma rays are in fact a form of electromagneti[c] radiation, travelling with the speed of light, and with wavelength[s] shorter than those of X-rays. Unlike beta particles, the gamma ray[s] emitted by a particular source can have only certain sharply-define[d] energies. For example, cobalt-60 emits 1.2 and 1.3 MeV gamma rays.

Sources and activities

Sources of radioactivity used in schools are relatively weak; they hav[e] low activities of about 18×10^4 Bq (5 µCi) or less. Typical sources are:

Radium-226 for alpha and beta particles, and gamma radiation.
Americium-241 for alpha particles (also some low-energy gamma radiation).
Strontium-90 for beta particles.
Cobalt-60 for gamma radiation.

The activity of a particular source depends on the quantity of radio-active material it contains, as well as the particular atomic species (nuclide). Thus a 2×10^5 Bq source of, say, strontium-90 contains twice as many ^{90}Sr atoms as a 1×10^5 Bq source.

Nuclear gunnery

Section F3

Many experiments in atomic and nuclear physics consist of bombarding one kind of particle with another. The ionization energy of atoms (the energy needed to remove one of the outer electrons) can be found by bombarding gas atoms with relatively slow-moving electrons 'boiled off' a hot wire. When accelerators are used to produce very fast-moving particles (speeds approaching that of light), much more damage is done: the atomic nucleus may break up and new particles can be created.

READING
'Radioactivity and the nuclear atom: a brief history' (page 373)

Some very important bombardment experiments crucial in the development of the nuclear model of the atom were carried out by Rutherford and his colleagues Geiger and Marsden in about 1913. They used alpha particles from radioactive sources to bombard metal atoms. To interpret their results it is essential to know the properties of alpha particles: their mass, electric charge, and energy.

The Rutherford model of the atom

The idea that matter ultimately consists of 'uncutable' atoms (particles which cannot be subdivided any further) is a very old one. Our present ideas of how atoms are made up of smaller particles began to develop early in the twentieth century. By 1900, J. J. Thomson had found that electrons could be removed from atoms, and in 1904 he proposed a so-called 'plum pudding' model for the atom: negative electrons embedded in a sphere of uniformly distributed positive charge. Some experiments in about 1910 produced surprising results which the Thomson model could not explain. When alpha particles were fired at a metal, a few were turned back or scattered through very large angles (figure F1). This suggested to Rutherford that the atom's positive charge was concentrated in a small but massive nucleus; the scattering was due to the electrical repulsion of the alpha particle by the central small nucleus. Geiger and Marsden were then set to work to test this model experimentally.

The experiment tested whether the scattering behaved as if the particles were acted upon by an inverse-square law force. The particles could not be followed along their paths so an indirect test had to be arranged. The number of particles recorded at particular angles (table F1) was compared with the number predicted if the force acting upon them were an inverse-square one. The experimental results agreed with the theoretical predictions.

A '1/r hill' is a physical model which has some of the properties of the nuclear atom. The hill is carefully constructed so that its height varies as $1/r$, r being the distance from the centre (figure F2). When a ball rolls on the hill its potential energy therefore varies as $1/r$ and the force on it as $1/r^2$.

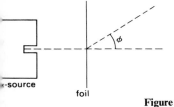

-source

foil

Figure F1

angle ϕ degrees	Number scattered at angle ϕ into fixed small area
150	33.1
135	43.0
120	51.9
105	69.5
75	211
60	477
45	1 435
37.5	3 300
30	7 800
22.5	27 300
15	132 000

Table F1

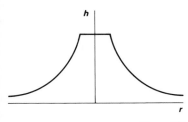

h

r

Figure F2

DEMONSTRATION F5
Qualitative test of the
gravitational hill model

Geiger and Marsden slowed down alpha particles by passing them
through thin sheets of mica. They found that slower-moving alpha
particles are more likely to be deflected; see table F2. Tests with the hill
show that the lower the initial kinetic energy of the ball, the larger the
angle it is deflected through.

Number of sheets of mica	Range of α-particles after leaving mica	Number of α-particles detected at fixed angle
0	5.5	24.7
1	4.76	29.0
2	4.05	33.4
3	3.32	44
4	2.51	81
5	1.84	101
6	1.04	255

Table F2

The fact that the vast majority of alpha particles either pass through
the foil undisturbed or are scattered at small angles is evidence for the
small size of the nucleus – the undeflected particles pass too far from
any nucleus to be significantly affected.

The size of the nucleus – defined by the distance of closest approach
of an alpha particle – is of the order of 10^{-14} m; that is, four orders of
magnitude smaller than the atom itself.

Section F2   EXPONENTIAL DECAY

Radioactive decay and half-life

Radioactive decay is a spontaneous process which cannot be controlled
and is not affected by chemical reactions, temperature, or pressure.
Radiation is emitted from the nucleus of an atom and the process does
not involve the atom's outer electrons. When a nucleus emits an alpha
or a beta particle the nucleus of a different atom, called the daughter
atom, is formed.

EXPERIMENT F7
The decay of radon and the
determination of its half-life

The half-life of a particular radioactive nuclide is defined as the time
it takes for half the nuclei originally present to disintegrate. Since
activity is proportional to the number of active nuclei present, the
activity due to a particular nuclide also falls to half its initial value in
one half-life.

Half-lives can vary from millionths of a second to millions of years.
Polonium-212 has a half-life of 3×10^{-7} s, whereas the half-life of
uranium-238 is 4.5×10^9 years. The activity of a radioactive substance –
the number of disintegrations per second – is directly related to the
number of undecayed nuclei present. The process is governed by
chance, like the throwing of a die. The chance of a given nucleus
decaying in, say, one second is fixed, but there is no way of being sure
whether that particular nucleus will decay during the second or not. If
many undecayed nuclei are present the decay curve will be quite
smooth, but when only small numbers are involved the statistical
fluctuations are much more noticeable.

QUESTIONS 15 to 18

The rate of radioactive decay depends on the number of undecayed nuclei present. The decay curve has the same shape as that for the decay of charge on a capacitor, where the rate depends on the amount of charge. Other processes grow at a rate which depends on the quantity of something present – for example, the growth of a population of bacteria. Such processes are called exponential decay or exponential growth, respectively. The characteristic of an exponential change is the constant ratio property: in equal intervals of time the quantity present always grows (or decreases) by a constant factor. Exponential growth has a constant doubling time; exponential decay has a constant half-life.

QUESTIONS 19 to 23

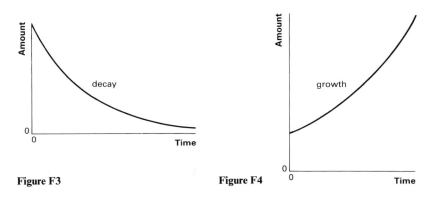

Figure F3

Figure F4

The exponential function

'About exponential changes' (page 394)

QUESTIONS 24 to 27

λ is the *decay constant*

In 10^2, 10^3, a^t, the quantities 2, 3, and t are called 'exponents'; they are also called 'powers' or 'indices'. This is where the word 'exponential' comes from. The growth equation $\Delta N/\Delta t = \lambda N$ can be solved numerically. The solution is of the form $N = a^t$. This gives a constant-ratio curve when values of N are plotted against t. When $\lambda = 1.0$ and $N = 1.0$ at $t = 0$, the value of a is 2.718... which is the natural number e. The general solution of the equation $\Delta N/\Delta t = \lambda N$ is $N = e^{\lambda t}$.

Similarly, the equation for decay, $\Delta N/\Delta t = -\lambda N$, has the solution $N = e^{-\lambda t}$. This assumes that $N = 1.0$ at $t = 0$, but this is not always true. The solutions become $N = N_0 e^{-\lambda t}$ for a decay and $N = N_0 e^{\lambda t}$ for a growth, where N_0 is the value of N at $t = 0$ and N is the value after time t (figures F5, F6).

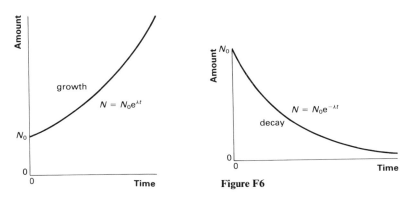

Figure F5

Figure F6

$$N/N_0 = e^{-\lambda t}$$
$$\Rightarrow \ln N = \ln N_0 - \lambda t$$

The constant-ratio property is a test of whether or not a curve i͏ exponential. If the change is exponential, a graph of $\ln N$ against tim͏ will be a straight line. The slope is positive for a growth curve, an͏ negative for a decay curve (figure F7).

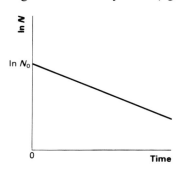

Figure F7

Half-life and decay constant

After one half-life ($t_{\frac{1}{2}}$)

$$N = N_0/2 = N_0 e^{-\lambda t_{\frac{1}{2}}}$$

from which it follows that

$$\lambda = \ln 2/t_{\frac{1}{2}} = 0.693/t_{\frac{1}{2}}$$

and

QUESTION 28 $t_{\frac{1}{2}} = 0.693/\lambda.$

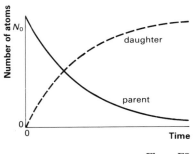

Figure F8

Radioactive decay and recovery

When alpha or beta radiation is emitted from a nucleus, a new nucleu͏ of a different substance is formed. This new nucleus is called th͏ daughter product. If the daughter nucleus is not radioactive, the number o͏ daughter nuclei present at any time is just the number of parent nucle͏ which have decayed: $N_0 - N$. But N is given by $N = N_0 e^{-\lambda t}$.

So $N_0 - N = N_0 - N_0 e^{-\lambda t} = N_0(1 - e^{-\lambda t}).$

Radioactive equilibrium

The daughter product may also be radioactive. If the daughter half-lif͏ is very much less than the parent half-life a state is soon reached wher͏ the daughter atoms are decaying as quickly as they are being produced͏ Thus the number of daughter atoms present in the sample is constan͏ This condition is called *radioactive equilibrium*. For example, th͏ amount of protactinium present in a sample of uranium quickly reache͏ an equilibrium value because the half-life of protactinium is rathe͏ short.

THE NUCLEUS

Proton number, nucleon number, and the Periodic Table

Z = proton (or atomic) number
(nucleon number) = number of protons
+ number of neutrons

$m_n = 1.675 \times 10^{-27}$ kg

$m_p = 1.673 \times 10^{-27}$ kg

QUESTION 29

Alpha-particle scattering experiments by Geiger and Marsden established Rutherford's model for the atom: a small, massive, positively-charged nucleus, surrounded by negative electrons. The amount of scattering depends, among other things, on the nuclear charge, that is, on the number of protons in the nucleus. In the Periodic Table the elements are arranged in order of increasing proton number, Z. The nucleon number of an atom, A, is usually about twice its proton number. This means that the mass of the nucleus is about twice the mass of the protons it contains. The extra mass is due to the uncharged neutrons, each of which has a mass approximately equal to the mass of a proton. Because neutrons are uncharged they do not cause ionization and can penetrate matter easily. This makes them difficult to detect. Although their existence was suggested by Rutherford in 1914, neutrons were not shown to exist until 1932. Then it was possible to propose a model for the nucleus containing A particles, Z of them protons and $(A - Z)$ neutrons. In the atom there are also Z electrons. All atoms with the same value of Z are chemically the same.

Isotopes

^A_ZX

isotope means 'same place'
(*i.e.*, in Periodic Table)

QUESTIONS 30 to 33

'Radioisotopes' in the Reader *Particles, imaging, and nuclei*

Atoms with the same Z but different A are *isotopes*: they are chemically the same, but have different masses. For example, seven isotopes of carbon have been identified. Most are radioactive. Some 98.9 per cent of naturally-occurring carbon is $^{12}_6$C, which is stable. On the other hand, $^{14}_6$C is a naturally-occurring radioactive isotope with a half-life of about 5000 years, and is the basis of radiocarbon dating. Radioactive isotopes can also be manufactured, often by neutron irradiation in a nuclear reactor (for example ^{23}Na + ^{1}n → ^{24}Na). Radioisotopes have many uses in science, medicine, and industry.

Electrons and ionization energy

QUESTION 34

QUESTION 35

DEMONSTRATION F9
Ionization by electron collision

Alpha particles can be detected by a Geiger–Müller (GM) tube because they ionize the gas inside the tube. The ions created by an alpha particle also act as condensation centres in a cloud chamber. Alpha particles have energies of about 5 MeV.

Beta particles (high-energy electrons) also cause ionization and can be detected with a GM tube. Experiments with electrons show that when they are accelerated through a low-pressure gas, energies of only 10–20 eV are sufficient to cause ionization. The energy of the bombarding electron is transferred to the gas atom which gains enough energy for one of its outer electrons to break free from the electric force which binds it to the nucleus. Since the bound system (the atom) has *less* energy than the ion plus free electron which results, the energy binding

the electron to the atom is taken to be *negative*, in the same way that th[e]
energy of a bound system such as the Earth and the Moon is taken to b[e]
negative.

Energy must be *added* to a neutral atom to free an electron, an[d]
ionization energy is conventionally taken as a *positive* quantity; it [is]
equal in magnitude to the electron's binding energy.

Ionization energy varies from element to element, in a way whic[h]
reflects the element's position in the Periodic Table (figure F9[).]
Elements in the same column of the Periodic Table have simila[r]
ionization energies. The addition of one more proton can change th[e]
ionization energy drastically: inert gases such as helium bind the[ir]
electrons firmly, whereas an alkali metal such as lithium has an oute[r]
electron which is weakly bound and has a low ionization energy. Usin[g]
our knowledge of ionization energies we can confirm that an atom['s]
outer electrons are about 10^{-10} m from the nucleus.

QUESTIONS 36 to 38

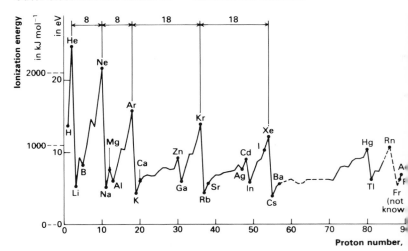

Figure F9
Ionization energies of the elements.

If the atom is bombarded by electrons with energies less than th[e]
binding energy, inelastic collisions may still occur. Some of the energ[y]
of the bombarding electron is absorbed by the atom but it is not enoug[h]
to free an atomic electron. This process is called *excitation*. The excite[d]
atom loses this extra energy – usually very quickly – by emitting light. [If]
the electron does not have enough energy to cause excitation then th[e]
collision will be elastic: the bombarding electron loses no energy, th[e]
atom gains none.

Transformation rules

An alpha particle is a helium nucleus emitted from the nucleus of [a]
decaying atom. A beta particle is an electron liberated when a neutro[n]
in the nucleus decays into a proton plus an electron (plus a massles[s]
neutral particle, the neutrino). The changes in Z and A which accom[-]
pany these decays are shown in table F3.

QUESTIONS 39 to 41

	change in proton number Z	change in nucleon number A
α decay	-2	-4
β decay	$+1$	0

Table F3

Unit L, 'Waves, particles, and atoms'

When a nucleus emits gamma radiation, neither Z nor A change. Gamma radiation is not a stream of particles, but is very short-wavelength electromagnetic radiation – shorter than X-rays. Atoms emit light when their electron energy levels change, and we can learn about these energy levels from the wavelength of light emitted. Gamma radiation is emitted when nuclear energy levels change, and analysis of the radiation gives information about nuclear energy levels.

What holds the nucleus together?

QUESTION 42

'he particles and forces of nature' in the
Reader *Particles, imaging, and nuclei*

'Lasers probe the atomic nucleus' in the
Reader *Particles, imaging, and nuclei*

A simple calculation shows that gravity is not nearly strong enough to hold the protons in a nucleus together against the electric repulsion between them. A new force, which acts on neutrons as well as protons, is at work. This force is evidently restricted to very small distances – its influence is not felt outside the nucleus.

Nuclear binding energy

This attractive force gives rise to a nuclear binding energy. Because a collection of neutrons and protons has *less* energy than the same particles have when far apart, the nuclear binding energy is *negative*. The larger the number of particles in the nucleus, the more negative the binding energy is. For each nucleon added the total binding energy decreases approximately by a further 8 MeV. Nuclear binding energies are much larger in magnitude than ionization energies.

However, the graph of nuclear binding energy (BE) against nucleon number A is not quite a straight line (figure F10), and a graph of BE/A (average binding energy per nucleon) against A shows some interesting features (figure F11).

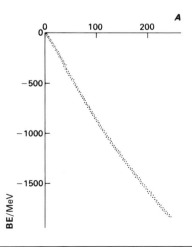

Figure F10

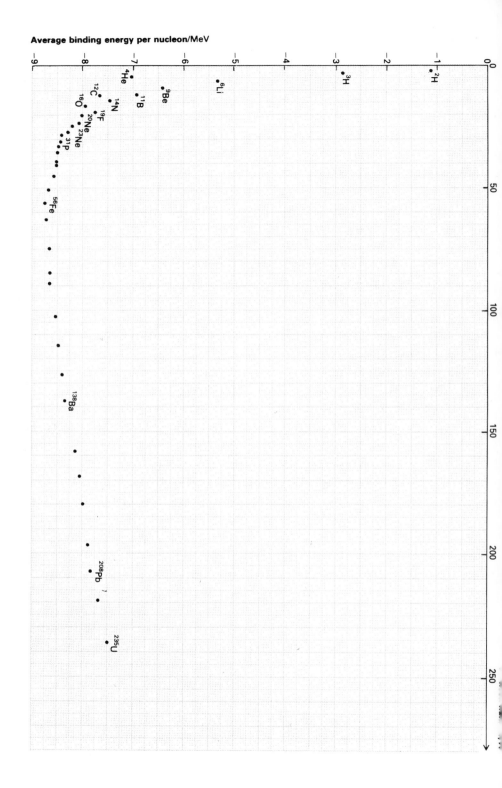

Figure F11
Average binding energy per nucleon as a function of number of nucleons.

The most stable nuclei are those with the most negative binding energies. The iron nucleus, ^{56}Fe, is the most stable. ^{4}He, ^{12}C, and ^{16}O are each more stable than their immediate neighbours. ^{235}U is considerably less stable than lighter nuclei. This suggests that if the 235 nucleons in ^{235}U could be rearranged into two or more lighter and more stable nuclei, energy would be released.

Calculation of nuclear binding energies

Unit H, 'Magnetic fields and a.c.'

The masses of atoms can be measured very precisely using a mass spectrometer. The masses of the elementary particles (proton, neutron, and electron) are also known with considerable precision. These very small masses are often expressed in unified atomic mass units (u). 1 u is one-twelfth of the mass of an atom of ^{12}C. It is a surprising fact that the mass of an atom – any atom – always turns out to be less than the value found by adding the masses of its constituent parts. The difference is biggest in the case of ^{56}Fe, less for ^{235}U, and considerably less for ^{6}Li. This mass loss (or *mass defect*) is a measure of the binding energy.

$1\,u = 1.66 \times 10^{-27}\,kg$

QUESTION 29

Einstein's theory of relativity linked the concepts of mass and energy. He showed that energy has mass. Experiments demonstrate that a very fast-moving electron is accelerated less by a given force than a slow-moving one: its mass is greater. The mass of a clock spring is greater when it is wound, and storing energy, than when it has run down; the mass of a beaker of hot water is greater than the mass of the same water molecules when cold. For these everyday examples the mass change is far too small to be detected. But the mass of an atom of ^{4}He is about 0.03 u (0.75 per cent) less than the mass of two protons plus two

QUESTIONS 44 to 46

neutrons plus two electrons.

This loss of mass means that the atom has less energy than its parts. Mass change and energy change are related by

$c = 3 \times 10^{8}\,m\,s^{-1}$ (speed of light)

$$\Delta E = c^2 \Delta m$$

The energy loss corresponding to 0.03 u is 28.3 MeV, so the average binding energy *per nucleon* for ^{4}He is $-28.3/4\,\text{MeV} = -7.1\,\text{MeV}$, as shown in figure F11.

Fission

'fission' means splitting

It is possible to cause certain heavy atoms (high nucleon number) to split into two roughly equal parts. A few free neutrons will also be produced. Although this can happen spontaneously, it is a very rare event. We can trigger this process by adding an extra neutron to the nucleus. A typical example is

$$^{235}_{92}\text{U} + ^{1}_{0}\text{n} \rightarrow ^{144}_{56}\text{Ba} + ^{90}_{36}\text{Kr} + 2^{1}_{0}\text{n}$$

The nucleus of ^{235}U 'captures' a low-energy neutron (slow neutron); the resulting nucleus is unstable and splits. (Note that the reaction shown is just one example of many possibilities.) The neutrons produced in the reaction can in turn be captured by other ^{235}U nuclei, and so the

F

chain reaction

Unit G, 'Energy sources'

reaction continues, perhaps at an increasing rate. Since the tota binding energy of the products is less (more negative) than the bindin energy of ^{235}U, energy is released in this reaction – about 200 MeV p nucleus. Most of this energy appears as kinetic energy of the fissio fragments. 200 MeV per atom is equivalent to about 80×10^6 MJ p kilogram – very much more than the energy available in a chemic reaction, for example, when one kilogram of oil or coal is burned (abo 30 MJ per kg).

Fusion

'fusion' means joining, uniting (or melting)

The 'energy well' shown in figure F11 slopes very steeply near th origin. If it were possible to move deeper into the 'well' by combinin two light nuclei on this part of the curve to form a single more massiv nucleus, the energy released per nucleon would be much greater than i fission.

Unit G, 'Energy sources'

Fusion is a difficult process to achieve because of the stron electrical repulsion between the nuclei when they are close to eac other. At extremely high temperatures (about 100 million K) the nucl have enough kinetic energy to overcome this repulsion. A possib fusion reaction involving deuterium (heavy hydrogen) is

$$^2_1H + ^2_1H \rightarrow ^3_2He + ^1_0n$$

This reaction releases 3.3 MeV, but the energy *per kilogram* is great than in fission.

'Our nuclear history' in the Reader
Particles, imaging, and nuclei

The Sun's energy source is a sequence of fusion reactions of this kir in which hydrogen is converted into helium.

READING

RADIOACTIVITY AND THE NUCLEAR ATOM: A BRIEF HISTORY

Our modern nuclear model of the atom derived, after a lapse of twenty years, from two remarkable discoveries. The first, by H. Becquerel in 1896, was of the effect we know as radioactivity; the second was the discovery by J. J. Thomson in 1897 that all atoms of all elements contained the same light, negatively-charged particles we now call electrons.

Becquerel's discovery

In 1896, Henri Becquerel was trying to see whether phosphorescent materials that glowed when light had shone on them emitted any penetrating rays like the X-rays that Röntgen had recently discovered. He reported in the journal of the French Academy of Sciences:*

'One of Lumière's gelatine bromide photographic plates is wrapped in two sheets of very heavy black paper so that the plate does not fog on a day's exposure to sunlight.
'A plate of the phosphorescent substance (uranium potassium sulphate) is laid above the paper on the outside and the whole exposed to the Sun for several hours. When later the photographic plate is developed, the silhouette of the phosphorescent substance is discovered, appearing in black on the negative.'

A little later, in the same volume, he wrote:

'Among the preceding experiments some were prepared on Wednesday the 26th and Thursday the 27th of February, and, as on those days the Sun only appeared intermittently, I held back the experiments that had been prepared and returned the plate holders to darkness in a drawer, leaving the thin layers of the uranium salt in place. As the Sun still did not appear during the following days, I developed the photographic plates on the first of March, expecting to find very weak images. To the contrary, the silhouettes appeared with great intensity. I thought at once that the action had been going on in darkness...'

Repeated tests soon identified the uranium as the cause of the effect.

J. J. Thomson and the electron

In 1897 Thomson published his classic paper† on his experiments with cathode rays which led him to propose that these rays consisted of

F

---

*ROMER, A. (ed. and trans.) *The discovery of radioactivity and transmutation.* Dover, 1964. From *Comptes rendus de l'Académie des sciences.* Volume **122**, 742.

†THOMSON, J. J. 'Cathode rays'. *Philosophical Magazine.* Volume **44** (5), October 1897, p. 293. (Reprinted in WRIGHT, S. (ed.) *Classical scientific papers – physics.*)

streams of negatively-charged particles of very small mass. These lat
became known as 'electrons'. The properties of these particles were ju
the same whatever gas was used in the tube and whatever metal w
used for the cathode.

Naturally Thomson tried to relate his discovery of the electron as
universal ingredient of atoms to what was then known about aton
and, in particular, to the perplexing problem of the periodiciti
recognized in the Periodic Table. Of these, the most important was t
arrangement of the elements in groups containing 2–8–8– etc., elemen
with properties repeating from group to group. Thomson wrote:‡

'We have seen that corpuscles§ are always of the same kind whatev
may be the nature of the substance from which they originate; this,
conjunction with the fact that their mass is much smaller than that
any known atom, suggests that they are a constituent of all atoms; tha
in short, corpuscles are an essential part of the structure of the atoms
the different elements. This consideration makes it important
consider the ways in which groups of corpuscles can arrange themselv
so as to be in equilibrium. Since the corpuscles are all negativel
electrified, they repel each other, and thus, unless there is some for
tending to hold them together, no group in which the distances betwe
the corpuscles is finite can be in equilibrium. As the atoms of t
elements in their normal states are electrically neutral, the negati
electricity on the corpuscles they contain must be balanced by a
equivalent amount of positive electricity; the atoms must, along wi
the corpuscles, contain positive electricity. The form in which th
positive electricity occurs in the atom is at present a matter about whi
we have very little information. No positively-electrified body has be
found having a mass less than that of an atom of hydrogen. All t
positively-electrified systems in gases at low pressures seem to be ator
which, neutral in their normal state, have become positively-charged l
losing a corpuscle. In default of exact knowledge of the nature of t
way in which positive electricity occurs in the atom, we shall consider
case in which the positive electricity is distributed in the way mo
amenable to mathematical calculation, i.e., when it occurs as a sphere
uniform density, throughout which the corpuscles are distributed. Th
positive electricity attracts the corpuscles to the centre of the spher
while their mutual repulsion drives them away from it; when
equilibrium they will be distributed in such a way that the attraction
the positive electrification is balanced by the repulsion of the oth
corpuscles.'

Thomson went on to consider how these corpuscles (electron
might be arranged within the sphere of positive charge and he was ab
to suggest a whole series of arrangements which, as one proceed
through the Periodic Table, broke up into sets of concentric rings.

This is the Thomson model of the atom, which is often referred to
a 'plum-pudding' model. It accounted for the way in which t

‡THOMSON, J. J. *The corpuscular theory of matter.* Constable, 1907.

§'corpuscle' = electron.

radiations from radioactive substances passed so readily through atoms in their path, for nowhere did they encounter anything either massive enough or with a big enough charge to deflect them much.

In spite of weaknesses, Thomson's model proved to be a very useful one in the early days, but by 1910 it was clear that something else was needed.

Rutherford and alpha particles

We now return to the other side of the story – radioactivity. Soon after Becquerel's discovery, it was found that there were three different kinds of radiation from radioactive substances – and these were named the alpha, beta, and gamma radiations.

It was the alpha particles that happened to be the tool which, almost by accident, gave Rutherford the clue which led him to propose the nuclear model for the atom. Before this could happen Rutherford had found out a great deal about alpha particles. Some of his reports are reprinted in *Classical scientific papers – physics*. In 1902 he discussed the causes of radioactivity. In 1903, he reported the deflection of alpha particles by electric and magnetic fields. In 1909, Rutherford and Royds collected the gas formed when alpha particles were trapped in a tube and showed that it was helium. This indicated that alpha particles were charged helium atoms. But Rutherford had thought that alpha particles and helium were connected long before his experiment in 1909. In a general review of the state of knowledge about radioactivity which Rutherford and Soddy published in 1903, they made a number of rough (but quite good) estimates of the charge, mass, energy, and speed of alpha particles. Without such estimates, the behaviour of alpha particles would have been much harder to understand and the alpha particle scattering experiment which gave the clue to the nucleus would have been unintelligible.

Rutherford proposed the nuclear model of the atom to account for the surprising results obtained by Geiger and Marsden in which they bombarded metal foils with alpha particles (1909). If most of the mass of the atom is concentrated in a very small positively-charged nucleus surrounded at a distance by a distribution of negative electrons, then most of the alpha particles would be expected to pass through the foil target unaffected. Those few that came close enough to the nucleus would be deflected, the angle of deflection or scattering depending on how close the alpha particle passed. He assumed that the force between nucleus and alpha particle obeyed an inverse-square law and calculated how the fraction of incident particles that are scattered through an angle ϕ should depend on ϕ. Geiger and Marsden's paper describing their tests of Rutherford's nuclear model was published in 1913. (*Philosophical Magazine.* Volume **27** (6), 1913, p. 604. (Reprinted in WRIGHT, S. (ed.) *Classical scientific papers – physics*.)

F

Questions

a In 1904, J. J. Thomson, the discoverer of the electron, suggested a model for the atom which is frequently called the 'plum-pudding model'. Describe this model and indicate how it incorporated the facts known at the time.

b How was radioactivity discovered and how were the radiations sorted out into the three different types?

c What is the evidence for the identification of
i an alpha particle with a helium ion, and
ii a beta particle with an electron?

d (Additional question)   How does a Geiger–Müller (GM) tube work and what are its limitations?

LABORATORY NOTES

Handling radioactive substances

Radioactive substances should be handled with the same care and treated with the same respect as concentrated acids. You may find yourself using naturally-occurring radioactive substances, such as compounds of uranium or thorium, or specially prepared 'sealed sources' (americium-241, strontium-90, or cobalt-60). Such sources must always be handled with tongs or a special source holder and never with the fingers. They should be pointed away from the body and, indeed, held well away from it. When not in use, even temporarily, they should be returned to their lead-lined storage boxes. Under no circumstances should you probe inside such sources or allow them to come into contact with any substance that might attack or dissolve the source or its container.

When handling the salts of uranium or thorium, you must ensure that they cannot be taken into the body, nor be dispersed around the laboratory. They should be handled above a suitable spill tray, lined with absorbent paper; you should wash your hands both before and after the experiment and, if necessary, cover any cuts or scratches, and wear protective clothing. Keep any object which may have been contaminated away from the mouth; and keep your papers and books out of the way.

Used properly, these radioactive sources, which are approved for use in schools by the Department of Education and Science, will cause no harm.

The use of Geiger–Müller tubes

The thin end-window tube is very easily damaged, take great care not to allow any solid object to come into contact with the end-window. This is normally covered with a plastic mesh cap. Some of these tubes contain ferromagnetic material and, if you use one in conjunction with a magnet, take great care not to allow the tube to be attracted so that it moves towards the magnet.

Geiger–Müller (GM) tubes work over a limited range of voltages. You should ask what voltage to use. Too high a voltage can damage the tube; at too low a voltage it will work ineffectively or not at all.

EXPERIMENT

F1 The deflection of beta radiation in a magnetic field

scaler
GM tube holder
thin window GM tube
beta particle source
source holder
mild steel yoke
2 Magnadur magnets
2 retort stand bases, rods, bosses, and clamps
2 lead blocks
plotting compass

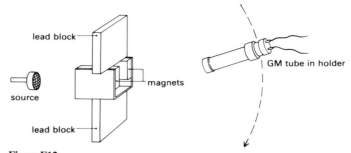

Figure F12
Arrangement of apparatus for the magnetic deflection of beta particles (seen from above).

Adjust the scaler to record beta particles arriving at the GM tube. Then arrange the apparatus so that you can examine the effect of allowing beta particles to pass through a magnetic field. Hence determine the sign of the charge on the beta particles.

EXPERIMENT

F2 Measuring the ionization current to find the number of ion pairs produced by an alpha particle and an estimation of the energy of the alpha particles emitted

picoammeter, 10^{-9} or 10^{-11} A
or
electrometer with $10^9\,\Omega$ or $10^{11}\,\Omega$ input resistance, and output meter

alpha particle source
ionization chamber
e.h.t. power supply
source holder
leads

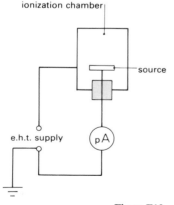

Figure F13
Measuring ionization current with a picoammeter.

DO NOT connect the e.h.t. supply directly across the meter! The range required (10^{-9} A or 10^{-11} A) depends on the source provided.

The ions resulting from the passage of alpha particles through the air in the ionization chamber move in the electric field between the source and the walls of the ionization chamber. This current can be measured and it is then possible to estimate the number of ions present.

The ionization current is very small – perhaps 10^{-9} A or 10^{-11} A. It can be found from the p.d. across a very large resistance through which it flows. An electrometer will probably be used for this measurement. Why would a standard moving-coil meter not be suitable?

Apply a small p.d. between the source and the chamber and carefully increase this until all the ions which are being produced are being collected. (How will you know when this is happening?)

Measure (or calculate) the ionization current.

Assuming the charge on a single ion to be the same as that on an electron, use your value of the ionization current to calculate the number of ion pairs produced per second.

Each ionizing event creates a pair of charged particles – say an electron and an ion with a single positive charge – which move in

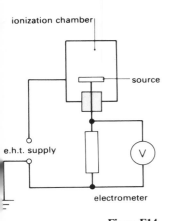

ionization chamber

source

e.h.t. supply

V

electrometer

Figure F14
Measuring ionization current by
measuring p.d. across a resistance.

opposite directions in the electric field. Why is the charge transferred by such a pair equivalent to *one* single charge moving right across the ionization chamber?

From the known activity of the source, find the number of disintegrations per second and hence find the number of ion pairs produced per alpha particle.

To find the energy of the alpha particle you will need to know the average energy required to produce each ion pair. The ionization energy of nitrogen (air) is 14 eV. The collision between an alpha particle and a molecule is a complex process and the average energy lost by the particle cannot be less than 14 eV per ionization. Experiments suggest an average of about 30 eV. The energy of the alpha particle is the product of the energy needed to produce one ion pair and the number of pairs produced. Calculate the energy of an alpha particle produced by the source ($1 \, \text{eV} = 1.6 \times 10^{-19} \, \text{J}$).

EXPERIMENT

F3 The penetrating power of alpha, beta, and gamma rays

alpha, beta, and gamma sources
source holder
GM tube and holder
scaler
stopwatch or clock
2 retort stand bases, rods, bosses, and clamps
metre rule
set of absorbers
Vernier callipers or micrometer screw gauge

This experiment offers a number of possibilities:

a Range of alpha particles in air.

b Range of beta particles in aluminium.

c 'Half-thickness' of lead for gamma radiation.

d The relation between the thickness of the absorber and the radiation transmitted for gamma and beta radiation.

e The relation between intensity and distance for gamma radiation.

In all cases, a count will be detected even in the absence of a source. You must make a correction for this *background count*.

F3a Range of alpha particles in air

Alpha particles are readily absorbed and it is necessary to use a very thin end-window GM tube or an ionization chamber. In the former case you will have to make an allowance for the effect of the end-window (the window of the MX 168/01 tube is equivalent to about 17 mm of air, and that of the MX 168 tube to about 30 mm).

Determine the range in air of the alpha particles from the source provided. Figure F15 shows the range in air of alpha particles with different energies. Use this graph to find the energy of the alpha particles emitted by the source used.

What thicknesses of different materials will just absorb the particles?

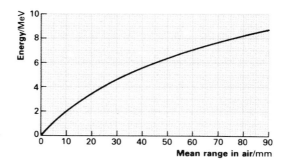

Figure F15
The mean range in air of alpha particles of different energies.

F3b Range of beta particles in aluminium

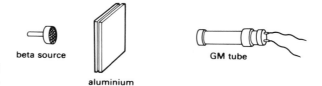

beta source

aluminium

GM tube

Figure F16
Range of beta particles in aluminium.

It does not follow that beta particles have more energy than alpha particles even though they may travel further in air.

Find the maximum energy of the beta particles from the source provided as follows. Keep the distance from the source to the detector fixed (between 1 and 2 cm), and interpose varying thicknesses of aluminium to determine the range of the radiation in aluminium (figure F16). Then use figure F17 to find the maximum energy of the beta particles from the source. Remember that the beta particles have travelled through air and the end-window of the GM tube as well as through the aluminium.

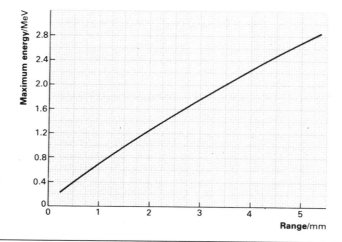

Figure F17
The range–energy curve for beta particles in aluminium.
*From KATZ, L. and PENFOLD, A. S.
'Range–energy relations for electrons and the determination of beta-ray end-point energies by absorption'. Rev. Mod. Phys.
24, 28, 1952.*

F3c 'Half-thickness' of lead for gamma radiation

Find the energy of the gamma radiation from the source provided. Keep the distance from the source to the detector fixed at about 2 cm, and by interposing varying thicknesses of lead find the thickness which will reduce the count-rate to one-half of its original value ('half-thickness'). Use figure F18 to determine the energy of the gamma radiation used.

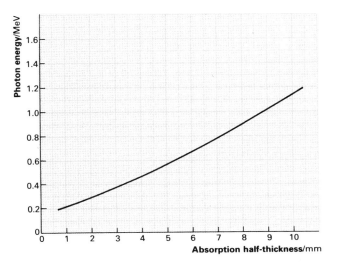

Figure F18
The half thickness–energy curve for gamma radiation in lead.
From DAVISSON, C. M. and EVANS, R. D. 'Gamma-ray absorption coefficients'. Rev. Mod. Phys. **24**, *79, 1952.*

F3d The relation between the thickness of the absorber and the radiation transmitted for gamma and beta radiation

Plot a graph of count-rate against the thickness of absorber introduced between the detector and the source, keeping the distance from detector to source fixed. Try both gamma and beta radiation. Suggest a relationship between count-rate and absorber thickness in both cases.

F3e The relation between intensity and distance for gamma radiation

Try to determine how the gamma ray count-rate varies with distance from the source. Plot graphs to determine whether the relationship is inverse, or inverse-square, or exponential, or...

It is difficult to measure the distance without involving a zero error so think carefully about what might be the best graph to plot. For example, when testing for an inverse-square law, is it better to plot count-rate against $1/(\text{distance})^2$, or distance against $1/\sqrt{(\text{count-rate})}$? If you choose correctly you will be able to estimate the zero error from your graph.

EXPERIMENT
F4 Photographic detection of radiation

radium source

either
dental X-ray film
or
bromide paper with developer and fixer

X-ray film or photographic printing paper is sensitive to the ionizing
radiation emitted by radioactive sources.

With dental X-ray film you do not need to use a darkroom. Place
the source face down on the unopened film pack for 20 to 30 minutes.
One type of film comes in a pod which contains its own developer and
fixer: follow the manufacturer's instructions. Another type is protected
by a built-in filter and can be processed in subdued light (but avoid
fluorescent lighting).

The same exposure time should be sufficient for bromide paper; but
the source must be held directly over the sensitive surface, so this
exposure must be made in a darkroom. To work in the light you can
wrap the sheet of bromide paper in black paper, but a much longer time
will be needed – hours or even days.

Which radiations are responsible for affecting
i the unwrapped bromide paper, and
ii the wrapped paper or film?

DEMONSTRATION
F5 Qualitative test of the gravitational hill model

alpha particle scattering analogue
drawing board

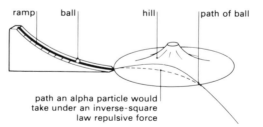

Figure F19
Gravitational analogue of alpha particle
scattering.

Where must the ball be aimed for it to be turned through the largest
angle?
How does the 'scattering angle' for the ball vary with the aiming error
('impact parameter')?
How does the scattering angle for a given aiming error depend on the
ball's speed?
List similarities and differences between this situation and an experiment in which alpha particles are scattered by a thin metal foil.

EXPERIMENT
F6 Decay and recovery of protactinium

scaler
GM tube holder
thin end-window GM tube
small polythene bottle, 50 cm³, containing prepared solution of uranyl(VI) nitrate
retort stand base, rod, boss, and clamp
stopwatch or clock
tray lined with absorbent paper

The decay chain involved is:

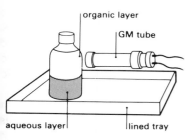

organic layer
GM tube
aqueous layer
lined tray

Figure F20
The decay of protactinium.

$$\text{uranium-238} \xrightarrow{\;\alpha\;} \text{thorium-234} \xrightarrow{\;\beta\;} \text{protactinium-234} \xrightarrow{\;\beta\;}$$

^{234}Pa decays quickly enough for its decay to be observed in a short experiment. It has, however, to be separated from its parent thorium by an organic solvent. The solutions necessary are contained in a small polythene bottle. You should perform the experiment in a spill tray.

Under no circumstances should the end-window of the GM tube be allowed to touch the bottle. This ensures that the GM tube does not become contaminated with the radioactive solutions.

First measure the background count-rate in the absence of the bottle. Shake the bottle for 10 to 15 s. The organic solvent containing the protactinium will float to the top. Position the bottle beside the GM tube so that the end-window is opposite the organic layer. Once the layers have separated, the beta particles can be counted. Take a ten-second count each half minute.

Having made allowance for the background count-rate, plot a graph of the count-rate against time and determine the half-life of protactinium from the graph. Do this by starting at several points and measure how long it takes for the count-rate to fall to one-half of the initial value chosen. The half-life is the mean of these times.

Also plot a graph of ln (activity) against time. If this should prove to be a straight line, you have added confirmation that the decay is obeying an exponential law.

Estimates of half-life based on low values of the count-rate may be unreliable. Why is this?

The thorium from which the protactinium has been removed remains in the aqueous layer in the lower half of the bottle. Protactinium will immediately start to reappear there as the thorium decays and the growth curve of protactinium can be plotted. To observe this recovery, repeat the experiment with the GM tube window opposite the lower (aqueous) layer. Plot a graph of count-rate against time.

F

EXPERIMENT

F7 The decay of radon and the determination of its half-life

picoammeter, 10^{-11} A (*e.g.*, electrometer with 10^{11} Ω input resistance and output meter)
ionization chamber
radon generator
2 cell holders with four cells
leads

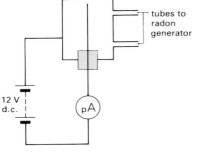

Figure F21
The decay of radon.

The decay chain involved in this experiment is

thorium-232 → radium-228 → actinium-228 → thorium-228 → radium-224 → radon-220–

Radon-220 is a gas and an alpha particle emitter. The materials whic
give rise to it must be kept in a closed container. It is convenient to stor
them in a plastic bottle fitted with two tubes and valves. These tubes f
on to tubules attached to an ionization chamber, the whole forming
closed circuit with no risk of the radon leaking into the atmosphere.

Report any accidental spills to your teacher.

As in experiment F2, the alpha particles emitted by the radon caus
some ionization in the chamber, which gives rise to an ionizatio
current when a p.d. is applied across the chamber. The ionizatio
current is very small: it can be calculated from the p.d. set up when
flows through a very high resistance, say 10^{11} Ω. An electrometer wi
probably be used for this measurement. Why could a standard moving
coil meter not be used?

Apply a p.d. of 12 V d.c. across the ionization chamber, and select a
appropriate current range (that is, an appropriate input resistance fc
the electrometer).

Pass ^{220}Rn into the chamber by *gently* squeezing the plastic bottl
until a full-scale reading is obtained. Record the ionization currer
every ten seconds and plot a graph of meter readings against time. Us
your graph to determine the half-life of radon.

EXPERIMENT

F8 A radioactive decay analogue

100 dice
graph paper
tray

This experiment is about events which happen by chance. If eac
member of a collection of objects has a fixed chance of 'decay' in a give
time interval, then the number remaining will decrease with time. Th
question concerns the pattern of this decay; does it have the same forr
as the decay of the radioactive substances?

In radioactive decay we do not know when any particular atom i
going to decay. But, given enough atoms, we can say what proportio

of them will decay in a given time interval. This fraction will vary from substance to substance.

If you throw an ordinary die, you have one chance in six of throwing, say, a five. If you threw one die many times (or many dice just once) what fraction of the throws (or of the dice) would result in, say, a five?

Given enough throws (or enough dice) you can predict that fraction with reasonable certainty. So throwing dice is analogous to radioactive decay in the sense that both processes lead to a fraction which is fixed for specific dice or for specific atoms. In neither case can we predict which of the atoms will decay or which of the dice will show a five next.

If we remove this fraction (that is, the 'fives') as if they have decayed, and then throw the remaining dice, and so on, the process imitates radioactive decay to some extent.

Throw the dice and count those that fall with a five upwards. Remove these dice. Throw the remaining dice, count and remove the fives, and so on. Continue for some ten throws. Plot graphs of the number 'decaying' and of the number remaining against the number of the throw. The shapes of the two graphs will be familiar but may look rather 'bumpier' than other examples you have met. Why is this? How could you improve the shape of the curves?

The graphs are typical of situations in which the rate at which something decays depends on the quantity or the number present. What tests can you apply to check whether or not the graph for the dice is exponential in form? Try one at least of these. Estimate the 'half-life' of a die.

OPTIONAL DEMONSTRATION
F9 Ionization by electron collision

Experiments in which electrons are accelerated in a gas at low pressure gave information about the structure of atoms. For each gas there is a characteristic electron energy which will ionize the gas atoms. The onset of ionization is shown by a rapid increase in current, as the number of charge carriers increases.

In an *elastic* collision between an electron ($m_e \approx 0.000\,5\,u$) and a stationary neon atom ($m \approx 20\,u$), will the energy transferred to the atom be a large or a small fraction of the electron's kinetic energy?

Must energy be added to a neutral atom to ionize it, or is energy released when the ion plus free electron is formed? (*Hint:* is the potential energy of the atom greater or smaller than that of the two charged particles, ion and electron, when they are separated?)

Careful experimentation may show evidence of inelastic collisions between electrons and gas atoms when the electron's kinetic energy is less than that required to ionize the atom. The atom may be excited – one of its electrons is promoted to a higher energy level. More precise information about ionization and excitation energies comes from careful measurement of the wavelengths of light in an element's absorption or emission spectrum.

QUESTIONS

The nature and properties of alpha, beta, and gamma radiation

1(E)a Describe how you would distinguish between alpha, beta, and gamma radiation by experiment.

b What happens to the kinetic energy of alpha and beta particles when passing through air?

c When asked to outline how he would distinguish between the signs of the charges carried by alpha and beta particles, a student drew a rough sketch (figure F22).

Figure F22

The Geiger tube can be moved in any direction. The student was able to detect the deflection of beta particles, but not of alpha particles.
i Why could he not detect the deflection of alpha particles?
ii What changes would you make to the apparatus to improve the chances of detecting the deflection of alpha particles?

(From Physics (Nuffield) O-level examination, Paper 2, November 1979)

2(I) Figure F23 is a sketch made by someone trying to design an arrangement that will detect the deflection of beta particles by an electric field. The idea is to have parallel plates, length *l*, spacing *d*, down the centre of which the beta particles travel. After travelling distance *l* they have been deflected sideways by a distance *s*. From there on, they travel more or less in a straight line and have been deflected by a distance *y* after going a further distance *L* to a detector.

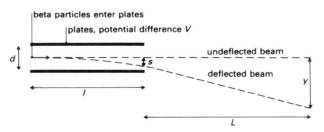

Figure F23

a What is the electric field, *E*, between the plates if the potential difference between them is 2 kV and *d* is 2 mm?

b What is the sideways force, *F*, on an electron, charge $e = 1.6 \times 10^{-19}$ C, between the plates?

c A fast-moving beta particle's mass, m, is about 10^{-29} kg (larger than its rest mass because of the relativistic increase of mass at high speed). What will be its acceleration, a, in the field?

d A beta particle travels at a speed approaching that of light, 3×10^8 m s^{-1}. If l is 0.1 m, for what time does this sideways acceleration last?

e Find the sideways distance, s, through which the beta particles are deflected by the field in time t under the acceleration calculated in **c**.

f If the experiment is done in air, L cannot be much more than 0.1 m or so. Estimate the deflection y.

3(P) In an experiment to determine the energy of an alpha particle using an ionization chamber and a very sensitive ammeter, the following results were obtained:

Ionization current $= 3 \times 10^{-11}$ A
Activity of the source $= 3.7 \times 10^3$ Bq

Given that the average energy required to produce an ion-pair in air is 30 eV, find:

a the number of ions produced per second;

b the number of ions produced by each alpha particle (on average);

c the energy of each alpha particle.

4(P) The graph (figure F24) shows how many ion-pairs are formed per millimetre by an alpha particle at each point on its track. Note that the graph places the *end* of the track at the origin.

a Suggest why the ionization per millimetre rises as the particle approaches the end of the track.

b Estimate the total number of ion-pairs formed by an alpha particle that produces a 50 mm track.

c If 30 eV is the average energy required to produce one ion-pair, what is the total average energy of a single alpha particle?

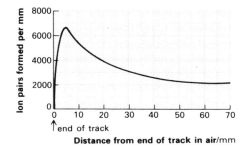

Figure F24
Number of ion pairs per mm of track versus distance for an alpha particle.
After GENTNER, W., MAIER-LEIBNITZ, H., *and* BOTHE, W. An atlas of typical expansion chamber photographs. *Pergamon, 1954.*

5(L) This question follows an argument set out by Rutherford and Soddy in 1903. The estimates are similar to those of Rutherford and Soddy but have been expressed in SI units.

a If each alpha particle from radium has energy 10^{-12} J, and there are 10^{21} radium atoms in a gram of radium, what total energy is emitted if each atom emits one particle?

b Each atom emits at least five particles. How does that alter the answer to **a**?

c Is the answer to **b** a minimum estimate of the energy emitted or a maximum estimate?

d The total electric current from ions produced by alpha particles emitted from one gram of pure radium in air is about 10^{-3} A. Each ion has a charge of about 10^{-19} C. How many ions are being produced each second?

e The least energy needed to produce an ion in air is about 10^{-18} J. How much energy is emitted in one second by one gram of radium?

f In one second, what fraction is emitted of the total energy which will be given out ultimately by the radium?

g Make a rough estimate of the lifetime of a sample of radium.

6(P) A lump of material containing radium is always slightly warmer than its surroundings. Burning 1 mole of radium to radium oxide releases 525×10^3 J.

a How much energy is that, *per atom of radium?*

b When radium atoms disintegrate, each emits, on average, five alpha particles each of energy about 5 MeV. How much is this in joules?

c How much energy is emitted when one mole of atoms disintegrates?

d How does the energy released in the radioactive decay of radium compare with the energy released on oxidation?

e Why are materials containing radium not very hot?

(Half-life of ^{226}Ra $= 1600$ years.)

7(E) You will have noticed the following precautions being observed when radioactive sources are being handled.

a The sources are handled only with tongs or special source holders.

b The sources are handled so that the radiation is not directed towards people.

c When not in use, sources are put in a box with lead shielding and kept in a locked cupboard with a warning notice on it.

Suppose a local councillor has heard that you are doing experiments with radioactive sources and is concerned lest you be exposed to radiation. Write a letter explaining the reasons for the precautions taken. (The councillor may not realize the existence of 'background' radiation.)

Alpha particle scattering and evidence for the nucleus

8(I) How big is an atom of gold?

(The density of gold is $19.3 \times 10^3 \, \text{kg m}^{-3}$ and its mass number is 197. The Avogadro constant is $6 \times 10^{23} \, \text{mol}^{-1}$.)

9(L) You are now asked to estimate the size of an atomic nucleus. Geiger and Marsden's experiments suggest that about 1 alpha particle in 8000 is turned back through a large angle by a gold foil 6×10^{-7} m thick. If such a foil has n layers of atoms then $1/n$ of this number would be turned back by one layer; that is, $1/8000n$ of the particles would be turned back.

All the alpha particles would be turned back if the cross-sectional area of the layer were entirely filled with nuclei. Half would be turned back if the layer were 'half-filled' with nuclei. Since $1/8000n$ is turned back, then only $1/8000n$ of the layer is filled with nuclei. See figure F25.

If we can assume that there are only small gaps between atoms, then the nuclei represent $1/8000n$ of the total area 'seen' by approaching alpha particles. Hence

$$\frac{\text{target area of nucleus}}{\text{cross-sectional area of atoms}} \approx \frac{1}{8000n}$$

The number of layers, n, in a given thickness of the foil will depend on the diameter of the atoms:

$$n \approx \frac{\text{thickness of foil}}{\text{diameter of atom}}$$

Use your result from question **8** to make a rough estimate of the diameter of a gold nucleus.

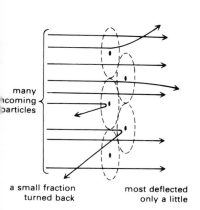

many incoming particles

a small fraction turned back

most deflected only a little

Figure F25
The 'target' area of the nuclei decides the fraction of alpha particles turned through a large angle.

10(I)a Suppose an alpha particle travelling at $10^6 \, \text{m s}^{-1}$ hits another stationary helium nucleus 'head on' (figure F26).

alpha particle

He nucleus at rest

Figure F26

How will each particle move after the collision?

b Now suppose the stationary target is a proton (hydrogen nucleus) also struck 'head on'. At what speed will each particle be travelling after the collision? You may assume that the collision is elastic. What else *must* be true about the collision?

c In collisions of alpha particles with gold nuclei, it is usually assumed that the gold nucleus is not moving after the collision. Is this a fair assumption to make?

11(L)   How large is a nucleus? If it is true that inverse-square law electrical forces act right down to the closest distance an alpha particle comes to a nucleus, then a nucleus must be smaller than that distance.

Consider a 5 MeV alpha particle approaching a gold nucleus 'head on'.

a   What is its kinetic energy in joules? What happens to this kinetic energy as the particle approaches the nucleus?

b   At some distance *r* from the nucleus, the speed of the alpha particle will be zero. Assume that a gold atom has a charge of +79*e* and an alpha particle has a charge of +2*e*. How much potential energy is stored in the system at this moment?

c   Assuming that the massive gold nucleus will acquire very little kinetic energy from this store, equate the two values for energy to find a value for *r*.

d   What is the electric potential at this distance?

e   What relationship does the electric potential bear to the original kinetic energy of the alpha particle? Why is this?

f   What 'turn around' force does the gold nucleus exert on the alpha particle at this distance of nearest approach? Before doing the calculation, guess – is it equal to the weight of an atom, of a cell of your body, of an eyelash, of a pencil, of a book, of a person?

12(P)   Describe the forces between the object A (at rest) and the bombarding particle P, that would explain the paths shown in figure F27.

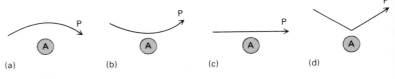

(a)   (b)   (c)   (d)

Figure F27

13(R)   In figure F28, A is an alpha particle being deflected by a charged nucleus N, *r* metres from A. Another alpha particle B happens to be 2*r* metres from N. They come from distant points A′, B′.

a   Why do the continuations (broken lines) of the paths of the alpha particles from A and B curve upwards on the sketch?

b   If the forces on A and B from N are electrical, how does the force on A compare in size with that on B? (Ignore forces between A and B.)

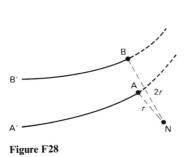

Figure F28

c   In what direction(s) are the forces on A and B?

d   Copy figure F28 and add arrowed lines (vectors) at A and B representing the direction of the *velocity* of each particle.

e   In **d** you will not have been able to know for certain how big the velocities are. But is the particle's velocity at A larger, smaller, or about the same size as its velocity at A′?

f On your diagram, sketch the path of an alpha particle deflected through about 90° by a nucleus. Mark in the position of the nucleus. Mark the point where the speed of the alpha particle would be least. Explain.

g If the speeds are equal at A′ and B′ in figure F28, how do the speeds at A and B compare?

h Suppose you could lay down a flat surface (a very thin sheet of card, for example) so that the incoming alpha particles travelled along the flat surface, and the scattering nucleus also lay on the surface. Would the alpha particles leave the surface as they were deflected by the nucleus? Explain.

Random events

14(E)a The arrival of particles at a Geiger tube from a radioactive substance is said to be 'random'. Do you think that the following are sensible descriptions of what 'random' means in this context?
i The particles arrive at a basically steady rate, which is disturbed by small variations.
ii In each second there is a fixed chance that a particle will arrive, but just when they do so is quite unpredictable.
iii The time between the arrival of the particles may be 0.1 s, 1 s, 5 s, or any other value: each is equally likely.

b In an earlier Unit, the word 'random' was used to describe the motion of the molecules in a gas. Give a sensible description of what 'random' means in that context.

Radioactive decay

15(P)a The half-life of ^{90}Sr is 27 years. The half-life of ^{24}Na is 15 hours. Which of two samples, one containing 10^{20} atoms of ^{90}Sr, and the other containing 10^{20} atoms of ^{24}Na, would have the highest activity?

b Sketch graphs showing how the number of atoms of ^{90}Sr and ^{24}Na changes with time.

c (*Harder*) After roughly how long would the activities become equal?

16(E) In exponential decay the number of atoms remaining after a given interval of time is always the same fraction of the number present at the beginning of the time interval. Therefore, someone might say, however long you wait there will always be some left. A sample of radioactive atoms will last for an infinite time. Is this reasonable?

17(L) No ordinary place on the Earth's surface is free from 'background radiation'. In making a measurement of this background a student set up a GM tube and scaler and recorded the scaler reading every 30 s. The results are shown in table F4.

Time/s	0	30	60	90	120	150	180	210	240
Count	0	13	31	48	60	75	88	102	119

Table F4

Determine the background count per minute. What would you expect the total count to be after a further 30 seconds?

18(P) A radioactive source is set up in front of the GM tube of question **17** in the same laboratory. Scaler readings of the activity of the source are taken for 1 minute at time intervals of 1 hour. These are shown in table F5.

Time/hours	0	1	2	3	4	5	6	7	8	9	10
Activity/counts minute^{-1}	828	510	320	202	135	95	70	51	41	38	37

Time/hours	11	12	13	14	15
Activity/counts minute^{-1}	31	33	34	29	35

Table F5

a What action should be taken about the background count (see question **17**)?

b Plot a graph to show how the activity of the radioactive substance changes with time and derive from it three different values of the half-life. Calculate the mean of these.

c Worried by the obvious fluctuations in the final hours of the count, one student suggests that it might be wise to continue counting for at least as long again. Is this a good idea? Explain.

d There is a slight increase in the count rate towards the end of the experiment. Is this significant? Explain.

e Suggest any steps you might take to improve the experiment.

Testing for exponential growth and decay

19(P) A student says she knows what an exponential curve is and offers to illustrate it by taking some drinking straws, cutting one in half, then one of the halves into half again, then one of the quarters into half again, and so on. She then places the straws side-by-side as shown in figure F29(a).

A second student says he has thought of a similar method but instead of using halves each time, he has cut thirds – figure F29(b).

a Are both models exponential?

b How could you check to see if a graph of one variable (y) against another (x) is 'exponential'?

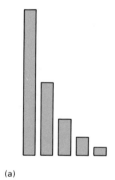

(a)

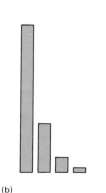

(b)

Figure F29

20(P) Refer to your table of the activity of the source (less background) in question **18**.

a Test whether the data fit a mathematical model of exponential decay by
i checking the constant-ratio property and saying what that ratio is, and
ii plotting a graph of ln(activity) against time.

b Given that $N = N_0 e^{-\lambda t}$ where N_0 and N are the number of radioactive atoms at time zero and time t respectively, and λ is the decay constant, use your graph to find the decay constant and hence the half-life.

21(P) Exponential changes are not confined to physics. They occur so widely that they have great importance in other studies, for example economics, industry, population statistics.

Table F6 gives data for the capacity of the electrical generating plant installed in Great Britain for each year from 1951 to 1980.

Year	51	52	53	54	55	56	57	58	59	60
Capacity/GW	16.2	17.7	19.2	20.6	22.5	24.6	26.6	28.0	30.0	31.9

Year	61	62	63	64	65	66	67	68	69	70
Capacity/GW	33.9	37.2	39.3	40.0	43.9	46.2	50.0	53.6	55.1	60.5

Year	71	72	73	74	75	76	77	78	79	80
Capacity/GW	65.6	68.8	71.1	72.1	71.8	69.9	69.7	69.9	71.6	70.9

Table F6
Data from Central Statistical Office Annual Abstract of Statistics *(1954–82).*
By permission of the controller, HMSO.

F

a By plotting a suitable graph, find out whether any part or parts of this development were exponential.

b There was an economic recession in the early 1960s. How was this reflected in the growth of installed generating plant about that time? What happened to this industry during the 1970s?

c What was the 'doubling time' during:
i the 1950s
ii the 1960s?

22(P) Table F7 shows the number of cars in private ownership in Britain from 1950 to 1980. Does the number rise exponentially?

Year	Cars in millions	Year	Cars in millions
1950	2.26	1966	9.51
1952	2.51	1968	10.8
1954	3.10	1970	11.5
1956	3.89	1972	12.7
1958	4.55	1974	13.6
1960	5.53	1976	14.0
1962	6.56	1978	14.1
1964	8.25	1980	15.1

23(P) Many estimates have been made of the growth of the World's population. Show that the growth suggested by the data in table F8 indicates a rise even 'faster' than an exponential.

Year	Estimate
1650	0.51×10^9
1750	0.72×10^9
1800	0.91×10^9
1850	1.13×10^9
1900	1.59×10^9
1920	1.81×10^9
1930	2.01×10^9
1940	2.25×10^9
1950	2.51×10^9
1960	3.03×10^9
1970	3.68×10^9
1980	4.42×10^9

Table F8
Estimates of the World's population.
The figures for the years from 1650 to 1900 are means of several estimates; subsequent figures are based on UNO data.

About exponential changes

The argument that follows is made up of a number of learning question and a commentary. Our concern is with the shape and meaning of the mathematical equations that describe exponential changes.

Changes are called exponential if the rate of change of something i in proportion to the quantity of that something already present.

But, before investigating the mathematical form of exponentia change two facts have to be recalled:

Logarithms The logarithms of two numbers are added to find the logarithm of their product. So

$$\lg AB = \lg A + \lg B.$$

(Note that $\lg x = \log_{10} x$.)

Powers Powers are added when numbers expressed as powers are multiplied.

$$10^2 \times 10^5 = 10^7; \qquad 2^3 \times 2^{0.2} = 2^{3.2}.$$

In 10^2, 2^3, a^t, the quantities 2, 3, and t are called 'exponents' as well as 'powers' or 'indices'. This is where the word 'exponential' originates.

24(L) Consider an equation of the form $\Delta N/\Delta t = kN$ where t is time and k is a constant. Suppose N stands for the number of families which have video recorders in their homes.

a What does ΔN stand for?

b What does $\Delta N/\Delta t$ mean?

It might be plausible to think that a family would not consider buying a video recorder unless they knew of other families like themselves who had done so, and who recommended it.

c Taking this view, suggest why a mathematical model like $\Delta N/\Delta t = kN$ might be appropriate for the way in which the number of families having video recorders changes.

d If 'keeping up with the Joneses' were the only factor, the simple mathematical model might be appropriate. What complications might make it less so?

e Think now about the spread of a fashion in, say, clothes. If such a mathematical model were appropriate for the period when a fashion was spreading, how might the existence of television in a society alter the value of k?

f Fashions do not continue to spread indefinitely. Sketch graphs of the change with time of the number of people following a fashion which starts with only a few people
i if pretty well everyone adopts the fashion in the end, and
ii if the fashion becomes unfashionable and more and more people reject it.
  Mark on your graphs the points beyond which the model $\Delta N/\Delta t = kN$ is no longer appropriate, supposing that it serves quite well for the initial spreading period.

To summarize, $\Delta N/\Delta t = kN$ is a simple recipe for how N will change in the next time interval Δt. Although it may be an exact recipe for some problems, for others it might be just nearly right or perhaps only roughly right. To be useful, a simple model such as this must apply, in part at least, to a wide variety of situations.

A solution to a differential equation
The equation $s = \frac{1}{2}at^2$ tells us what the displacement s will be at some time t. The equation $\Delta N/\Delta t = kN$ does not do anything so definite as that. It says that if N has *this* value *now*, a short time, Δt, later it will differ from this value by ΔN, and it gives a recipe for working out ΔN. Such equations are called 'differential' equations.

But one still needs to know what N will be at some time t. If N_0 is the number of yeast cells in a culture of yeast now, the number to be expected after several hours may be needed. Sometimes an equation to do this can be found, sometimes not. If it can, it is called a solution of the differential equation. It is 'a' solution, not 'the' solution because it will differ according to the starting point. The recipe for change doesn't

say how many yeast cells there are to start with; but the number of cel[l]
at time t does depend on how many were there in the first place.

The next question shows you why $N = a^t$, where a is some numbe[r]
makes a reasonable choice as a solution of $\Delta N / \Delta t = kN$.

25(L)a   If $N = 2^t$ how would you use the x^y key of a calculator (or the log key, o[r]
logarithm tables) to find N when $t = 7$? (We know you could do it
without, of course.)

b   Use the same method to find the value of $N = 2^{0.5}$.

c   Plot a graph of $N = 2^t$ from $t = 0$ to $t = 1.0$. Why does the graph start a[t]
$N = 1.0$? Use axes which allow N to go from 1.0 to at least 3.

d   Sketch the graph of $N = 1^t$. Also, *either* sketch roughly the graph of
$N = 3^t$ on the same axes, getting the exact values at $t = 0$ and $t = 1.0$;
or plot a graph of $N = 3^t$.

e   Consider the graph of $N = 2^t$. When t goes up in equal steps from 0 to
0.2, then to 0.4, to 0.6, and so on, the same thing happens each time to
the value of N. Say carefully what that same thing is, perhaps by
inspecting your graph.

f   Find $N = 2^{0.2}$ from your graph. Knowing that $2^3 = 8.0$, what is $2^{3.2}$?
(Remember that $2^{(a+b)}$ means $2^a \times 2^b$.)

g   Look at this table:

a^t
$a^t \times a^t = a^{2t}$
$a^t \times a^t \times a^t = a^{3t}$
etc.

Use it to explain why, when t increases in equal steps, a^t is multiplie[d]
by a constant factor.

Exponential changes have, as we have seen, the 'equal ratio' property i[n]
common. The changing quantity alters by a constant *factor* when tim[e]
(or another variable) increases in equal steps. The equation $N = a[^t]$
shares this property.

Next we must ask what value of a in $N = a^t$ goes with a particula[r]
value of k in $\Delta N / \Delta t = kN$. In particular, what value of a fits the cas[e]
when $k = 1$?

26(L)a   Using the same scale as you used for your graph in question **25**, plot
out step by step the growth of N, starting with $N = 1.0$ when $t = 0$, if
$\Delta N / \Delta t = +1.0N$. Here N might be the number of bacteria in a colony,
for example. To start you off, when $N = 1.0$, $\Delta N = 1.0 \times \Delta t$. Taking time
intervals Δt as 0.1 second, ΔN begins by being 0.1. See figure F30.

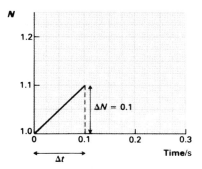

Figure F30

After 0.1 s, N has become 1.1. In the *next* 0.1 s, ΔN is a little larger than 0.1. Why?

After ten steps you reach $t = 1.0$, when N will have risen to about 2.6 or 2.7. Stop at this point.

b Repeat this process of curve-drawing for the growth represented by $\Delta N/\Delta t = 0.7N$, starting with $N = 1.0$ at $t = 0$. Why is this growth slower than the earlier one?

Now compare these curves with those from question **25**.

c What value will N reach at $t = 1$ if $N = 2^t$?

d What will N be at $t = 1$ if $N = a^t$?

e If the curves drawn in **a** and **b** above can be represented by equations such as $N = a^t$, what values of a would be suitable for each curve (approximately)?

A value for a in $N = a^t$

There is now reason to think that the curve you have drawn for $\Delta N/\Delta t = +1.0N$ in your answer to question **26a** has the form $N = a^t$, because this equation has the proper 'constant ratio' property. You can check that your curve has this property too, by looking at it. $\Delta N/\Delta t = 1.0N$ is a recipe for growth with 'constant ratio'. But what value has a?

Whatever the value of a is, it is bigger than 2, because your curve lies above the graph of $N = 2^t$. However, a is less than 3 because your curve lies below the graph of $N = 3^t$. Your curve, drawn from the recipe $\Delta N/\Delta t = 1.0N$, could be represented by the equation $N = a^t$, if a were between 2 and 3. Your graph may have given a value around 2.6; more accurate calculations give 2.718... This number has a mathematical symbol, e. So the solution to the recipe $\Delta N/\Delta t = 1.0N$ if N starts at 1.0 when $t = 0$, is $N = e^t$.

A less special case

The recipe $\Delta N/\Delta t = 1.0N$ is a very special case; we must move towards the much more general case $\Delta N/\Delta t = kN$ with k not equal to unity. The next question uses the graph drawn in question **26b** for which k was 0.7, to find out how to cope with the more general problem.

27 Question **26e** suggests that the solution to $\Delta N/\Delta t = 1.0N$ (the first curve) can be written $N = e^t$, where e is the number 2.718...

a Use a calculator or logarithms to find $e^{0.7}$, or read the value at $t = 0.7$ from the first curve $\Delta N/\Delta t = N$.

   The value of $e^{0.7}$ should be nearly 2.0, and in question **26b** you should have seen that the solution to $\Delta N/\Delta t = 0.7N$ is approximatel $N = 2.0^t$. Instead of writing $N = 2.0^t$ we can write $N = (e^{0.7})^t$ since $e^{0.7} \approx 2.0$.

   So a solution to $\Delta N/\Delta t = 0.7N$ is $N = e^{0.7t}$.

b Now make a speculation. If the equation $\Delta N/\Delta t = kN$ has a solution $N = a^t$, is the value of a likely to be e^k or ke or e/k?

A solution in general

The lesson from question **27** is that if we wish to solve $\Delta N/\Delta t = kN$, th solution is like that for $\Delta N/\Delta t = 1.0N$ which is $N = e^t$, except that th constant k will appear multiplying t.

   When $N = 1.0$ at $t = 0$, the equation that tells us the value of N time t, if $\Delta N/\Delta t = kN$, is $N = e^{kt}$.

   We have been thinking about growth because that is easier to thin about than decay. But this result can cope with decay, too. The recip for change, when N *decreases*, as in radioactivity decay, is jus $\Delta N/\Delta t = -kN$. The negative sign arranges that the changes ΔN are a *subtracted* from the previous value of N. This is like the growth versio but with a negative value of k. Following the same rule (N is e to th power t times k) gives

$$N = e^{-kt}.$$

   We have seen that e^{kt} is multiplied by a constant factor when t rise in equal steps. So $1/e^{kt}$ (which is the same as e^{-kt}) is *diminished* by constant factor when t rises in equal steps. And that is just what a exponential decay curve that obeys the recipe $\Delta N/\Delta t = -kN$ does.

   Finally, what about not requiring N to be unity when $t = 0$, which i after all, not very realistic?

   Suppose there are initially $N_0 = 10\,000$ radioactive nuclei in sample at some moment $t = 0$, and that one-fifth of them, 2000, decay i the first second, leaving 8000. In the next second we should expect one fifth of $8000 = 1600$ to decay, leaving 6400; and so on. Clearly, the actua numbers left, N (8000, 6400, 5120,...) and the fraction of nucl remaining, N/N_0 (0.8, 0.64, 0.512,...) diminish by the same proportion themselves in each second: N/N_0 obeys the same recipe for change as N But the fraction remaining, N/N_0, must be equal to unity at $t = 0$. So th solution to $\Delta(N/N_0)/\Delta t = -k(N/N_0)$ is simply $N/N_0 = e^{-kt}$. In the cas of radioactive changes the letter k is usually replaced by λ, the deca constant.

   In general, if $\Delta N/\Delta t = kN$, and $N = N_0$ when $t = 0$, then

$$N = N_0 e^{kt}$$

   This is the exponential equation. It gives the number of active nucl left in a sample of radioactive material after any time, or the number bacteria in a growing colony. It has uses, often as an approximation, in studying the growth and decay of many things in the sciences, economics, resources, populations, and so on.

28(R) A scientist wished to find the age of a sample of rock in which he knew that radioactive potassium ($^{40}_{19}$K) decays to give the stable isotope argon ($^{40}_{18}$A). He started by making the following measurements:

Decay rate of the potassium in the sample = 0.16 disintegrations s^{-1}
Mass of potassium in the sample   = 0.6×10^{-6} g
Mass of argon in the sample   = 4.2×10^{-6} g

a Show how he could then calculate that for the potassium:
  i the decay constant (λ) was 1.8×10^{-17} s^{-1}
  ii the half-life was 1.3×10^9 y.

b Calculate the age of the rock, assuming that originally there would have been no argon in the sample. Show the steps in your calculation.

c Identify and explain a difficulty involved in measuring the decay rate 0.16 s^{-1} given earlier in the question.

(Short answer paper, 1981)

Unified atomic mass unit u

29(L) The unified atomic mass unit (u) is defined as one-twelfth of the mass of an atom of carbon-12. The molar mass of ^{12}C atoms is 0.012 kg mol^{-1}. The Avogadro constant, L, is 6.022×10^{23} mol^{-1}.

F

a What is the mass of one atom of ^{12}C?

b Calculate the value of 1 u, in kilograms.

The mass of a proton is 1.673×10^{-27} kg and the mass of a neutron is 1.675×10^{-27} kg.

c Express these two masses in u.

d The mass of an electron is 9.11×10^{-31} kg. What is its mass in u?

Radioactive isotopes and their applications

30(R) Heavy hydrogen, ^{2}H, often called deuterium, was first separated from a mixture which was mostly ^{1}H by evaporating several litres (dm^3) of liquid hydrogen very slowly, until only about 1 cm^3 remained.
  In this residue ^{2}H was detected. Why should this process tend to concentrate ^{2}H in the residue? Would the method be as effective with liquid neon, containing mostly ^{20}Ne and a little ^{22}Ne, to separate the two?

31(P) The molar mass of lead, as tabulated in data books, is 207.21 g mol^{-1}. Table F9 gives the molar masses of lead taken from several different minerals.

Mineral	Origin	Molar mass/$g\,mol^{-1}$
Cleveite	Norway	206.08
Broggerite	Norway	206.01
Pitchblende	W. Africa	206.05
Kolm	Sweden	206.01
Thorite	Sri Lanka	207.8
Thorite	Norway	207.9

Table F9

What do these values suggest to you?

32(R) The radioactive isotope of carbon ^{14}C, is used for the dating of some archaeological finds. ^{14}C emits beta particles, and has a half-life of 5570 years. This isotope is present in small amounts in the carbon present as carbon dioxide in the atmosphere.

The ^{14}C is made continually by the interaction of cosmic rays with the nitrogen in the upper atmosphere. There is evidence that the rate of production of ^{14}C and its concentration in the atmosphere have remained fairly steady for at least the last 10 000 years. All living organisms have a small content of radioactive carbon when they are alive.

When an animal or plant dies, it stops exchanging carbon dioxide with the atmosphere. The ^{14}C in wood or bones then decays, reducing in amount and activity year by year.

a When an organism is alive, it contains about one atom of ^{14}C to every 10^{10} atoms of ordinary ^{12}C. In one second, a fraction 4×10^{-12} of the atoms of ^{14}C present may be expected to decay.

Estimate the activity in decays per second of 1 g of carbon from a living organism.

b What activity would you expect from 1 g of carbon taken from the bones of a bison eaten by Man some 11 000 years ago, when the great ice sheets covered much of Europe and North America?

c About how far back in time do you think the ^{14}C clock might be of use to archaeologists?

33(R) This question is about explaining ideas in physics.

Write an explanation of radioactivity suitable for a friend studying A-level physics who missed the teaching of this particular subject, giving a careful explanation of the topics: decay, radiation, isotopes.

Show also how your explanations could help your friend to understand **one** everyday application of the subject.

(Long answer paper, 1980)

Ionization

34(I) The Avogadro constant L is chosen (defined) to be the number of ^{12}C atoms in a sample of that isotope with a mass of exactly 12 g.

a What, approximately, is the mass of an alpha particle?

b What speed has an alpha particle which has a kinetic energy of one million eV?

35(I)a Calculate the energy (in joules) gained by an electron when it is accelerated through a potential difference of 10 V.

b Write down the kinetic energy (in J) of:
i a 5 eV electron
ii a 20 eV electron.

c Calculate the speeds of:
i a 5 eV electron
ii a 10 eV electron
iii a 20 eV electron.
(Use $m_e = 9.1 \times 10^{-31}$ kg.)

36(P) Figure F31 shows the energy needed to remove one electron from an atom, leaving it ionized, for atoms up to proton number about 90. The energies are given in electronvolts, and, for comparison, in kJ mol^{-1}.

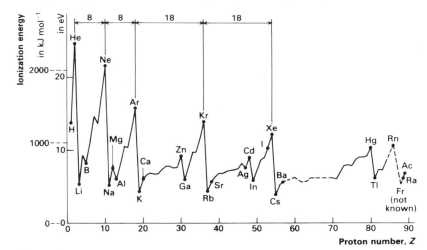

Figure F31
Ionization energies of the elements.

a The elements helium, neon, argon, krypton, and xenon come at peaks in figure F31. What chemical properties do these elements share?

b The elements lithium, sodium, potassium, rubidium, caesium, come at the lowest points of figure F31. What kind of chemical properties do these elements share?

c Look where sodium, Na, comes in figure F31. Now find the element neon, Ne, which has only one electron less than sodium. Is a sodium atom's 'last' electron tightly held to the atom?
    Find a similar pair of elements about which similar remarks might be made.

d Do calcium, strontium, and barium occupy similar positions on the graph? Are they chemically similar?

e Radium belongs to the same chemical family as calcium, strontium, and barium. The ionization energy of francium, the element one before radium, is not known. Predict the value, from the overall shape of the graph.

37(L) Figure F32 shows the radii of atoms of most elements, and the radii of the ions of a few.

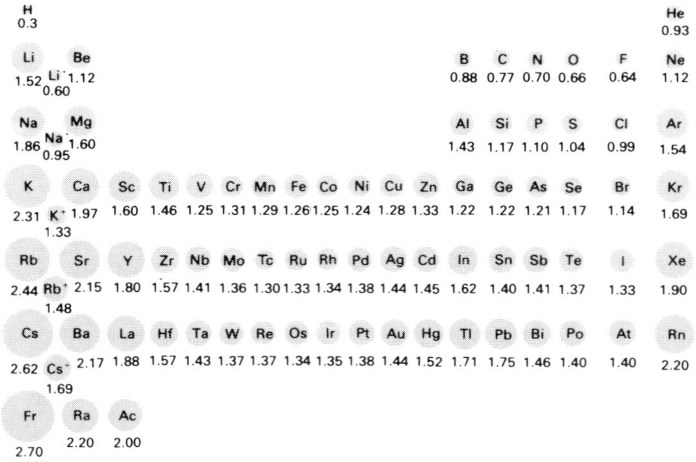

H 0.3 He 0.93

Li 1.52, Li⁺ 0.60 Be 1.12   B 0.88 C 0.77 N 0.70 O 0.66 F 0.64 Ne 1.12

Na 1.86, Na⁺ 0.95 Mg 1.60   Al 1.43 Si 1.17 P 1.10 S 1.04 Cl 0.99 Ar 1.54

K 2.31, K⁺ 1.33 Ca 1.97 Sc 1.60 Ti 1.46 V 1.25 Cr 1.31 Mn 1.29 Fe 1.26 Co 1.25 Ni 1.24 Cu 1.28 Zn 1.33 Ga 1.22 Ge 1.22 As 1.21 Se 1.17 Br 1.14 Kr 1.69

Rb 2.44, Rb⁺ 1.48 Sr 2.15 Y 1.80 Zr 1.57 Nb 1.41 Mo 1.36 Tc 1.30 Ru 1.33 Rh 1.34 Pd 1.38 Ag 1.44 Cd 1.45 In 1.62 Sn 1.40 Sb 1.41 Te 1.37 I 1.33 Xe 1.90

Cs 2.62, Cs⁺ 1.69 Ba 2.17 La 1.88 Hf 1.57 Ta 1.43 W 1.37 Re 1.37 Os 1.34 Ir 1.35 Pt 1.38 Au 1.44 Hg 1.52 Tl 1.71 Pb 1.75 Bi 1.46 Po 1.40 At 1.40 Rn 2.20

Fr 2.70 Ra 2.20 Ac 2.00

Figure F32
The Periodic Table showing atomic, and some ionic, radii/10^{-10} m.
After CAMPBELL, J. A. 'Atomic size and the Periodic Table'. J. Chem. Educ., **23**, 525, 1946.

(atomic radii determined from covalent bond distances)

a The atoms of Na, K, Rb, Cs, Fr are large. They all have small ionization energies. Suggest a link between these two facts.

b The ions of Na, K, Rb, Cs, are a good deal smaller than the atoms of these elements. Suggest a reason, based on your answer to **a**.

c The lithium atom has a nucleus with three protons, and contains three electrons within a radius of 1.52×10^{-10} m. The gold atom has a nucleus with 79 protons, and contains 79 electrons within a radius of 1.44×10^{-10} m. What general, rough rule can you give for the order of magnitude of the radius of atoms of an element with Z protons and Z electrons? Is it a surprising rule?

38(L) The binding energy of the electron in the hydrogen atom is -13.6 eV.

a Express this in joules.

The electron and proton both carry a charge of magnitude e.

b Write down an expression for the electrical potential energy when an electron is at a distance r from a proton.

c Use your answers to **a** and **b** to estimate the proton–electron distance.

This question implies that the total energy of the hydrogen atom is the potential energy of the proton–electron pair. In fact the electron has some kinetic energy.

d How does this affect your answers?

Radioactive transformations

39(P) The isotope ^{235}U decays into another element, emitting an alpha particle. What is that element? (Use a Periodic Table.)

This element decays, and the next, and so on, until a stable element is reached. The complete list of particles emitted in this chain is:

$$^{235}_{92}U \rightarrow [\alpha, \beta, \alpha, \beta, \alpha, \alpha, \alpha, \alpha, \beta, \alpha, \beta] \rightarrow X$$

What is the stable element X? (You could write down each element in the series, but there is a quicker way.)

40(R) $^{226}_{88}$Ra is a naturally-occurring isotope of radium and has a half-life of 1622 years. The radioactive decay is represented by the equation

$$^{226}_{88}Ra \rightarrow X + \alpha + \gamma,$$

where X is a daughter nuclide, α is an alpha particle, and γ is a gamma-ray photon. The proton number of X is

A 86 **B** 88 **C** 89 **D** 224 **E** 227   (Coded answer paper, 1980)

41(P) Use tables of nucleon (mass) numbers and proton (atomic) numbers to complete reactions **a, b, c** below.

Each item should have the nucleon number at the top left (superscript) and the proton number at the lower left (subscript). For example, for uranium of nucleon number 238 write $^{238}_{92}U$; for an electron write $_{-1}^{0}e$.

a $^{212}Pb \rightarrow {}^{212}Bi +$

b $^{212}Bi \rightarrow \quad + {}_{-1}^{0}e$

c $\rightarrow {}^{208}Pb + {}_{2}^{4}He$

Nuclear forces

42(L) The magnitudes of the electrical and gravitational forces between two particles are given by

$$F_e = \frac{1}{4\pi\varepsilon_0}\frac{Q_1 Q_2}{r^2} \quad \text{and} \quad F_g = G\frac{m_1 m_2}{r^2} \quad \text{respectively.}$$

a For each force say whether it is attractive or repulsive.

b Obtain an expression for the ratio F_e/F_g.

c This ratio is independent of r. What is the *physical* basis for this?

d Evaluate the expression obtained in **b** for two protons.

e Can the gravitational force be responsible for holding protons together in the nucleus?

The force which does hold protons (and neutrons) together in the nucleus is called the *strong nuclear force*. It is clearly stronger (within the nucleus) than the electrical force. But it is vanishingly small at distances greater than about 3×10^{-15} m.

f Suggest some evidence for the short-range nature of the nuclear force. (Think about what would happen if the nuclear force were still greater than the electrical force at a distance of about 10^{-10} m, *i.e.*, the distance between atoms.)

Nuclear binding energy

43(L) Use the curve for average binding energy per nucleon against nucleon number A (figure F11) to answer the following questions.

a Which nucleus is more stable, ^{4}He or ^{12}C?

b Arrange the following nuclei in order of increasing stability:

^{235}U, ^{12}C, ^{6}Li, ^{4}He, ^{2}H, ^{208}Pb, ^{56}Fe

c (*Harder*) In the Sun, helium is formed by the fusion of hydrogen. Suggest why the fusion of helium to form heavier elements does not occur. More extreme conditions of temperature and pressure (for example, those found in red giant stars or supernovae explosions) are needed for the formation of heavier elements.

d From figure F11 suggest elements which are likely to be formed in large quantities in the evolution of the heavier elements.

Mass and energy

44(P) The specific heat capacity of water is $4.2 \, \text{kJ kg}^{-1} \text{K}^{-1}$.

a How much energy is needed to raise the temperature of 5 kg of water from 20°C to 70°C?

b Use $\Delta E = c^2 \Delta m$ to estimate the increase in mass of the water when the temperature is raised by this amount.

c Express the mass increase as a fraction of the mass of the water.

45(P)a Estimate the mass of a rubber band.

b Estimate its force constant and obtain a value for the energy stored when the band is stretched to twice its original length.

c Use $\Delta E = c^2 \Delta m$ to estimate the increase in mass of the rubber band when it is stretched.

d By what percentage has the mass increased?

46(L) Atomic masses can be determined very precisely by mass spectrography. For example, the mass of a neutral atom of $^4_2$He is 4.002 6 u.

a Compare the mass of this neutral atom with the total mass of its component parts ($m_p = 1.007\,3$ u, $m_n = 1.008\,7$ u, $m_e = 0.000\,55$ u). Which is greater, and how large is the difference?

b Use $\Delta E = c^2 \Delta m$ to estimate the total nuclear binding energy for $^4_2$He:
i in joules (J)   *ii* in mega electronvolts (MeV).

c Calculate the average binding energy per nucleon. Your result should be comparable to the value given in figure F11.

Unit G
ENERGY SOURCES

Maurice Tebbutt
Faculty of Education, University of Birmingham

Contributors:

Michael Carrick, North Bromsgrove High School;
Bev Cox, Streetly School, Sutton Coldfield;
John Gardner, Shenley Court School, Birmingham;
Tom Gregory, Codsall School, Staffordshire;
John Minister, Bishop Walsh School, Sutton Coldfield;
Richard Spiby, King Edward VI Camp Hill School for Boys,
Birmingham

G

INTRODUCTION

This Unit is about energy sources, which means that it is also about energy itself. As you work through it you will meet familiar ideas, new ideas, and ideas which will be developed further in later Units. You will also need to try to make sense of data and graphs which are less straightforward than many others in physics. Many of the variables involved are not those normally found in physics, but this does not mean that you cannot think about them as a physicist, and find how far this gets you. Physics cannot give all the answers, but it does have its own contribution to make to the debate on energy supply. The Unit cannot cover all aspects of the topic, but will only lay the groundwork. Nor is it intended to lead you to adopt any one particular view.

Energy supply has been a topic of concern throughout the World since the early 1970s at least. This concern has been voiced in terms such as 'crisis', and expresses the fear that the World does not have enough energy for its present or future needs. But some might argue that it has too much for its own good.

It is worth looking first at some features of energy supply and demand.

THE BACKGROUND TO ENERGY SUPPLY AND DEMAND

SOME BASIC IDEAS

Energy is conserved

QUESTION 1   The conservation of energy is a basic principle of physics and features explicitly or implicitly at many places in science. If energy is conserved, how could we ever be 'short of energy'? The problem is that much of the energy from fuels heats the surroundings, spreads out, and becomes less and less useful.

QUESTION 2   **Sankey diagrams**
Many of the processes which are examined in this Unit are concerned with energy transfers and are often complicated. Sankey diagrams, like the one in figure G1, are frequently used to represent these complex situations.

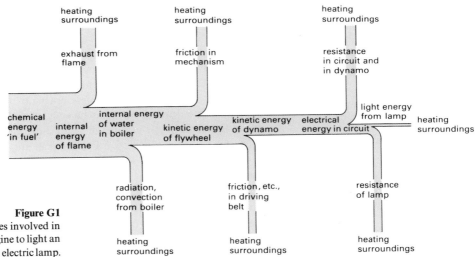

Figure G1
The energy transfer processes involved in using a steam engine to light an electric lamp.

The transfer of energy within the system is generally represented in the diagram as being like a stream flowing from left to right, with inputs and outputs usually shown vertically until they join or leave the main stream. The width of the stream represents the quantity of energy involved at that stage. (The idea of flow is used a great deal, both in this and other branches of physics. It is worth considering carefully what, if anything, flows in each case; or whether flow is just a useful way of visualizing a complex process.)

Figure G1 gives a general indication of the energies involved. Further data would be needed to make the width of the 'streams' directly proportional to the energy which each 'stream' represents.

Efficiency

Efficiency indicates what proportion of the energy input to a system which does a particular job is actually used to do that job.

$$\text{Thus efficiency is } \frac{\text{useful energy output}}{\text{total energy input}}.$$

It is expressed either as a fraction or as a percentage. What counts as 'useful' in the definition depends on the purpose of the system: whether it is being used for heating or for working, for instance. This point is taken up again in 'Energy converters', below.

'Energy slaves'

QUESTION 3

QUESTIONS 4 and 5

The human race has been very ingenious in developing machines which can do work for it, and because these machines can be called upon to help when required they can be regarded as 'energy slaves'. Although this idea usefully illustrates how our society depends on machines, 'energy slave' is not a precisely defined term and is used in various ways.

Fuels

If machines are to do work they need to be provided with energy in a form which they can use. Materials such as coal or oil, which, when burned enable machines to work for us, are called fuels. The fuel is consumed, though energy is conserved. In addition, burning occurs at high temperatures, and some energy inevitably goes to heating up the surroundings unproductively.

The term 'fuel' is often used in a broader sense than this. See, for example, 'Primary, end-use, and functional energy' on page 410.

Fuel consumption varies considerably from country to country. The people of developed countries have a range of energy slaves to draw upon for industry, agriculture, and domestic needs while the most primitive societies still depend on human and animal labour.

Energy converters

QUESTION 6

Table G1 shows the efficiencies of a range of energy transfer processes. Those in the first group, which all involve a chemical change (burning fuel) and the thermal transfer of energy to heat something up can be quite efficient. The second group involves mechanical change (for example, water running downhill) and electromagnetic processes. These too can be efficient. The members of the third group are all heat engines; all involve burning fuel and thermal transfer of energy in devices designed to do mechanical work. These processes are necessarily less efficient than the others. Efficiency depends on what you want: burning petrol heats things up well, but it is less efficient at making cars move.

Device	Efficiency/%	Processes
Space or water heaters		
Large commercial boiler	90	
Domestic boiler		chemical change
gas	75	and thermal transfer
oil	70	
coal	60	
Motors and generators		
Hydro-electric turbine	90	
Large electric motor	90	mechanical change and
Large electric generator	90	electromagnetic processes
Small electric motor	70	
Heat engines		
Steam turbine	45	
Diesel engine	40	chemical change,
Car engine	25	thermal transfer,
Steam locomotive	10	mechanical work

Table G1

Efficiencies of energy converters.
These values are indicative of efficiencies which can be obtained. The actual performance of the devices can often be worse than shown because of poor maintenance, for example.

Unit K, 'Energy and entropy'

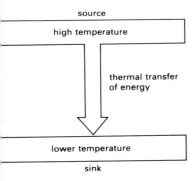

Figure G2
Thermal transfer of energy.

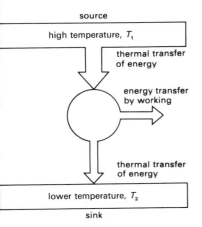

Figure G3

These differences are more than chance ones and reflect a fundamental law. The treatment given here is by no means comprehensive.

A body at a high temperature (called a source) seems to be able to transfer internal energy readily to a body at a lower temperature (a sink), through the process we call thermal transfer of energy. (Thermal transfer of energy goes from hot to cold.) In practice, some energy is transferred elsewhere owing, for example, to poor insulation, but there is no fundamental reason why all the energy transferred from the source should not go to the sink (figure G2).

In figure G3 we are making the system do work. In order to cause any energy transfer from the source it must be in contact with the sink. If, in some way, the work done by the system were equal to the energy transferred from the source, none would be transferred to the sink and it might as well not be there, hence removing the condition for thermal transfer to occur at all. In fact some energy *must* be transferred to the sink and this quantity of energy depends on the temperatures of the source and the sink.

The maximum efficiency turns out to be

$$E = \frac{T_1 - T_2}{T_1} \quad \text{(temperature measured in K)}$$

The turbines in a modern power station take in steam at, say, 530 °C (about 800 K), and reject it at 30 °C (300 K). Thus the maximum efficiency is 63 per cent, but practical efficiencies are likely to be considerably less than this.

Primary, end-use, and functional energy

Fuels are not necessarily used in the form in which they are originall
obtained. In this original form they are called primary fuels; the tota
energy that they have the potential to provide is called their primar
energy.

QUESTION 7 Crude oil, a primary fuel, is usually converted to various secondar
fuels, such as petrol, or fuel oil. In some cases the secondary fuel is in th
form in which it is ultimately used to provide energy; in these cases th
energy that the secondary fuel could provide (the secondary energy
may also be referred to as the end-use energy. However, the secondar
fuel may undergo one or more further conversions, for example into
electrical energy, which is the end-use energy in this case.

The amount of energy actually available to the final user i
sometimes called functional energy and depends on the efficiency of the
final energy conversion process. Functional energy is usually less than
the end-use energy which is a measure of the energy as it was delivered
to the final process. Many terms like these used in connection with
energy are not used precisely or consistently, but table G2 may help to
sort out common usage.

PRIMARY STAGE			SECONDARY STAGE			END-USE STAGE			
Primary fuel	Primary energy/J	Converter	Secondary fuel	Secondary energy/J	Converter	End-use fuel	End-use energy/J	Converter	Functiona energy/J
Example 1 Crude oil	1000	Refinery	Fuel oil	≈ 950	Power station	Electricity	≈ 300	Heater	≈ 300 in internal energy
Example 2 Crude oil	1000	Refinery	Petrol	≈ 900	none	Petrol	≈ 900	Car	≈ 200 in kinetic energy
Example 3 Coal	1000	Power station	Electricity	≈ 330	none	Electricity	≈ 330	Electric motor	≈ 260 in kinetic energy

Table G2
Examples of energy conversion to illustrate the use of the terms primary fuel and energy; secondary fuel and energy; end-use fuel and energy; and functional energy. The values given for the energies involved are illustrative rather than exact figures.

HOME EXPERIMENT GH1
The cost of a 'cuppa'

Table G3 and figure G4 (between pages 430 and 431) show how the
primary and end-use energies compare for various fuels in the U.K.

	Coal, etc. /10⁹ therms	Oil /10⁹ therms	Natural gas /10⁹ therms	Electricity /10⁹ therms	Total /10⁹ therms
Primary energy	29.0	28.1	18.0	3.7	78.8
End-use energy	7.6	23.2	16.6	7.5	54.9

Table G3
U.K. energy consumption 1981. (Note that electricity as primary energy is nuclear and hydro electricity, whereas electricity as end-use energy includes all electricity however generated.)
From: United Kingdom Energy Statistics, 1982. *Government Statistical Service.* United Kingdom Energy Statistics. HMSO, 1982.

Energy units

The fuel industry uses a variety of units other than the joule for measuring energy. Some of these, like kWh (kilowatt-hour), Btu (British thermal unit), and therm are comparable to the joule, that is, they are physical units which are defined, can be measured, and have values which are constant.

QUESTION 4

Other units, such as the megatonne of coal equivalent (Mtce but often written, incorrectly, as mtce) are not defined in the same way. This unit of energy is that which would be obtained if one million tonnes of coal were completely burned. The problem is that different kinds of coal can provide different amounts of energy if burned, so that it is common to use an average or a conventional value. Not everyone agrees on the convention which should be adopted and a degree of confusion results. For this reason, the conversion factors given in table G4 (see next page) may not agree with others which may be found, and the variations are such that it does not seem sensible to quote the values in this table to more than two significant figures.

Changes in fuel supply

Figure G5 shows that World demand for fuel did not grow steadily between 1925 and 1981, but that the rate of growth increased (apart from one or two relatively short periods).

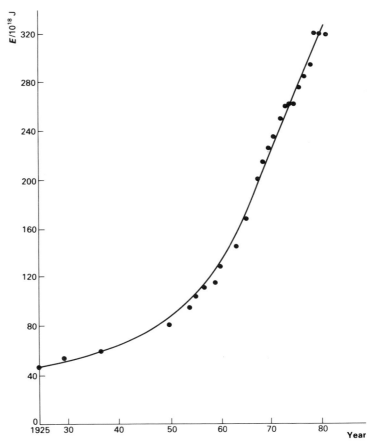

Figure G5
World demand for fuel since 1925.
Data from: Yearbook of World energy statistics, *United Nations, 1981, and* BP statistical review of World energy, 1982, *British Petroleum Company P.L.C., 1983.*

Multiply by factor to convert unit below into unit on the right	Mtce	Mtoe	m³	Therm	Quad	J	kWh	kcal	Btu	bbl
Megatonne of coal equivalent, Mtce	1	0.69	750×10^6	280×10^6	28×10^{-3}	29×10^{15}	8.1×10^9	6.9×10^{12}	28×10^{12}	4.8×10^6
Megatonne of oil equivalent, Mtoe	1.4	1	1.1×10^9	400×10^6	40×10^{-3}	42×10^{15}	12×10^9	10×10^{12}	40×10^{12}	7.0×10^6
Cubic metre of natural gas, m³	1.3×10^{-9}	0.92×10^{-9}	1	360×10^{-3}	36×10^{-12}	39×10^6	11	9.2×10^3	36×10^3	6.4×10^{-3}
Therm	3.6×10^{-9}	2.5×10^{-9}	2.7	1	100×10^{-12}	110×10^6	29	25×10^3	100×10^3	18×10^{-3}
Quad	36	25	27×10^9	10×10^9	1	1.1×10^{18}	290×10^9	250×10^{12}	10^{15}	180×10^6
Joule, J	34×10^{-18}	24×10^{-18}	26×10^{-9}	9.4×10^{-9}	0.95×10^{-18}	1	0.28×10^{-6}	240×10^{-6}	950×10^{-6}	0.17×10^{-9}
Kilowatt-hour, kWh	120×10^{-12}	86×10^{-12}	93×10^{-3}	34×10^{-3}	34×10^{-12}	3.6×10^6	1	860	3.4×10^3	0.60×10^{-3}
Kilocalorie, kcal	140×10^{-15}	100×10^{-15}	110×10^{-6}	40×10^{-6}	4.0×10^{-15}	4.2×10^3	1.2×10^{-3}	1	4.0	0.70×10^{-6}
British thermal unit, Btu	36×10^{-15}	25×10^{-15}	27×10^{-6}	10×10^{-6}	1.0×10^{-15}	1.1×10^3	290×10^{-6}	0.25	1	180×10^{-9}
Barrel of oil, bbl	210×10^{-9}	140×10^{-9}	160	57	5.7×10^{-9}	6.0×10^9	1.7×10^3	1.4×10^6	5.7×10^6	1

Table G4

Conversion factors for some units commonly used in connection with energy. For example, $10\,000\,\text{m}^3$ of natural gas $= (10\,000) \times (39 \times 10^6)\,\text{J}$.

Figure G6, in which the World demand for fuel is plotted on a logarithmic scale against time, has three distinct linear sections, showing that in each section the growth in demand was exponential. A characteristic of exponential decay curves is constant half-life and the equivalent idea for exponential growth is doubling-time. Hence the three sections of figure G6 have different doubling-times.

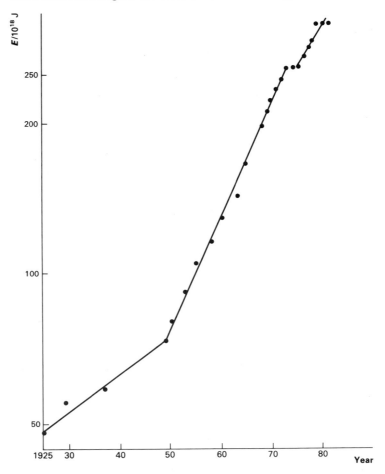

Figure G6
World demand for fuel (logarithmic scale) since 1925.
Data from: Yearbook of World energy statistics, *United Nations, 1981, and* BP statistical review of World energy, 1982, *British Petroleum Company P.L.C., 1983*.

The real world (of fuel supply) may not be quite so easily described by mathematical relations as the more restricted and controlled world of physics, but such ideas (mathematical models) may be useful in gaining an understanding of different possible futures.

Making predictions

We can predict the future of a bacterial colony in a controlled environment when we know both the stock of food available for it to live on and its rate of growth in the past. If growth in the past has been exponential, and if the conditions remain the same, we can assume that growth will continue to be exponential and the food will be used up in a predictable time.

QUESTION 12

Although our assumptions are too simple and there are other factors to be taken into account (we return to these later), in principle the same process can be used for predicting the future use of fuels. There are far more uncertainties here than in the example above. For instance the exact size of the fuel resources is not known, since the resources first have to be discovered before they can be extracted.

There has been a tendency for initial estimates of oil reserves, for example, to be rather too small and to be increased later (figure G7). This helps to explain why, despite great increases in consumption, reserves have kept well ahead, as shown in figure G8. Also, as the price of fuel rises it becomes economical to use deposits which have been too small or too difficult to use hitherto.

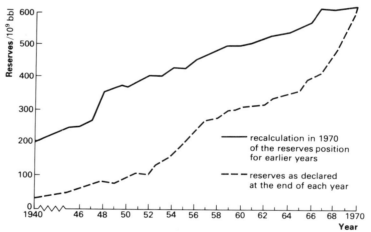

Figure G7
The growth of proven oil reserves over time. Proven reserves figures as declared at a particular time seriously understate the amount of oil which has actually been discovered
From: ODELL, P. R., Energy: needs and resources. *Macmillan, 1974.*

The pattern of demand in the future is also uncertain. The growth in demand in the past has not been uniformly exponential, so it is unlikely to be so in the future. Nevertheless, possible values for the lifetimes of resources can be calculated by making simple assumptions about the pattern of future growth. Which pattern actually comes about will depend on factors such as the growth of World population, the expectations of that population, and the way in which industry develops in both the industrialized and the less developed nations. All of these factors make forecasting difficult, and forecasts can be markedly inaccurate even in the short term, and even when made by leading experts. However, it is possible to put into perspective the estimates which are continually being made by a variety of people – expert and otherwise.

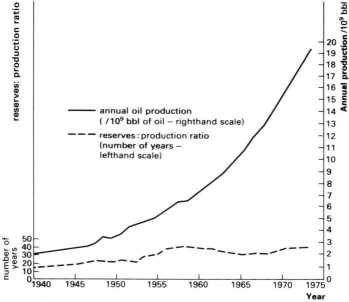

Figure G8
Proven oil reserves and annual oil consumption 1940–1975. The ratio of remaining known reserves to annual production has generally trended upwards (from only 15 years in 1940 to well over 30 years in the 1970s), in spite of rapidly rising use of oil over this period. *From: ODELL, P. R., Energy: needs and resources. Macmillan, 1974.*

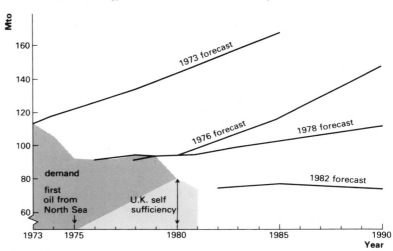

Figure G9
Variation of actual demand, and consequent variation in forecasts of demand, with time. *Source: Esso Magazine Supplement Winter 1982/83. Public Affairs Department, Esso Petroleum Company Limited.*

QUESTIONS 13 and 14   All of the discussion of growth hitherto has assumed that if the growth is exponential it will continue unrelentingly until the resource is suddenly found to be used up. In practice, growth is unlikely to remain

exponential and the growth rate will decrease as the resource become
more difficult to obtain and more expensive.

The supply of fuels is finite

Fossil fuels are being used up and therefore will eventually run ou
Figure G10 shows how the consumption of the World's stock of o
might look on a 2000 year time-scale centred on the present day. Thi
diagram could be repeated with somewhat different time-scales for a
the fossil fuels. Using fossil fuel is very like living off one's capital – i
runs out sooner or later depending on the rate at which it is used an
how much there was to start with.

QUESTION 15

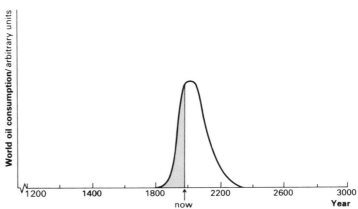

Figure G10
Variation in World oil consumption on a long time-scale.

Is there a crisis?

Whether there is a crisis in fuel supply depends on how soon th
fossil fuel 'capital' will be used up and whether there will be anythin
to replace it. One way of dealing with the problem would be to live o
one's income – that is, to use the renewable energy sources such as sola
power. There are also various kinds of nuclear energy source which ar
currently feasible. But nuclear fuels are also finite, so, although usin
them may delay the crisis, it will not, in the long run, avoid it.

QUESTIONS 16 to 18

The next part of this Section deals with some of these ideas in detai

MORE ABOUT FUEL USE

Distribution

The World's dependence on fossil fuels

Oil contributes roughly two-fifths of the World's fuel consumption, ga
one-fifth, coal almost a third, and less than one-tenth is provided b
hydroelectricity and nuclear power together. Given the projecte
lifetimes of the resources considered in the first part of this Section, th
World as a whole seems to be heavily, even dangerously, dependent o
those fuels which are likely to be depleted quickly. Figure G11 show
how this dependence has been growing.

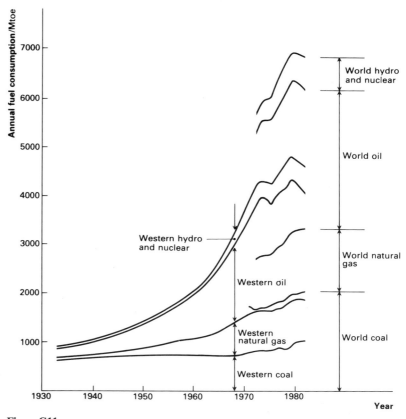

Figure G11

Variation of fuel consumption with time for the 'World', and for the World less U.S.S.R., Eastern Europe, and China – labelled 'Western'. Reliable data were not available for U.S.S.R., Eastern Europe, and China before 1971 hence 'World' graphs can only start at that point. Both sets of graphs are cumulative, that is, the data for gas are added to those for coal before being plotted.

Data from: BP statistical review of the World oil industry *and* BP statistical review of World energy. *British Petroleum Company P.L.C., 1963–1983.*

The dependence of different nations on fossil fuels

QUESTION 19   Some areas of the World are apparently even more dependent on fossil fuels than most, but how critical this is likely to be depends on the availability of resources to these areas. Table G11 (page 436) shows the consumption of various fuels by area, and table G13 (page 439) shows the production of fuels in these same areas. Although the fuel categories are a little different, it is possible by comparing the two tables to see which countries depend heavily on imports of particular fossil fuels.

Dependence of different peoples on fossil fuels

Table G11 shows that North America and Western Europe together consume about half the World's fuel each year. The fuel consumption *per capita* enables comparisons to be made without variations in population causing confusion.

The inhabitants of some areas of the World consume up to thirt
times as much primary fuel as some others. In the U.K. the *per capit*
consumption of primary fuel is, on average, the equivalent of about 10
kW h per day. Of course, the energy actually available to the individua
user is likely to be much less than this because of the losses in the
conversion of primary fuels and the fact that much of this fuel is used by
industry on behalf of individual citizens.

QUESTION 20   Nevertheless the inhabitants of the developed world have man
more energy slaves available to them than those in the less develope
countries. Since at least some of the fuel is used in manufacturing good
for sale at home and abroad, it means that the developed countries ar
able to earn more for their efforts. Figure G12 is a common represen
tation of the connection between fuel use and earnings; however, there i
some dispute about whether displays like this are entirely fair.

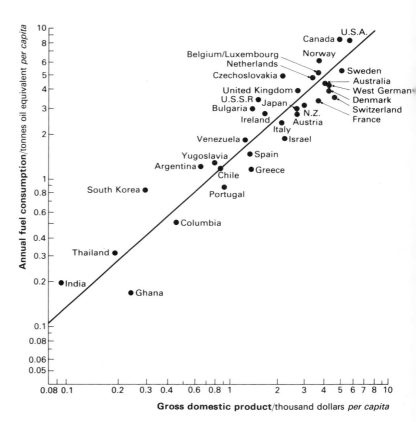

Figure G12
Fuel consumption against gross domestic product.
From: DARMSTADTER, J., DUNKERLY, J., *and* ALTERMAN, S. How industrial societies use
energy. *Johns Hopkins University Press, 1977.*

What do we use fuels for?

QUESTION 22 Notice that the asterisked categories in table G5 can be grouped together as 'low temperature space heating', which comprises 41 per cent of the total compared with 33 per cent of the total for the industrial production category which essentially involves 'high temperature' processes.

U.K. energy use by sector	Per cent
Industrial production	33
Domestic space heating*	18
Transport	16
Space heating industrial and commercial buildings*	13
Space heating public buildings*	10
Other domestic uses	10

Table G5
How energy is used in the U.K.

Some characteristics of fuels

Energy density

QUESTION 21 This term indicates that the energy in some fuels is more 'compressed' or intense than in others. For example, the energy density of coal is generally greater than that of peat; and, although vast quantities of energy fall on the Earth each day from the Sun, the energy density is relatively small. The energy density of a fuel determines the size of the converter which is needed to utilize it – a coal furnace is smaller than a solar furnace of similar power.

Transportability

QUESTION 23 Fuels are not generally found in the places where they are going to be used. Some, like oil or gas, are relatively easy to move about while others, like wave energy, are difficult to move in that form though relatively easy to move if converted to electricity.

Time

QUESTIONS 24 and 25 Fuels may not be available when they are needed either because they cannot be turned on on demand – like solar power or wind energy – or because the demand varies over time. Figure G13 shows how demand for electricity varies on a typical daily basis in summer and in winter. Figure G14 shows a somewhat atypical day. The gas industry has a similar pattern of demand (figure G15, page 421).

Although some storage is possible, both the gas and electricity industries are generally faced with the need to meet the demand as and when it occurs. If they fail, they may cause inconvenience or worse, with consequent public outcry.

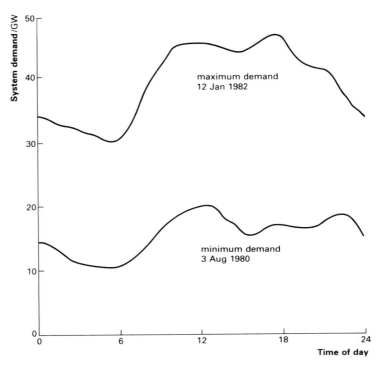

Figure G13
Variation in demand for power from Central Electricity Generating Board with time of day and time of year.
From: WADDINGTON, J. *and* MAPLES, G. C. *'The control of large coal and oil-fired generating units.'* CEGB Research, *February 1983.*

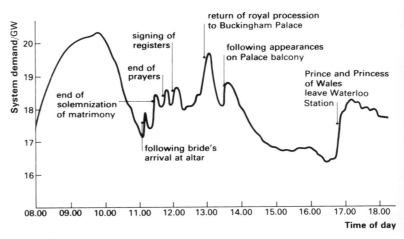

Figure G14
Variation in demand for power from Central Electricity Generating Board during an atypical day, 29 July 1981.
From: WADDINGTON, J. *and* MAPLES, G. C. *'The control of large coal and oil-fired generating units.'* CEGB Research, *February 1983.*

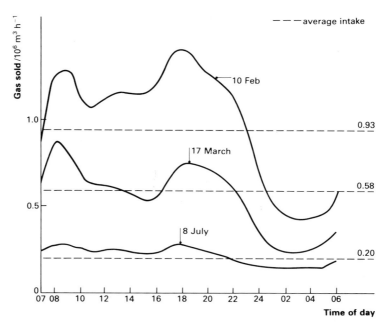

Figure G15
Gas sold in West Midlands Region of British Gas, for three days during a year.
Data: *British Gas Corporation.*

Feasibility

To use coal for heating does not require any elaborate apparatus, though maximizing the efficiency may involve some complication. On the other hand, using sunlight, waves, uranium, or deuterium may require complex devices which cost a great deal in terms of money and also of energy before they begin providing energy for use by our society.

A similar argument applies to any other form of energy conversion. The capital cost and the energy requirements mean that large energy converters cannot be built without a substantial amount of forward planning.

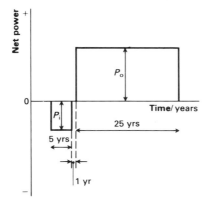

Figure G16
Schematic diagram showing how a power station requires energy while it is being built before it produces energy once it is running. P_i and P_o are the average input and output powers. The area under the graph is a measure of the energy involved.
From: CHAPMAN, P. Fuel's paradise. *Penguin Special, Copyright © Peter Chapman, 1975. Reprinted by permission of Penguin Books Ltd.*

Undesirable consequences of fuel use

One undesirable consequence of fuel use is pollution. Everyone know what this is, but it is paradoxically difficult to define in a way whic people will agree upon. It is generally taken to be the results of Man activities which affect the environment in an undesirable way. On problem is that most of our activities affect the environment in som way. Whether this is undesirable depends in part on whether th benefits of the activity which produces the alleged pollution ar considered to outweigh the disadvantages.

Thermal pollution

The Earth is continually receiving large quantities of energy from th Sun, and emitting energy by radiation. The rate at which energy radiated depends critically on the Earth's temperature. This chang until the rate at which energy is emitted by radiation is just equal to th rate at which energy is received (less the small proportion which 'fixed' as plant life or biomass). This is important because quite sma changes in temperature can produce substantial changes in weath patterns and the size of the polar ice caps, for instance.

The consumption of fuels adds to the energy input to the Earth an its temperature must rise to enable the extra energy to be radiate away. This might not be a problem for the Earth as a whole for som time. But population, and hence energy use, tend to be concentrated i relatively small regions and in such regions the energy input from fue can be a substantial fraction of the solar input.

QUESTION 26

Other sources of pollution

The burning of fossil fuels produces combustion products which ca combine with water in the atmosphere to produce acid rain, which ma well fall on a country other than that of its origin; some of th combustion products are radioactive because the fuel contains sma quantities of radioactive material; quantities of solid matter may b produced as smoke or ash. Extraction of fossil fuels may also damag the environment.

Nuclear power stations require the manufacture and treatment c fuel and the disposal of waste which are polluting to different extents.

This is an area of continuing concern which will be considere further in Section G4.

Risks in fuel production

Just as human activity may be regarded as polluting, it also entails som elements of risk. Where the consequences of an activity frequently lea to injury or damage it is possible to determine statistically the ris involved to any individual. Where the activity almost never 'goe wrong' or where it is novel, it is possible to estimate what the risk might be, but this determination is different from the one describe above. It is also necessary to consider whether the person at risk ha knowingly agreed to take the risk. Figures G17 and G18 show the risk involved in a variety of activities. The risks attached to air crashes ar

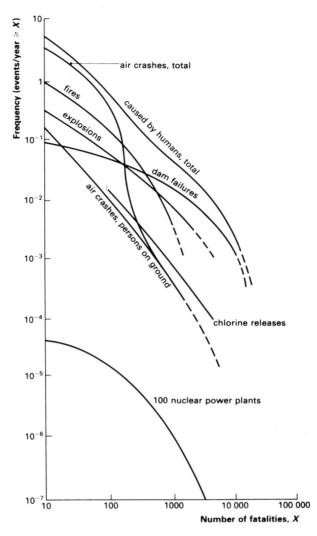

Figure G17

Probability of accidents, causing more than some number, X, of fatalities, against X for a variety of causes of accidents.

From: The Rasmussen Report, 1975. The reactor safety study. *NUREG 74/014. U.S. Nuclear Regulatory Commission, Washington.*

presumably calculated on actual evidence. Those who fly in aircraft probably accept the risk, but do all of us knowingly accept the risk of having an aircraft fall on us?

Nuclear power stations have been notably free from fatalities attributable to them, therefore the curves relating to their operation must have been calculated using the theory of risk analysis. There is therefore more doubt about whether these figures are reliable. Moreover, the population at large has no choice about whether it should be exposed to the risks involved in nuclear power stations, any more than it has about the risks involved in fossil-fuelled power stations.

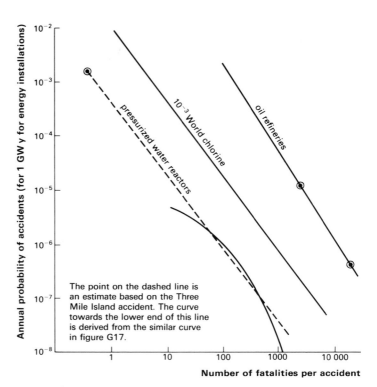

Figure G18
Probability of accidents of varying severity for energy installations. GW y means gigawatt year. Accident probabilities resulting from 1/1000 of the World's production of chlorine are included for comparison.
Data: Professor John Fremlin.

The point on the dashed line is an estimate based on the Three Mile Island accident. The curve towards the lower end of this line is derived from the similar curve in figure G17.

MORE ABOUT GROWTH

In the earlier part of this Section growth of fuel demand was examined and predictions were made based on some simple and probably naive assumptions – basically that demand would either remain static or would continue to grow exponentially at the average rate which has applied over the last ten years or so.

Here are some of the factors which may affect growth in demand.

Population growth

QUESTION 27 If World population continues to grow exponentially, then the demand for fuel will increase exponentially for this reason alone, assuming that fuel consumption per head remains at today's level.

Increasing expectations

QUESTIONS 28 to 30 It may be argued that the distribution of fuels is unfair and that more fuel ought to be available to each inhabitant of the less developed world (and that we in the industrialized nations ought to have no more, or perhaps less than at present).

It can be argued that centralized systems of both fuel consumption and production of goods may not be appropriate to the less developed countries. There is, however, evidence that fuel demand *per capita* is rising faster in the less developed world than elsewhere and this could produce a rapidly rising demand for fuel resources.

Reducing reserves

All of this discussion must be set against a background of increasing difficulty of extracting fuel and hence increasing cost. This is likely to affect the poorer countries more than the rest, and may possibly lead to an even greater imbalance than at present.

Making savings

It is easy to make savings in fuels. After all, in the past humans survived by their own and their animals' labour, and many people in the World may be said to do so today.

In reality, if we consumed considerably less fuel, the living standards of those in the industrialized nations would be substantially reduced. Whatever the rights and wrongs of this, it is not likely to be popular!

This is not to say that savings cannot be made, since much fuel is used wastefully. Many of the possible savings depend on changes of attitude by individuals or by governments; the problem is that attitudes are notoriously resistant to change. Other savings depend on developing techniques to ensure that resources are used less wastefully. Many of these changes take time. Table G6 summarizes some of the factors involved in planning energy savings.

	Savings made by:	
	Individuals	**Societies**
Time required to put idea into practice/Years	1	10
Relative cost	low	high

Possible ways of making savings	Domestic
	  heating
	  labour saving appliances
	  cooking
	
	Transport
	
	Communication
		Choice of central
		  or local power
		  stations
		
		Choice of conversion
		  method
		
		Choice of method of
		  distribution

Possible sources of energy	Electricity Gas, coal Solar	Fossil fuels Nuclear Solar, wind, wave

Other considerations	Living standard
	
	Domestic safety
	
	Amount saved
		Conformation to
		  legislation
		
		Choice of site:
		  environmental and
		  economic factors

Table G6
Controlling factors for energy planning.

More speculative fuel resources, for example nuclear fusion, would take longer to develop, cost more, and present more technical difficulties than those mentioned in the table.

POWER FROM NUCLEAR FISSION

ENERGY CAN BE RELEASED IN NUCLEAR REACTIONS

Unit F, 'Radioactivity and the nuclear atom' showed that the particle in an atom's nucleus (the nucleons) are strongly bound together by an attractive force. Since energy is required to separate the nucleons the binding energy is negative. Figure G19 shows how the average binding energy per nucleon depends on the number of nucleons and shows that if a massive nucleus were to split into smaller fragments (fission) the total binding energy would become more negative and so energy would be released. Similarly, if two light nuclei could be fused together to form one more massive nucleus, energy would again be released.

Fission – the basic ideas

Whilst fission can occur spontaneously, it is a very rare event and is more usually triggered by adding an extra neutron to a suitable nucleus. There are a number of possible nuclei but the one which is most commonly used is ^{235}U. Naturally-occurring uranium contains 0.7 per cent of ^{235}U, the rest being mainly ^{238}U. This can also undergo neutron-induced fission, as can ^{233}U and some plutonium isotopes. (Plutonium, proton number 94, does not occur naturally, nor does ^{233}U.)

A typical reaction is:

$$^{235}_{92}U + ^{1}_{0}n \rightarrow ^{236}_{92}U \rightarrow ^{144}_{56}Ba + ^{90}_{36}Kr + 2^{1}_{0}n + \text{energy}$$

but there are many possible products of the process as shown in figure G20.

Figure G19 shows that the energy released is about 1 MeV per nucleon, or a total of some 200 MeV. This gives a very high value of specific energy release (energy released per unit mass) compared, for example, with burning coal or other similar chemical reactions.

Most of the energy which is released is in the form of kinetic energy of the fission fragments. These are brought to rest in the fissionable material thus increasing its internal energy. This energy can be extracted by using a coolant.

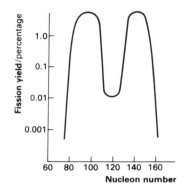

Figure G20
Yield of different fission products as a function of their nucleon number. Each fission produces two product nuclei, that tend to be of different size.
WINTERTON, R. Thermal design of nuclear reactors. Pergamon, 1981.

QUESTION 31

Table G7
Energy released per thermal fission of ^{235}U.
Source: Physics (Advanced) Notes for the guidance of teachers on the Nuclear physics option. *GRIFFITH, J. A. R., and TEBBUTT, M. J. J.M.B., 1977.*

Source	Energy/MeV
Kinetic energy of fission products	165
Kinetic energy of neutrons	5
γ-rays (immediate)	7
β-particles and γ-rays (from later decay of fission fragments)	13
Neutrinos/antineutrinos from decays	10
Total	200

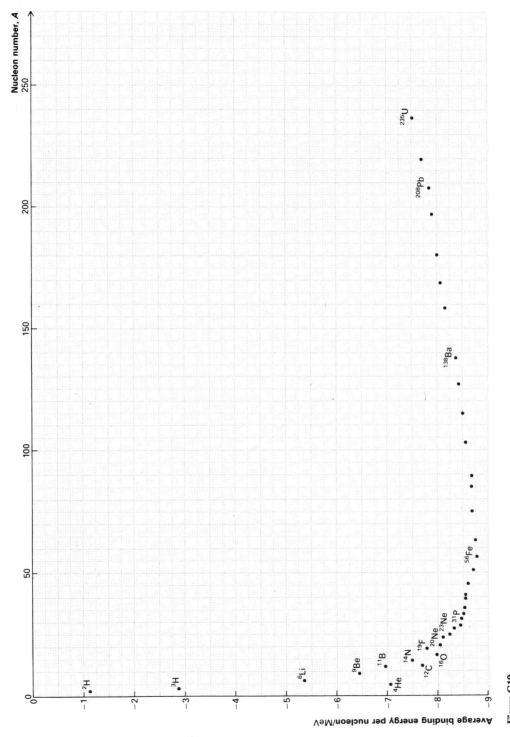

Figure G19
Average binding energy per nucleon against nucleon number.

Although the energy released in one fission event is very lar[g] compared with, say, the energy released when one carbon atom bur[ns] (combines with oxygen), many fissions are needed per second to provi[de] power on the scale supplied by power stations. These fissions a[re] induced by neutrons, and neutrons are emitted when the nuclei split. [A] rather neat arrangement in which the neutrons emitted induce enoug[h] further fissions to produce a reaction which is at the least self-sustainir[g] is called a chain reaction.

Fission – more detailed ideas about getting it to work

Critical mass

The neutrons emitted from a fission will not induce further fissions [if] they are absorbed by non-fissionable material or lost from the surface [of] the fuel. As the size of the block of fuel increases so does the chance [of] absorption by fissile material; the chance of loss decreases as th[e] surface-to-volume ratio decreases.

Thus, as a fuel assembly is made up, a stage is reached when a cha[in] reaction can occur. This happens when a so-called 'critical mass' of fu[el] has been assembled. The precise value of this quantity depends o[n] factors such as the fuel being used and the shape of the assembly.

Cross-section

QUESTION 32

Nuclei are about 10^{-14} m in diameter and are more or less spherical i[n] shape. A typical cross-section is therefore about 10^{-28} m^2. But not a[ll] particles passing through this area will necessarily interact with th[e] nucleus. For example, some nuclei appear to allow neutrons to pa[ss] through them relatively unimpeded – it is as though they were sem[i-] transparent to the neutrons. We say that the cross-section for absor[p-] tion of neutrons is less, in this case, than the geometrical cross-sectio[n.] Similarly nuclei may have cross-sections for some interactions whic[h] are larger, perhaps much larger, than the geometrical cross-section[.] The actual value depends on the nature of the nucleus and th[e] interacting particle, and on the latter's energy.

The cross-section for an interaction is a useful way of indicating ho[w] likely that interaction may be, but it should not be thought that th[e] nucleus undergoes some peculiar swelling and shrinking process.

Obtaining fission in uranium-235

If the fuel for a nuclear reactor consists of natural uranium there [is] about 150 times as much ^{238}U as there is ^{235}U. Neutrons induce fissio[n] of ^{235}U, but ^{238}U absorbs them (generally without inducing fission). [If] the cross-sections for these reactions were the same there would be on[ly] a one in 150 chance of a particular neutron inducing fission.

QUESTIONS 33, 34

Figure G21 shows how the cross-sections vary with the neutro[n] energy. The fission cross-section of ^{235}U must be at least 150 times th[e] absorption cross-section of ^{238}U for the reactions to be equally likel[y.] This only occurs for low energy neutrons. (In fact the detailed picture [is] much more complicated than this.) The energy of such neutrons is of th[e] same order as the kinetic energies of the surrounding atoms – henc[e] they are called thermal neutrons.

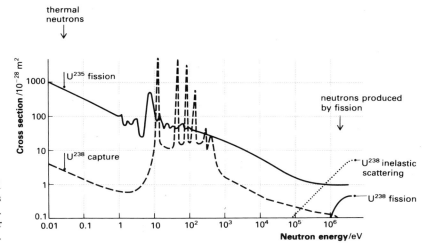

thermal
neutrons

Figure G21

Principal neutron reaction cross-sections
for uranium.

WINTERTON, R. Thermal design of nuclear
reactors. *Pergamon, 1981.*

QUESTIONS 35 to 39

Moderation

Neutrons emitted from the fission process have high kinetic energies (about 2 MeV). In order to take part in the fission process described above they must be slowed down by being made to collide with the nuclei of a material called a moderator. But if too many are absorbed by ^{238}U or the moderator and other parts of the structure (or lost from the surface of the assembly) the chain reaction will not take place.

Enrichment and reprocessing

QUESTION 40

The likelihood of establishing a chain reaction is increased if the proportion of material capable of undergoing fission is artificially increased. Such material is called enriched fuel and may have, say, 3 per cent of ^{235}U instead of the naturally-occurring 0.7 per cent.

Once the reaction starts, this proportion is reduced as the ^{235}U nuclei split. The chance of maintaining the chain reaction decreases. In addition the fission products often absorb neutrons very strongly which further inhibits the chain reaction. For this reason, fuel may have to be withdrawn and reprocessed to eliminate the unwanted material. The extraction and treatment of this waste material is a matter of some concern which will be examined in Section G4, part C, Nuclear safety.

Control

QUESTION 39

Once a chain reaction is achieved it is unlikely to continue at just the required rate. In practice the reaction rate is set too high, and the process is controlled by absorbing enough neutrons to keep the chain reaction steady. Rods of a material with a large neutron absorption cross-section are moved in and out of the reactor to do this.

Different kinds of reactor

There are many different kinds of thermal fission reactor, in addition to fast fission reactors. Section G4, part A, Nuclear power stations, provides an opportunity to examine some of these in detail. The process of nuclear fusion and possible fusion reactors is studied in Section G4, part B, Nuclear fusion.

QUESTIONS

Energy is conserved

1(I) Figure G22 shows a familiar arrangement used to demonstrate energy conversions. When the mass is released it accelerates, the generator turns, and the lamp lights. Usually at some stage the mass and generator reach a steady speed and the lamp gives a steady output. This continues until the mass reaches the floor when the generator slows to a stop and the lamp no longer lights.

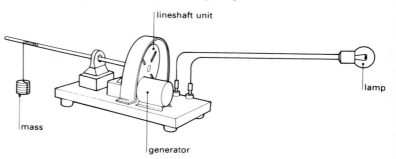

Figure G22
An energy conversion experiment.

Figure G23 attempts to show in a simple way how energy is transformed as it is transferred through the system. Notice the convention which has been used here: the part of the system being considered is represented by a labelled box and the 'form' in which the energy might be thought to exist is added as a label underneath each box.

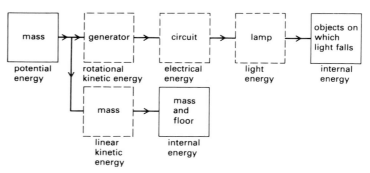

Figure G23
Energy transformations in the experiment shown in figure G22.

a *i* What is meant by 'the potential energy of the mass'? Is it strictly a property of the mass alone?
ii What is the process by which energy is transferred from the first stage in figure G23 to the second?
iii What is meant by saying 'the generator has a certain quantity (say 15 J) of rotational kinetic energy'?

A Publication of the
Government Statistical Service

Energy Flow Chart 1983

United Kingdom

DEPARTMENT OF ENERGY

JUNE 1984

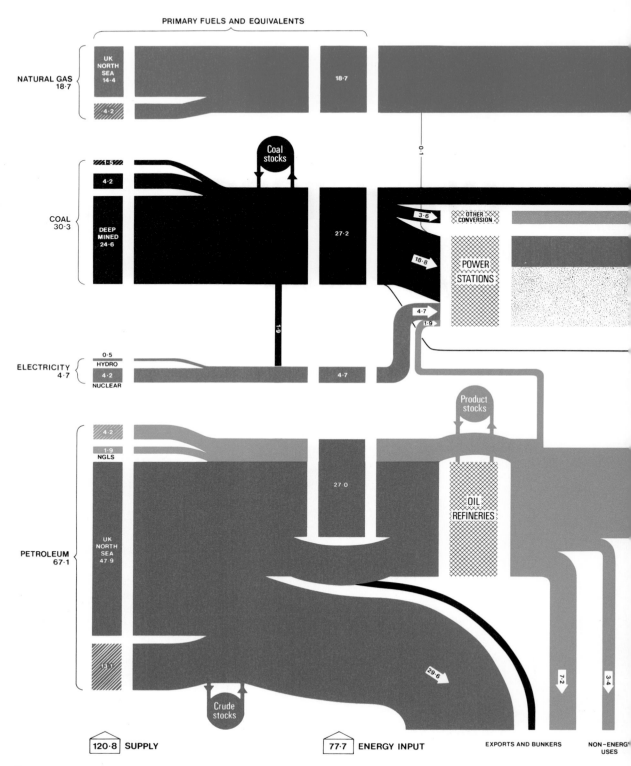

UK ENE

(THOU

PRIMARY FUELS AND EQUIVALENTS

NATURAL GAS
18·7

UK NORTH SEA 14·4

4·2

18·7

0·1

Coal stocks

COAL
30·3

1·5

4·2

DEEP MINED 24·6

27·2

3·6

OTHER CONVERSION

18·8

POWER STATIONS

4·7

1·9

1·9

ELECTRICITY
4·7

0·5 HYDRO

4·2 NUCLEAR

4·7

Product stocks

PETROLEUM
67·1

4·2

1·9 NGLS

27·0

UK NORTH SEA 47·9

OIL REFINERIES

13·1

29·6

7·2

3·4

Crude stocks

120·8 SUPPLY

77·7 ENERGY INPUT

EXPORTS AND BUNKERS

NON-ENERGY USES

Figure G4

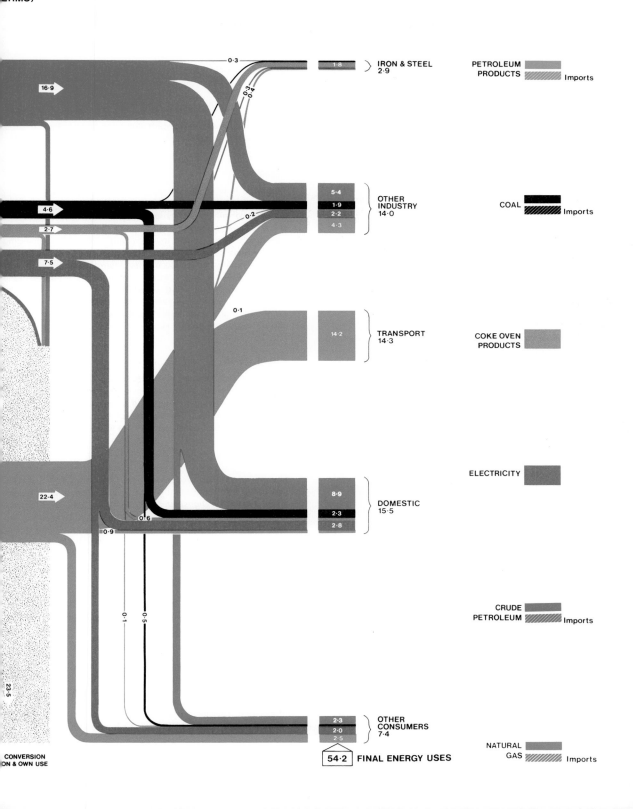

IRON & STEEL
2.9

1.8

OTHER
INDUSTRY
14.0

5.4
1.9
2.2
4.3

TRANSPORT
14.3

14.2

DOMESTIC
15.5

8.9
2.3
2.8

OTHER
CONSUMERS
7.4

2.3
2.0
2.5

54.2 FINAL ENERGY USES

PETROLEUM
PRODUCTS   Imports

COAL   Imports

COKE OVEN
PRODUCTS

ELECTRICITY

CRUDE
PETROLEUM   Imports

NATURAL
GAS   Imports

16.9

4.6

2.7

7.5

22.4

0.3

0.3
0.0

0.2

0.1

0.6

0.9

0.1

0.5

23.5

CONVERSION
ON & OWN USE

Energy Flow Chart 1983
United Kingdom

The chart illustrates the flow of primary fuels from the point at which they become available from (on the left) home production or imports to their eventual final uses (on the right), either in their original state or after being converted into different kinds of energy by the secondary fuel industries.

All flows are measured in thousands of millions of therms and the widths of the bands on the chart are roughly proportional to the absolute sizes of the flows they represent. Stocks of coal and petroleum are represented by circles. (The circles are not related to the size of the stocks — and they do not show whether there has been a stock rise or stock fall.)

Primary supplies and primary fuels

The chart is similar to the previous issue relating to 1980. Primary consumption of petroleum is the sum of consumption of petroleum products at power stations and gas works, deliveries for other uses, refinery fuel and refinery losses. Petroleum products derived from crude oil which are used for non-fuel purposes (eg as a raw material for the manufacture of chemicals and plastics, as bitumen for road making etc) are not included as energy consumption.

As can be seen, most of our primary fuel supply is not finally consumed in the original state in which it is produced or imported. Crude petroleum is refined to produce petroleum products (eg petrol, fuel oil, gas/diesel oil, jet fuel etc). The largest proportion of coal flows to power stations where it is transformed into electricity.

Nuclear and hydro electricity are often referred to as primary electricity to distinguish them from that generated at conventional power stations burning fossil fuels, ie coal, petroleum or natural gas. There are many ways in which the output of nuclear and hydro electricity can be measured. In the chart and in all related statistics the electricity generated by these means is expressed in terms of the notional amount of fossil fuels that would have been needed to generate the same amount of electricity at contemporary conventional steam power stations.

Secondary fuels

The principal secondary fuels are petroleum products, electricity and coke (which in the chart includes other manufactured solid fuels). Secondary fuels are in the main required for specific purposes for which the use of primary fuels is inappropriate. For many uses there is no practical alternative to electricity as a fuel and coke is an essential material for the iron and steel industry.

Losses

This large flow (in dotted grey) shows those losses that occur between primary supplies and deliveries to final users. Each fuel industry consumes energy in the course of its operations and some is lost during its subsequent distribution. Electricity generation in particular involves large losses in converting primary fuels to electricity. The chart does not show the further losses of energy which occur after energy is supplied to final consumers which result principally from the inefficiencies in the multitude of energy using appliances, eg domestic fires and boilers, cars, lorries, aircraft, central heating plant etc. It is estimated that these latter losses could in total amount to almost half of the energy supplied to final consumers.

Final uses

This section of the chart illustrates how energy consumption is distributed between the main final consuming sectors and how the different kinds of primary and secondary fuels are shared between the sectors. The figures for coal and petroleum are deliveries as actual consumption data is not available.

Statistics

The chart has been prepared by the Economics and Statistics Division of the Department of Energy and is based on statistics taken from the "Digest of United Kingdom Energy Statistics 1984". (Table 6) 'Energy balance for the United Kingdom'. The flow chart is a simplification of these figures and some of the terms used in the chart are not used in the Table. Table 2 of "Energy Trends" (Supply and Use of Fuels) is an abbreviated version of the energy balance table. Due to rounding the sum of constituent items may not equal totals.

The "Digest of United Kingdom Energy Statistics 1984" is prepared by the Economics and Statistics Division of the Department of Energy and published by Her Majesty's Stationery Office.

"Energy Trends", a Statistical Bulletin, which is also prepared by the Economics and Statistics Division of the Department of Energy, is published monthly, and is available on annual subscription. Details about subscriptions may be obtained from Information Division, Room 1397, Department of Energy, Thames House South, Millbank, London SW1P 4QJ.

b Diagrams such as figure G23 are intended to show the main features of the conversion process, but they can easily include too much and become too complex, or they can miss out important points.

Some important boxes have been omitted which relate to energy 'losses'. Show where these boxes ought to be inserted and, following the convention used in figure G23, show what form the energy takes and where it is located.

c The initial period of fall involves all the boxes in the diagram. The later stage of steady state is characterized by the state of those parts of the system shown thus $----$ not changing. Those shown thus $------$ continue to change.

Which kind of border should the boxes you have inserted in answer to **b** have?

d What do you know about the total energy of this system? Why has the word 'losses' been printed in quotation marks in part **b** of this question?

e Construct your own diagram for another conversion chain with which you are familiar.

Sankey diagrams

2(L) Construct first a non-quantitative Sankey diagram for a car travelling at a constant speed.

Then make it as quantitative as you can using these data: At 90 km h^{-1} the car will use 7 litres of petrol per 100 km; in order to keep the speed steady the power output at the flywheel is 15 kW and at the wheels is 12 kW. The energy produced when 1 litre of petrol is completely burned is 10 kW h.

Efficiency

3(E) This question treats the human body as a machine and as a heater. This oversimplification is intended only to give you some idea of the quantities of energy involved.

Human beings derive their energy from food. Hence one measure of food input is the energy it can provide in appropriate circumstances. The input depends on many factors but is normally about 10 MJ per day in this country.

a Measure (or estimate or recall) the mechanical power output of the human body when operating a pulley, walking, or running upstairs. (These are simple school experiments which you may have met before.)

b Estimate
i the total energy you expend per day in mechanical tasks and locomotion.
ii how much this amount of energy would cost if you bought it from the local electricity board. How much does it cost to buy food which will provide 10 MJ? (Examples: rice or mutton provide about

15 MJ per kg. Look at a recent electricity bill to find the cost of one unit, *i.e.*, 1 kW h.)

iii how efficient the human body is as a machine.

iv the power rating of the body as a heater. (*Hint:* estimate the power dissipated by the body assuming that all the energy obtainable from the food input can be used for heating.) Estimate the power delivered in, say, a theatre, by the 500 people inside it. Do such buildings need heating systems? If so, what for?

c The fuel (food) which a mammal's body takes in is used in a number of ways: for instance, doing external work, keeping the parts of the body working, building body parts, and keeping warm – some is wasted, of course.

i Construct a Sankey diagram to illustrate this. It is likely to be mainly qualitative but some of the information needed to quantify it is available from other parts of this question.

ii The overall energy input and output of a body must be balanced. Use the Sankey diagram to consider what might happen to the system when one of the components is changed. Useful changes to consider are: increasing or decreasing the food intake (which relates to dieting in human beings), increasing the external work which is done (perhaps by exercise), or changing the temperature of the environment (which has implications for battery farming of animals).

Energy units

4(P) This question provides practice in using various units. You will need to use data tables to find information for some parts, and in some cases you will have to make estimates.

How much energy (in J) is provided by the following?

a A 1 kW heater running for one hour.

b Burning 1 litre of petrol (enthalpy of combustion of petrol $= 4.7 \times 10^7 \, \mathrm{J \, kg^{-1}}$, density $= 7400 \, \mathrm{kg \, m^{-3}}$).

c Burning 1 tonne of coal (10^6 tonnes of coal produce $8 \times 10^9 \, \mathrm{kW \, h}$).

d A 1000 MW power station in 1 hour.

e One kg of water falling through 100 m.

f A one tonne car in slowing from 80 km h^{-1} to rest.

g The total annual fuel consumption of the World (about 7000 Mtoe). (Use table G4, page 412.)

h A nucleus with a kinetic energy of 100 MeV coming to rest.

i One Btu (British thermal unit), which is the quantity of energy needed to increase the temperature of 1 lb of water by 1 °F (5/9 K). (You are expected to work this out, not just look up the answer in table G4.) (1 lb = 0.45 kg.)

5(L) This question tries to make clear the various ways in which the term 'energy slave' is used, taking domestic appliances as examples. Some of these use fuels just to do work, some for heating, and some for both.

a Make a list of all the machines used in your home for doing work (ignoring for the moment those just concerned with heating). How many are there? This is one use of the term – the actual number of machines which are available regardless of their power or how much they are used. Why may it be misleading to use this number?

b The second use of the term is to compare their power with that of a person. Add the power ratings of the machines (in W). How many 'slaves' is this equivalent to – assuming the average power output for useful work calculated in question **3**? How could this be misleading?

c The third use of the term is to compare the energy 'consumption' over a certain time with the useful work which can be done by a person in that time. Estimate the average daily energy 'consumption' of all these machines. How many slaves does this amount to? (The word 'consumption' is in quotes to show that it is not quite correct. What is wrong about it and what would be a better way of putting it?)

d Many of the appliances which use fuel in the home are used for heating as part or all of their function (for example, washing machines or boilers). Repeat part **c** after adding devices which are used for heating and compare your previous answer with that from part **c**.

Efficiency

6(L) According to the data in figure G4 (between pages 430 and 431):

a What is the *operating* efficiency of power stations in the U.K.?

b What is the efficiency of the electrical *transmission* system?

c What is the *overall* efficiency of the electrical supply system?

d What is the *overall* efficiency of oil refineries?

e What is the efficiency of the gas supply system?

f Many of the machines you listed in your answer to question **5** are electrical. What, roughly, is the average daily consumption of primary fuel needed to run these machines?

Primary, end-use, and functional energy

7(L)a Householders are concerned about the cost of the fuel which they need to buy in order to provide each kW h for heating their home. How much more expensive can electricity be than gas, for example, for the cost of each kW h of internal energy to be the same? (Use table G1, page 409.)

b On the other hand, society ought to be concerned about the quantity of primary fuel needed to provide the kW h of internal energy. Estimate this quantity for electricity and gas.

Growth

8(P)a Check whether the population of the World (table G8) has been growing exponentially and, if so, determine its doubling-time.

b Table G9 gives data for World fuel consumption over the same period. Is the growth in population sufficient, on its own, to account for the growth in fuel consumption? If not, what deduction can you draw?

Year	Population/10^9
1925	1.890
1950	2.505
1955	2.726
1960	2.990
1965	3.281
1968	3.484
1970	3.609
1971	3.678
1972	3.747
1975	4.258
1981	4.528
1984	4.763
1985	4.837
1986	4.917

Table G8
World population 1925–1984.

Year	Fuel consumption/10^{18} J	Year	Fuel consumption/10^{18} J
1925	47	1972	236
1929	55	1973	248
1937	59	1974	250
1949	73	1975	250
1950	81	1976	264
1953	92	1977	273
1955	105	1978	282
1958	118	1979	292
1960	131	1980	290
1963	146	1981	289
1965	165	1982	289
1968	192	1983	293
1969	206	1984	304
1970	217	1985	312
1971	226	1986	319

Table G9
Total World Fuel consumption since 1925.
Source of data: Based on BP Statistical Review of World Energy 1987 *for recent figures (since 1965), and various sources before that.*

9(L) Imagine a bacterium which breeds by dividing each hour. A colony of bacteria needs a volume of nutrient proportional to the number of bacteria present if all are to survive. One bacterium is placed in a jar of nutrient at the beginning of the day.

a Sketch graphs, using scaled axes, to show how the population and the breeding rate change with time.

b If the bacteria fill the jar after 24 hours, when will it have been half filled?

c If the bacteria were sensitive and intelligent when might they become alarmed at the idea that nutrient was becoming scarce?

d If they wish to continue to breed at the same rate after the first 24 hours, how much extra nutrient would be required for one hour's further life? And for another hour after that?

e If the original volume of nutrient had been underestimated, and in fact there was twice as much nutrient as had been thought, what difference would this make to the potential lifetime of the colony?

Questions **10** and **11** can be tackled by the incremental method and graph plotting technique used elsewhere in the course, or by means of a calculator or computer. If you have it available, the 'Dynamic modelling system' enables you to set up the equations and solve the problem very rapidly.

10(I) If you borrow money the interest charges are added to the original debt and the whole lot then attracts interest. If you neglect to pay back either the original debt or the interest the overall debt rises more rapidly than would be expected at first sight. Thus at a rate of interest of 2 per cent per month (a typical figure for a credit card) you might expect the doubling-time to be 50 months. How long is it in fact?

11(L) It is useful to know how the doubling-time (t_D) is related to the growth rate (g). If you are mathematically inclined you can find the relation quite easily by analytical methods.

Alternatively, repeat question **10** for growth rates of 4, 6, 8, and 10 per cent. (This is likely to be tedious to do unless a computer or calculator is available or the labour is shared.) Now process the results appropriately to find the relationship. (*Hint:* either plot a suitable graph or guess a relationship and check it by calculation.)

Making predictions

12(L) The term 'resources' is generally taken to mean the total of all the fuel stocks which could possibly exist. Since some of these are as yet undiscovered this figure must be an estimate. 'Reserves' or 'proved reserves' are the fuel stocks whose existence is based on much firmer evidence, such as geological surveys, drilling, etc. Reserves (or proved reserves) are thus a subset of resources. 'Recoverable reserves' are again a subset of proved reserves, since it may not be possible to extract all the proved reserves. Notice that the categories that are used in table G10 do not all correspond exactly with these.

G

Fuel	Quantity/ millions of tonnes	Equivalent/ Mtoe
Bituminous coal/ anthracite		
Proved reserves in place	775 470	519 565
Recoverable	488 333	327 183
Additional resources	3 928 222	2 631 909
Sub-bituminous coal/lignite		
Proved reserves in place	544 958	179 836
Recoverable	394 074	130 044
Additional resources	5 994 062	1 978 040
Peat		
Proved reserves in place	57 022	14 256
Recoverable	15 819	3 955
Additional resources	261 618	65 405
Oil		
Crude reserves	80 633	80 633
Oil shale	41 137	41 137
Bituminous sands	40 001	40 001
Natural gas	($/10^9$ m^3)	
Reserves	77 109	69 398

Table G10
Fuel resources and reserves.
Source: U.N. Yearbook of World energy statistics, 1980

a If the World were to continue to use the various fuels at the most recent rate listed in table G11, for how long could we be sure to have each fuel available? (For the purpose of this question you can add the recoverable reserves for the coals and for peat together.)

b Is there any evidence to suggest that consumption will continue at a constant rate?

13(L) Table G11 shows how the demand for different fuels changed over a 10-year period.

Consumption/Mtoe per year

Region	Oil	Natural gas	Coal	Hydro	Nuclear	Total 1986	Total 1976	% Change	(1986) Population /10⁶	(1986) Consumpt per capita /toe
N. America	813.9	459.5	471.1	157.2	129.7	2031.4	2009.8	+1.1	266	7.6
Latin America	215.4	71.2	22.1	82.2	1.5	392.4	263.4	+49.0	414	0.95
America – Total	1029.3	530.7	493.2	239.4	131.2	2423.8	2273.2	+6.6	680	3.6
United Kingdom	77.0	49.1	67.1	1.4	12.7	207.3	207.4	0	56	3.7
West Germany	119.6	40.2	77.1	5.0	27.3	269.2	259.6	+3.7	61	4.4
W. Europe – Total	585.1	192.5	250.0	101.1	138.6	1267.3	1209.6	+4.8	353	3.6
USSR	445.0	505.3	376.2	52.5	35.2	1414.2	1016.7	+39.1	281	5.0
China	99.2	12.1	531.2	28.2		670.7	487.9	+37.5	1069	0.63
Japan	204.4	36.4	70.2	19.6	41.4	372.0	344.7	+7.9	121	3.1
Africa	81.7	28.5	66.2	16.1	1.0	193.5	117.6	+64.6	572	0.34
Rest of World – Total	1266.6	783.9	1565.9	178.9	102.9	3898.2	2813.3	+38.6	3884	1.0
World Total 1986	2881.0	1507.1	2309.1	519.4	372.7	7589.3			4917	1.5
World Total 1976	2894.0	1145.7	1785.9	364.0	106.5		6296.1			
% Change	−0.4	+31.5	−29.3	+42.7	+250.0			+20.6		

Table G11
World primary energy consumption.
Note: in this table America (total) is equivalent to North America and Latin America; but Western Europe (total) is more than j UK and West Germany. Similarly. 'Rest of World' is more than the countries actually listed.
Source of fuel data: Based on BP Statistical Review of World Energy 1987.
Source of population data: See table G8.

a If the annual growth rate was one-tenth of the growth in ten years, what would be the doubling-time for the rate of consumption of each fuel?

b Is it reasonable to assume that the annual growth rate is the average rate over ten years? What are the problems in taking a shorter period'?

c Use the growth rates for various fuels from **a** to calculate how long the recoverable reserves are likely to last if growth continues at these rates

d What difference will the additional resources shown in table G10 mak to your answers? Assume that all the additional resources are proved but of course not all of these will be recovered. (Estimate this by assuming that the same proportion is recovered as in the table.)

e What, in general terms, would happen to the figures for recoverable reserves if the price of the fuel were to rise; or fall?

14(L) In question **9** it was assumed that the organism continued breeding at the same rate regardless of the impending lack of nutrient. In practice, the increasing scarcity of food would become steadily more apparent. This would increasingly inhibit the breeding rate.

The population–time graph might be something like the one in figure G24.

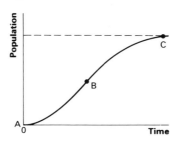

Figure G24

a What does the corresponding graph of breeding rate against time look like?

b What is the time corresponding to point B? (Conditions as in question **9**.) Is it 12 hours, somewhere between 12 and 22 hours, between 22 and 23 hours, or more than 23 hours? Why?

c What is the time corresponding to point C?

d What is the effect of the increasing shortage of food on the organism's life style or its 'intrinsic rights'?

Is there a crisis?

15(L) As fuels become more difficult to extract, the rate at which consumption grows will decrease.

a Sketch a graph showing how consumption (quantity of fuel used per unit time) varies with time for a constant growth rate of 2 per cent per annum. (Use a total time scale of about 4–5 doubling periods; it may help to refer back to question **10**.)

b On the same axes show how consumption varies with time when allowance is made for extraction becoming more and more difficult.

c What does the area under the consumption–time graph represent? It is possible to plot a range of such graphs with varying values of initial growth rate. What do you know about the total area under each one?

d If the maximum consumption is high what is the consequence for the lifetime of the resources?

e If we wish the limited resources of fossil fuels to last as long as possible when should we start reducing consumption?

G

16(E) At one time the number of horses in London was increasing exponentially and there were predictions that in a short time the whole city would be submerged in dung! Speculate first on the possible reasons why this did not happen, and then on whether there are parallels which can be drawn between that crisis and the present fuel crisis.

17(E) Make rough calculations of the following quantities, and answer any subsidiary questions:

a The area of the Earth's surface. (Radius of Earth ≈ 6370 km.)

b About two-sevenths of this is land – what is this area?

c What area of land would each person in the World have had in 1981 if it had been evenly shared out? (See table G8, page 434.)

d In fact, about one-fifteenth of the land area is cultivated. Grassland and forest account for a further quarter each. What accounts for the rest of the land area? What area would have been available per person in 1981 for food growing? Substantial numbers of people did not have enough to eat in 1981. This might suggest that the area available for food production is too small. Is this necessarily a reasonable conclusion? Why?

e What is the doubling-time for the World population? (See question **8**.) Which year is one doubling period after 1981; two doubling periods?

  What would be the area available for food growing per person on these dates? Eventually this is likely to be too small, with obvious effects. It is not possible to decide when this will be without further data but the trend should be clear.

18(E)a 'Developed agriculture' seems to produce more food per unit area than less developed agriculture. Which columns of table G12 provide this evidence? Suggest two reasons for this.

Column	1	2	3	4
Crop	Output/ $kW\,h\,ha^{-1}\,y^{-1}$	Ratio $\dfrac{energy\ output}{energy\ input}$	Output/ kg protein $ha^{-1}\,y^{-1}$	Input/kW h (kg protein)$^-$
Rice, Philippines	797.3	16.5	11.4	3.17
Taro-yam, Tsembaga	388.9	16.4	5.6	4.25
Corn grain, U.S.A.	21 390.6	2.6	481.0	17.25
Wheat, U.K.	15 612.4	3.4	400.0	11.67
Bread (in shops)	12 223.2	0.53	350.0	67.51
Potatoes, U.K.	15 806.8	1.57	376.0	26.67
Pigs in gardens, Tsembaga	861.2	2.1	62.0	11.39
Battery eggs, U.K.	1 944.6	0.14	137.0	98.06
Milk (average), U.K.	2 778.0	0.37	129.0	57.78

Table G12
Efficiency and productivity of some food producing systems. ('ha' = hectare = 10^4 m^2).
Source: FOLEY, G. and NASSIM, C. The energy question. Penguin, 1981.

b Calculate, roughly, the energy input per hectare-year for rice growing, for growing corn in the U.S.A., and for battery farming. Are your answers consistent with those to part **a**?

c Is it reasonable to say that developed agriculture converts oil to food rather than converting sunlight to food? If so what are the implications for food production of depleting fuel resources?

Fuel distribution and use

19(E) Examine the data for the U.K., North America, U.S.S.R., and Japan in tables G11 (page 436) and G13.

Region	Production/Mtoe					
	Oil	Natural gas	Coal	Hydro	Nuclear	Total
N. America	569.5	469.4	529.6	157.2	129.7	1855.4
Latin America	330.4	67.7	18.1	82.2	1.5	499.9
America – Total	899.9	537.1	547.7	239.4	132.2	2356.3
United Kingdom	128.6	38.8	65.8	1.4	12.7	247.3
West Germany	4.1	11.1	80.4	5.0	27.3	127.9
W. Europe – Total	198.2	155.5	199.0	101.1	138.6	792.4
USSR	613.0	617.4	377.8	52.5	35.2	1695.9
China	130.6	11.8	539.0	28.2		709.6
Japan	0.6	1.9	10.6	19.6	41.4	74.1
Africa	250.8	42.8	99.1	16.1	1.0	409.8
Rest of World – Total	1834.7	875.8	1601.4	178.9	101.9	4592.7
World Total	2932.8	1568.4	2348.1	519.4	372.7	7741.4

Table G13
Production of fuels by area 1986.
Source of fuel data: Based on BP Statistical Review of World energy 1987.

a Is each of these regions self-sufficient as far as total fuel requirements are concerned?

b What is the situation as far as individual fuels are concerned?

c What can be done in the short term to deal with any shortfall in supply?

d How is a continuing shortfall likely to affect decisions about fuel options?

20(E) Figure G12 (page 418) shows the relation between fuel use *per capita* and gross domestic product (GDP) *per capita*. It is sometimes claimed that the fuel use quoted ignores the fuel (food) consumed by the body and wood used for heating. The GDP allegedly ignores the food which is often the only product of less developed communities and which is ignored since it is used within these communities.
 Make sensible, but quick and rough, estimates of:

a the energy equivalent of the food consumed per head per year;

b the energy provided *per capita* per year by burning wood (assume that the energy equivalent for burning wood is one-third of that for coal);

c the value of the food produced per head per year; and predict how figure G12 would be changed if these quantities were taken into account. (Your answers to question **3** will give a reasonable guide to some of these answers.)

Characteristics of fuels

21(E) Consider a family (of four people) in the U.K. whose domestic fuel requirement is 10^{11} J per year.

a Assume for the moment that the requirement is entirely for heating.
Estimate what quantities of the following resources are required to provide for their needs. Would each be a feasible solution for the need of the U.K. population of, say, 55 million?
i Forest, assuming that temperate forests produce 700 tonnes of dry material suitable for fuel per km^2 per year. The energy equivalent of wood varies widely: assume it is $3 \, kW \, h \, kg^{-1}$.
ii Miners, assuming that each miner produces 500 tonnes of coal per year.
iii Area of solar panels, assuming that the average insolation (the measure of solar radiation falling on a unit area per unit time) is $10 \, MJ \, m^{-2} \, day^{-1}$ in the U.K. and the efficiency of solar panels is 40 per cent.

b Assume that it is feasible to use the resources in **a** to generate electricity. What difference, if any, would be made to your answers if the 10^{11} J were provided by electricity?

22(E) In which sector(s) in table G5 (page 419) could fuel most easily be saved? Would it be possible to save 20 per cent of the total?
If 20 per cent could be saved overall in the industrialized nations, what would be the effect on global fuel consumption, and on the lifetimes of fuel resources? (You will have to make some assumptions about the other changes which will be occurring at the same time – it is easiest to start by assuming that fuel consumption would otherwise stay constant.)

23(P) Many fuels are produced far from where they will be used. Table G14 shows the cost of moving them by various means.

Calculate the percentage energy loss in the following situations:

a transporting oil by pipeline from the North Sea to the U.K. (100 km);

b transporting coal from the Siberian coal fields to Moscow by heavy train (3000 km);

c transporting oil by supertanker from the Middle East, round the Cape to Western Europe (25 000 km);

d transporting oil by small tanker from the Middle East (via the Suez Canal) to Western Europe (15 000 km);

e transporting coal by small train from the U.K. coalfields to small, local power stations (80 km).

Method	Energy consumption/ kW h tonne^{-1} km^{-1}	Average speed/ m s^{-1}
Average pipeline	0.05	5
Barge	0.12	6
Cargo vessel	0.06	7
Supertanker	0.026	6
Local road delivery (petrol)	1.74	19
Road: 5 tonne diesel	0.81	25
Road: 20 tonne diesel	0.53	22
Road: 38 tonne diesel (articulated)	0.41	21
Rail: light train (1000 tonne)	0.25	33
Rail: heavy train (15 000 tonne)	0.10	21
Air: Boeing 707	6.86	225
Air: Boeing 747	4.88	230

Table G14
Energy needs in freight transport.
Source: RICHMOND, P. E. *and* FLOYD, T. (*eds*) Energy data sheets, *Southern science and technology forum, 1980.*
Original source: BOULADON, G. Institute Batelle, 1974.

24(L) Power stations are normally most efficient when running under full load. The variation in demand over a day means that there must be capacity to meet peak demand – but much of this will be out of use for most of the day. This is wasteful of capital equipment when it is standing idle and of the fuel needed to run the station up and down.

Assume that the demand shown in figure G13 (page 420) (upper line) is to be met from 'base load' stations which run for most of the time plus hydroelectric stations which provide for peak demand. The water for the hydroelectric stations is pumped into reservoirs at times when the base load station output is greater than demand. This is called a 'pumped storage system'.

a Estimate the 'base load' capacity which would allow this system to meet demand. How do you do this? What is the power output of the hydroelectric stations which will be required?

b If the water moves through 100 m difference in level, what volume of water would need to pass through the turbines each second to provide the peak power output? What area of lake 30 m deep would be required to provide the energy which the base load stations cannot provide? Is this a feasible arrangement?

c (*Hard*) The system considered in parts **a** and **b** would require more than half the 'base load' stations to be shut down in the summer. Would it be feasible to have a pumped storage system which kept base load stations running all the year round? (You will need to construct a possible demand curve over one year. For ease of calculation it should be a very simple shape. The values for system demand in figure G13 (page 420) can be assumed to apply to any year.)

25(P) Figure G15 (page 421) shows how demand for gas in the West Midlands Region of British Gas varied during three days at different times of the year. The North Sea gas producers require the supply rate to be nearly constant. Storage facilities must therefore be provided so that when the supply rate is greater than the demand, gas is stored, to be released when the opposite condition prevails. Some of this capacity is provided in a large storage ring main round the region consisting of 142 km of 0.6 m diameter pipe and 36 km of 0.75 m pipe. The minimum operating pressure is about 25×10^5 Pa ($N\,m^{-2}$) and the maximum normal operating pressure is about 55×10^5 Pa.

a What is the volume of the ring main?

b If gas is extracted from the main without being replaced what will happen to the main pressure?

c What volume of gas, measured at one atmosphere (10^5 Pa), would be needed to raise the pressure of the main to its minimum operating level?

d The uppermost line in the graph shows the most severe conditions for the system. From this graph determine roughly what volume of gas is available to be stored during the period when the supply rate is greater than demand.

e What pressure increase would storing this gas produce in the main? Is this acceptable? If not what proportion of the stored gas must be stored elsewhere?

26(E) Estimate or find out the fuel consumption of London and its area. What proportion of the insolation is this? (See question **30**.) What effect do you think the fuel consumption may have on the temperature in London? Does the evidence bear this out? Compare the ratio of fuel consumption to insolation for London with the average ratio for the Earth as a whole calculated in question **30**.

Population growth and rising expectations

27(E) How does the growth rate in population (question **8**) compare with the growth rate in fuel demand (question **13**)? Is it likely that the increase in fuel demand is due simply to rising population?

28(E) By what factor (approximately) would fuel demand rise on a global scale if *per capita* fuel consumption were to rise to Western European standards?

What would be the annual growth rate in demand due to this effect alone if this increase were to be achieved over 30 years?

29(E) (*Hard*) Using your answers to question **27** and **28**, comment on the combined effect of increasing population and increasing expectations on the growth in demand for fuels in the past. Is there evidence in table G11 (page 436) for expectations increasing faster in developing rather than developed countries?

30(E) The solar energy flux in space near the Earth is $1.4\,\text{kW m}^{-2}$. Absorption in the atmosphere reduces this on average by 50 per cent.

Estimate the total solar energy delivered at the Earth's surface in one year. You need to take account of the day–night effect and of the fact that not all of the Earth's surface is at right angles to the radiation. (This can be done without using complicated methods.) (Radius of Earth $\approx 6370\,\text{km}$.)

Compare your answer with the total global fuel consumption in the same time.

This may suggest that solar power is the solution to all our energy problems. Section G4 may 'throw some light' on this.

Energy from fission

31(L) Some idea of the energy available from the fission of a single nucleus of uranium can be obtained by considering either the interchange of mass and energy (using $\Delta E = c^2 \Delta m$) or the change of electrical potential energy as the fragment nuclei move apart.

This is just one example of fission:

$$^{235}_{92}\text{U} + ^{1}_{0}\text{n} \rightarrow ^{236}_{92}\text{U} \rightarrow ^{96}_{37}\text{Rb} + ^{138}_{55}\text{Cs} + 2^{1}_{0}\text{n} + \text{energy}$$

a Look up the atomic masses of ^{235}U and the neutron, preferably expressed in unified mass numbers, u. (Atomic mass of ^{96}Rb is 95.932 58 u; atomic mass of ^{138}Cs is 137.919 50 u.)
i Nuclear masses are needed for the calculation. Is it necessary to take into account the masses of the electrons in this case? (*Hint:* how many electrons would feature on each side of the equation?)
ii Find the change of mass per fission in kg.
iii Use $\Delta E = c^2 \Delta m$ to obtain a value for the energy released by this process.

b Assume that the total kinetic energy of the two fission fragments is equal to the loss of electrical potential energy as the two nuclei move apart.
i Write down the expression you would use to calculate the value of the electrical potential energy when the two nuclei are the diameter of a uranium nucleus apart (take this as $1 \times 10^{-14}\,\text{m}$).
ii Assume that they move apart to a large distance so that all this potential energy is transformed to kinetic energy. Calculate the kinetic energy of the fission fragments.
iii Speculate on possible reasons for the difference between this answer and that for part **a***iii*.

c *i* Taking the value of 5×10^{-11} J per fission, calculate how much energy would be released by the fission of 1 kg of uranium-235. (You will need to calculate how many nuclei there are in one kilogram and to assume that they all disintegrate.)

ii How does this value compare with that for burning coal if about 1×10^{14} J are released by burning 4 000 tonnes of coal?

iii The 'natural abundance' of ^{235}U is 0.7 per cent. How much natural uranium metal would contain 1 kg of ^{235}U?

iv Assume that 0.1 per cent of uranium ore is uranium. How much ore would have to be mined to produce 1 kg of ^{235}U?

v Compare the masses of material which have to be mined to produce equivalent quantities of energy from fission of ^{235}U and burning coal.

Cross-section

32(L) Some idea of what is meant by 'nuclear cross-section', and how it might be measured for a particular element, can be obtained from the following simplified argument.

a Imagine a tube of the reactor core containing only one neutron, moving parallel to the tube axis, and one uranium target nucleus located somewhere in the tube (figure G25). Write down the chance of the neutron hitting the nucleus. (Assume that the neutron itself is of negligible size.)

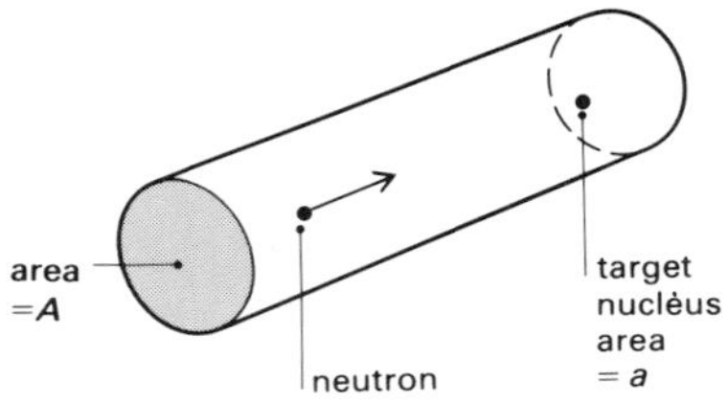

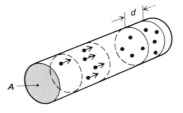

Figure G25
One neutron's chance of hitting a uranium nucleus.

area = A

neutron

target nucleus area = a

b If the tube now contains n neutrons per cubic metre, travelling at speed v m s^{-1} how many neutrons could reach the target per second?

c If a thin slice of the tube, thickness d, contains N target nuclei per cubic metre, how many targets are there – and what is their total cross-sectional area? (Assume there is no 'shadowing'.)

Figure G26
Many neutrons' chances of hitting target nuclei in a 'slice' of metal.

d

A

d What is the chance of a neutron hitting one of these target nuclei in one second?

e If R is the number of interactions (collisions) per second per unit volume of the target slice, obtain an expression for R in terms of a, n, N, and v.

f Now write an expression for the effective target area of a nucleus, a, usually called the nuclear cross-section.

g What sort of measurements would you need to make in order to calculate the nuclear cross-section for an element?

Fission of ^{235}U

33(R) The energy of a molecule in a gas at temperature T (kelvin) is, according to the kinetic theory of gases, $3kT/2$, where k is Boltzmann's constant. What is this energy in eV for a gas at 800 K? This is also the energy of a 'thermal neutron' at this temperature. How does it compare with the energy of a neutron released in fission of ^{235}U? (See table G7, page 426.)

34(L) Figure G21 (page 429) shows how cross-sections of two nuclei vary with the energy of the neutrons which interact with them. This question is intended to show that the best conditions for fission of ^{235}U occur when the neutron energy is below about 0.1 eV.

a The neutrons arising from fission have energies of about 2 MeV. This is marked on the graph. (Notice that the neutron energy scale is logarithmic.)
i Read from the graph the cross-sections for neutrons with this energy for fission and for inelastic scattering with ^{238}U, and for fission with ^{235}U (notice that this scale too is logarithmic).
ii The figures give the relative probability of these interactions assuming that the nuclei are equally represented. For ease of calculation assume that there are one hundred times as many ^{238}U nuclei as ^{235}U nuclei. Obtain figures for the relative overall probabilities of the interactions.
iii In a similar way, find the cross-sections and the relative overall probabilities of fission of ^{235}U and capture by ^{238}U, for neutron energies of 10 keV, 100 eV, 1 eV, and 0.01 eV.
For which of these energies is a chain reaction most likely?
iv The uranium in thermal reactors has between 3 per cent (enriched) and 0.7 per cent (natural) of ^{235}U. What is the ratio of ^{238}U to ^{235}U for these extremes? Do these figures make any significant difference compared with the value of 100 used in parts *ii* and *iii*?

b While the neutrons are slowing down there is a substantial chance of them being captured by ^{238}U.
i How should the fuel elements and moderator be arranged in order to minimize the chance of capture by ^{238}U?
ii How might you explain the form of the ^{238}U cross-section curve between 10 and 200 eV? (*Hint:* critical potentials.)

Moderation and control

35(R) The neutrons emitted during the fission of ^{235}U have an energy of about 2 MeV and have to be 'slowed down' by a moderator for maximum probability of fission.

a Even if 3 neutrons are emitted (the average is about 2.5) this leaves most of the 200 MeV released on fission unaccounted for. What is the most likely form for the rest of the energy released?

b What happens to the highly-charged fission fragments and their energy?

c How are α-particles slowed down in a cloud chamber? Can we use that mechanism for neutrons? If not, why not?

d What process can be used to slow down the emitted neutrons from 2 MeV to about 0.02 eV needed for the best chance of fission?

(Questions **36** to **39** are about the details of this process.)

36(L) (*Hard*) A steel ball, A, of mass, m, and initial kinetic energy E_k has a head-on elastic collision with another ball, B, of mass M, initially at rest. Show that ΔE_k (the kinetic energy lost by A) is given by

$$\Delta E_k = \frac{4mM}{(M+m)^2} E_k$$

A more difficult problem is to show that this expression becomes

$$\Delta E_k = \frac{2mM(1-\cos\theta)}{(M+m)^2} E_k$$

if ball A is scattered through angle θ. Does this expression give sensible answers for glancing ($\theta = 0$) and head-on ($\theta = 180°$) collisions?

37(L)a For a neutron (mass m) colliding head-on with a nucleus of mass M assume that

$$\Delta E_k = \frac{4mM}{(M+m)^2} E_k$$

and derive an expression for the fractional energy loss of the neutron $\Delta E_k/E_k$.

b Taking the neutron's mass m to be 1 u, for what value of M does the neutron lose most energy?

c Draw up a table showing percentage energy lost by a neutron in collision with nuclei of masses 1, 2, 10, 12, 16, 112, and 238 u. Which elements are likely to have these nuclear masses?

d Which element seems to be the best moderator?

38(P) How many head-on collisions would be needed for a 2 MeV neutron to be slowed down to 0.05 eV (thermal energy) if graphite were used as a moderator?

(*Hint:* if the energy loss is 33 per cent then the energy of the neutron after one collision is $\frac{2}{3} \times 2 \times 10^6$ eV; after two collisions it is $\frac{2}{3} \times \frac{2}{3} \times 2 \times 10^6$ eV $= (\frac{2}{3})^2 \times 2 \times 10^6$ eV. After n collisions it is $(\frac{2}{3})^n \times 2 \times 10^6$ eV.)

The actual number of collisions required is about 120. Suggest why your answer is less than this.

39(E) Steel balls always separate after a collision, but this is not so for neutrons and nuclei. The neutrons can be absorbed or captured to become part of the nucleus. The probability of capture is expressed as the neutron capture cross-section. Similarly, the probability of 'bouncing-off' a nucleus is expressed as the scattering cross-section.

a What cross-sections should a good moderator have?

b Choose two candidates for good moderators from table G15 and explain your choice. (The data relate to thermal neutrons.)

Nucleus	H	D	B	C	O	Cd	U
Scattering cross-section $\sigma_s/10^{-28}$ m^2	20.436	3.390	3.6	4.75	3.76	5.6	8.90
Absorption cross-section $\sigma_c/10^{-28}$ m^2	0.332	0.00053	759	0.0034	0.00027	2450	7.59

Table G15
Scattering and absorption (capture) cross-sections of various nuclei for thermal neutrons.

c If your choice does not agree with your answer to question **37d** explain why.

d Which of these elements would make good control rods?

Enrichment

40(L) Energy is released in the core of a particular reactor at a rate of 1000 MW. If each fission of uranium 235 produces 200 MeV:

a Calculate the number of uranium nuclei disintegrating per second.

b What mass of uranium 235 is 'used' per second?

c Estimate the mass of uranium 235 in the core if one-third of the fuel is changed each year.

d Assume that new fuel is enriched to contain 3 per cent of ^{235}U and that spent fuel contains only 0.8 per cent of ^{235}U. Estimate the total mass of the uranium in the core.

e Give reasons why a nuclear power station with an output of 1000 MW to the national grid would have a larger core than your estimate.

G

SECTION G3
CONSERVING FUEL IN THE HOME

The word 'conservation' is used in two different but linked senses i connection with energy. One has already been used earlier in this Unit the physical law that says that energy cannot be created or destroye but can be transformed from one form to another.

The other meaning is more colloquial, using as little as possible of precious resource – in this case fuel. It is mainly this second meanin with which we will be concerned in this Section.

It is important to realize that we must keep on heating a house if w want the temperature indoors to remain higher than the temperatur outside, even when we have warmed it to the temperature that we wan There is a flow of energy (associated with the temperature differenc between inside and outside) through the walls, roof, windows, etc., an this energy has to be replaced in order to maintain the insid temperature. Once a steady temperature has been reached inside, th total rate of loss of energy from the house is equal to the rate at which i is being supplied inside. This energy is supplied by the central heatin system or fires and heaters, as well as by other sources, such as cooker lights, people, sunshine.

The following questions deal with a few of the things that must b considered when thinking about the flow of energy out of a house. Th first question shows how the flow equations can be built up.

Thermal resistance

41(L) Thermal flow of energy through a block of material can be thought of as similar to flow of electric charge through a wire. Two similar situations are shown in figures G27 and G28.

a What is the thermal equivalent of flow of charge?

b In figure G27 we might say that the potential difference drives current through the wire. What is it that drives energy through the block?

c The wire obeys Ohm's Law, which can be written as:

Potential difference $\propto$ rate of flow of charge (provided the conditions remain the same).

i Rate of flow of electric charge may of course be referred to as 'current'. Write down its units in two different ways.
ii What would be the analogous relationship for the block? Write down the unit for the righthand side of the relation in two different ways. What single word could replace the phrase on the righthand side of the relationship?

d What experiment could you do to try to justify the relationship for thermal flow of energy in c*ii*? In practice there are a number of

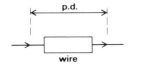

Figure G27

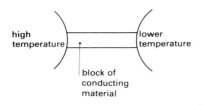

Figure G28

difficulties with this sort of experiment. What difficulties can you think of and how would you try to overcome them?

e For a wire we can write:

$$V = IR$$

where V is the potential difference across the wire, I is the current through it, and R is the wire's electrical resistance.

For the block the analogous equation is:

$$\Delta\theta = \phi\mathscr{R}$$

where $\Delta\theta$ is the temperature difference across the block (*cf.* potential difference), ϕ is the rate of thermal flow of energy (power, *cf.* current), and $\mathscr{R}$ is the thermal resistance of the block. What are the units of $\mathscr{R}$?

The idea of thermal resistance is important. The greater the thermal resistance of the walls and roof of a house, the smaller the rate of loss of energy from that house, and so the lower the rate at which energy must be supplied to maintain a given difference between the inside and outside temperatures.

f In order to compare the electrical resistance of different materials we use the resistivity, ρ, which is the constant of proportionality in the equation:

$$R = \frac{\rho l}{A}$$

where R is the resistance of a wire of length l, and cross-sectional area A.

What are the units of ρ?

g Sometimes electrical conductivity, σ, is used instead of electrical resistivity, ρ. The equation for electrical resistance can then be written:

$$R = \frac{l}{\sigma A}$$

For thermal flow of energy the thermal conductivity, k, of a material is often used rather than its resistivity. Since resistivity $= 1/$conductivity the equation for thermal resistance becomes:

$$\mathscr{R} = \frac{l}{kA}$$

What are the units of k?

The following questions use the equations:

$$\mathscr{R} = \frac{l}{kA}$$

$$\Delta\theta = \phi\mathscr{R}$$

G

42(P) A furnace wall, 0.5 m thick, is made from brick of thermal conductivity $0.8\,\mathrm{W\,m^{-1}\,K^{-1}}$. If the internal temperature is 850 °C and the outside temperature is 30 °C find:

a The thermal resistance of a section of wall of area one square metre.

b The power loss through this section of wall.

43(L) Consider a brick wall measuring 4 m by 3 m containing a single-glazed window 1 m by 1.5 m. If the thickness of the brick is 0.1 m and the thickness of the glass is 4 mm, find:

a The thermal resistance of the glass. ($k_{glass} = 1\,\mathrm{W\,m^{-1}\,K^{-1}}$.)

b The thermal resistance of the brick. ($k_{brick} = 0.6\,\mathrm{W\,m^{-1}\,K^{-1}}$.)

c If the outside temperature is 0 °C and the inside temperature is 20 °C what is
  i the power loss through the brick?
  ii the power loss through the glass?
  iii the total power loss through the wall and the window?

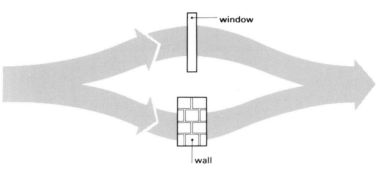

Figure G29
Paths of thermal flow of energy through a window in a wall.

Notice that some energy goes through the window and *other* energy goes through the wall. Hence the brick wall and glass window represent thermal conductors in parallel.

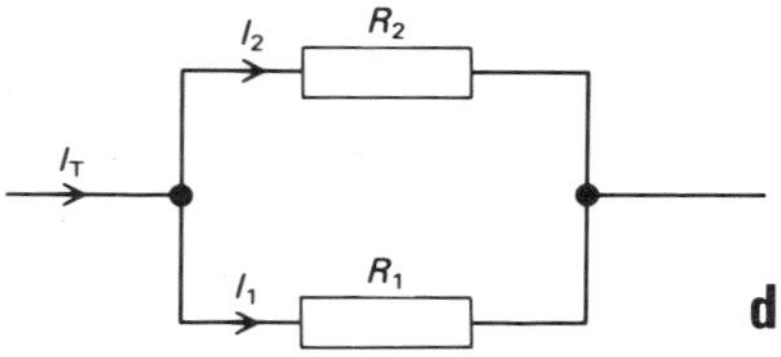

Figure G30
Electrical analogue of a window in a wall.

d The electrical equivalent would be as shown in figure G30.
  i What is I_T in terms of I_1 and I_2?
  ii How does this compare with your calculation of the total power loss through the wall and window in part **c**?

44(P) The brick wall in question **43** is now insulated with a 10 mm layer of fibre board of thermal conductivity $0.05\,\mathrm{W\,m^{-1}\,K^{-1}}$.

a Calculate the thermal resistance of the fibre board.

b Using the analogy with electricity calculate the new power loss through the brick and fibre board section of the wall.

45(P) A cylindrical hot water tank is 0.5 m in diameter and 0.8 m high and made from copper 3 mm thick.

a If the water inside is at 70 °C and the outside temperature is 20 °C calculate the rate of loss of energy from the tank. (Thermal conductivity of copper $= 385 \, \text{W m}^{-1} \, \text{K}^{-1}$.)

    (*Note.* The answer to this question will seem very large. The reason for this will become apparent later.)

b The tank is now insulated with a glass fibre quilt, 20 mm thick and of thermal conductivity $0.04 \, \text{W m}^{-1} \, \text{K}^{-1}$. Estimate the rate of loss of energy now.

c Since the thermal conductivity of air is $0.024 \, \text{W m}^{-1} \, \text{K}^{-1}$, which is less than that for the glass fibre quilt, suggest why it is beneficial to replace air with glass fibre. (Don't spend long on this now since it reappears later.)

Thermal resistance coefficient, X

Heating engineers find it convenient to use a quantity which is a property of unit area of a particular thermal conductor (for example, standard cavity brickwork or single-glazed window). One such quantity is the 'U-value' of the wall or window, which is the rate of thermal flow of energy per square metre for a temperature difference of one degree. If we think in terms of thermal resistance instead of thermal conductivity we can define an alternative property – the 'thermal resistance coefficient', X. This is the thermal resistance of unit area of the material. The area which matters in each case is, of course, the area perpendicular to the flow of energy. Because of the analogy between thermal and electrical resistance it is convenient to use thermal resistance coefficients rather than U-values in this course.

46(L)a If the thermal resistance coefficient of one square metre of a conductor is X, what is the thermal resistance $\mathscr{R}$ of 2, 3, or A square metres?

b Hence write down the relation between $\mathscr{R}$, X, and A.

c What are the units of X?

d By substituting for $\mathscr{R}$, obtain a relation between X, l, and k.

47(L)a What is the thermal resistance coefficient of a brick wall 0.1 m thick if the thermal conductivity of brick is $0.6 \, \text{W m}^{-1} \, \text{K}^{-1}$?

b What is the thermal resistance coefficient of a timber wall 20 mm thick if the thermal conductivity of timber is $0.15 \, \text{W m}^{-1} \, \text{K}^{-1}$?

c As far as thermal insulation is concerned which of the two walls above would be better?

48(P)a Calculate the thermal resistance coefficient for 4 mm thick glass and hence calculate the thermal resistance of a single-glazed window $2 \, \text{m} \times 1 \, \text{m}$ with glass 4 mm thick. ($k_{\text{glass}} = 1 \, \text{W m}^{-1} \, \text{K}^{-1}$.)

b What is the power loss through this window when the outside temperature is 0 °C and the inside temperature is 20 °C?

c Using your general knowledge of the heating requirements of rooms, say whether your value is about right, much too high, or much too low.

Surface resistance

If the power loss found in the previous question were correct our heating systems would need to be much more powerful than they are.

HOME EXPERIMENT GH2
Surface layer

As you may know from experience the surface of a window pane is not at the general temperature of the mass of air near to it. In fact, the temperature variation near the window is rather as shown in figure G31.

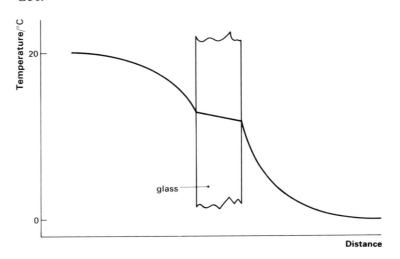

Figure G31
Temperature variation near a single-pane window.

Near each glass surface there is a surface layer of air with a temperature drop across it. This surface layer presents a resistance to the energy which must flow through it from the interior to the exterior. This resistance is called the surface resistance.

The way the temperature varies in the surface layer and the distance over which it varies depend on a number of factors, so it is usual to use measured values of surface resistances rather than to try to calculate them. Table G16 lists some values which are simpler than those which a heating engineer might use, because they ignore differences which might be produced by the textures of different surfaces (glass or brick), for example.

49(E) Suggest some factors which may affect the value of the surface resistance coefficient. Examination of table G16 should help with some of these, for instance compare **A** and **B**; **D** and **E**; and **C** and **D**.

		Surface resistance coefficient/m^2 K W^{-1}
A	Internal wall or window	0.13
B	External wall or window	0.06
C	External roof	0.05
D	Internal floor/ceiling (upward energy flow)	0.11
E	Internal floor/ceiling (downward energy flow)	0.15

Table G16
Some common values of surface resistance coefficients.

50(E)a Do the thermal resistance of a glass window and its surface resistances act in series or in parallel?

b Calculate the total thermal resistance coefficient for the window in question **48**. By what multiple, approximately, is the power loss calculated in question **48** changed now? *Note:* the values of surface resistance coefficients for walls and windows are not usually the same but you should assume that they are for the purpose of the question.

c You can now use the electrical analogy to calculate the temperatures of the surfaces of the glass. (This calculation assumes a linear variation of temperature with distance. Figure G31 shows that this is not strictly so. The answers to this and similar questions must therefore be approximations.)

Cavities

Cavities between layers of brickwork (cavity walls) have been used in buildings for some time, though cavities between layers of glass in windows (double glazing) are more recent. Double glazing has the express purpose of reducing energy flow. Cavity walls were introduced to prevent water penetration but may have the added benefit of reducing energy flow. These next questions examine how effective cavities are in doing this.

Key

Hor ↑ means a horizontal cavity with energy flow upwards.
Similarly Hor ↓ means a horizontal cavity with energy flow downwards.

Faced means that one face of the cavity is covered with aluminium foil.

Orientation	Surface	Resistance coefficient/m^2 K W^{-1} Width of cavity d/mm			
		5	20	60	100
Vert	unfaced	0.11	0.18	0.18	0.18
Hor ↑	unfaced	0.11	0.15	0.16	0.16
Hor ↓	unfaced	0.11	0.18	0.21	0.22
Vert	faced	0.18	0.57	0.57	0.57
Hor ↑	faced	0.18	0.35	0.42	0.43
Hor ↓	faced	0.18	0.63	1.45	1.58

Table G17
The thermal resistance coefficients of cavities for various orientations, surface finishes, and thicknesses.

51(L)a Using the same axes, plot graphs of thermal resistance coefficient against cavity thickness for the different cavity orientations and linings (table G17).

b On the same axes plot a graph showing how you would expect the thermal resistance coefficient of air to vary with thickness if it behaved like a slab of insulator ($k_{air} = 0.024$ W m^{-1} K^{-1}).

c What would be the thermal resistance coefficient of the cavity if it acted simply as two internal surfaces? (Use table G16.) How would this value vary with the thickness of the cavity? Plot this information on the same axes.

d Some patterns should now be apparent. Make some brief notes to describe them.

e Under what conditions does the cavity behave most like a slab of conductor?

f Compare the graphs for faced and unfaced cavities and say what effect lining one face of the cavity with aluminium foil has on the thermal resistance coefficient. Would this happen if the only mechanism for transferring the energy were conduction? Could another mechanism, such as convection or radiation, account for these differences in the graphs, and is this mechanism an important one? Explain your answers.

g Compare the graphs for horizontal cavities with upward and downward energy flows and suggest a mechanism of energy transfer which could account for the differences. How important is *this* mechanism?

h A cavity could be regarded as just two internal surfaces. Does a cavity behave like this?

i How do the measured values of thermal resistance coefficient compare with the value calculated for conduction only?

j If each mechanism of energy transfer is represented by its own resistance, how should the resistances be connected in order to produce this result?

Double glazing

It should now be clear that air in a cavity behaves quite differently from the same thickness of a solid, largely because it is a fluid in which convection and radiation take place as well as conduction.

Analysing such a system in detail would be very difficult because there are many factors which can affect the value of the cavity resistance. Fortunately, in practice, a limited number of types of cavity is used and it is possible to specify a definite cavity resistance for each situation.

Figure G32 summarizes the coefficients needed for calculations of thermal energy transfer across cavity systems. (There may, of course, be more than one cavity in the system, or none at all.)

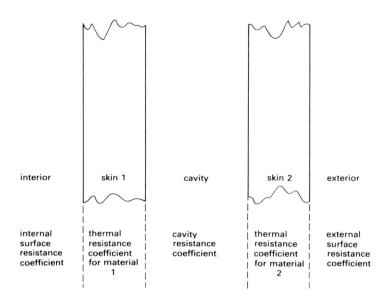

interior | skin 1 | cavity | skin 2 | exterior

internal surface resistance coefficient | thermal resistance coefficient for material 1 | cavity resistance coefficient. | thermal resistance coefficient for material 2 | external surface resistance coefficient

Figure G32
Summary of resistance coefficients required for calculating energy flows through cavity systems.

52(P) In questions **48** and **50** you calculated the power loss through a 2 m × 1 m glass window, first without taking surface layers into account, then with. Now imagine that the window is double-glazed. Assume that the cavity resistance coefficient for a 15 mm air gap is $0.14 \, \text{m}^2 \, \text{K} \, \text{W}^{-1}$.

a What other resistance coefficients need to be considered in describing this system?

b Calculate the power loss through the double-glazed window.

c Use the electrical analogy to find the temperature differences across all the resistances in the system.

d Draw a labelled graph, similar to figure G31 (page 452), to show how temperature varies with distance near this window.

e Is it permissible, using this analogy, to calculate the temperature at a distance, say, 3 mm from the outer surface of the window? Explain your answer.

53(E) Explain why a heavy curtain can increase the total thermal resistance of a single-glazed window by a significant amount.

54(E) Why do some householders have the cavities in the walls of their houses filled with glass fibre, foam, or insulating beads? Is this consistent with what you have learned?

You should now have a clearer understanding of why a glass fibre quilt is a better insulator than just the air around a hot water cylinder. (See question **45c**.)

55(E)a When answering question **49** did you suggest exposure as one of the factors which might affect the value of surface resistance coefficient?

b Using values you have already met, calculate the resistance coefficient you would expect for at least one of the types of window listed in table G18. Compare your answer with those in the table.

c The figures will probably not match exactly because you have ignored yet another factor – which? (*Hint:* does wood appear to be a better or worse conductor than glass?)

d If you were employed to advertise double glazing, how would you design a test window to produce results which, while correct, would present double glazing most favourably?

		Total thermal resistance coefficient/m^2 K W^{-1}		
		Exposure		
Window construction		*sheltered*	*normal*	*severe*
Single glazed	wood frame	0.26	0.23	0.20
	metal frame	0.20	0.18	0.15
Double glazed	wood frame	0.43	0.40	0.37
	metal frame	0.33	0.31	0.29

Table G18
The total thermal resistance coefficients (including surface and cavity resistances where appropriate) of single- and double-glazed windows under different conditions.

56(P) A cavity wall measures $2\,m \times 6\,m$ and has a double-glazed window in it measuring $1.5\,m \times 3\,m$. Use the following data to answer the questions below.

External temperature $= 2\,°C$
Internal temperature $= 18\,°C$
External surface resistance coefficient $= 0.06\,m^2$ K W^{-1}.
Internal surface resistance coefficient $= 0.13\,m^2$ K W^{-1}.
Cavity resistance coefficient for the window $= 0.14\,m^2$ K W^{-1}.
Thermal conductivity of glass $= 1\,W\,m^{-1}\,K^{-1}$.
Thickness of glass $= 4\,mm$.
Resistance coefficient for cavity brickwork of overall thickness 260 mm $= 0.67\,m^2$ K W^{-1}.

a Draw a diagram (in a similar way to an electrical circuit) showing all the thermal resistances, whether they are in series or parallel, and their values.

b Find the total thermal resistance of the brick section of the wall.

c Find the energy flow through the brick section.

d Find the temperature on the inside surface of the brick section.

e Find the total thermal resistance of the glass section of the wall.

f Find the energy flow through the glass section.

g Find the temperature on the inside surface of the glass section.

h Suppose the air inside the room became very humid. Use your answers to **d** and **g** to say whether you would expect condensation on the brick or glass first. Explain your answer.

57(P) Building regulations specify that the average value for the resistance coefficient of an external wall should be at least $1.6\,\mathrm{m^2\,K\,W^{-1}}$. (This includes surface resistance.) Consider a wall which just satisfies this requirement and comprises a brick section and a window section. The brick section has a total resistance coefficient of $2.0\,\mathrm{m^2\,K\,W^{-1}}$. The window is of a wooden-framed, double-glazed type which has a total resistance coefficient of $0.4\,\mathrm{m^2\,K\,W^{-1}}$ under normal conditions. If the total area of the wall including the window is $20\,\mathrm{m^2}$, what is the area of the window?

The heating of buildings

58(L)a Suggest suitable temperatures for points A and B in figure G33.

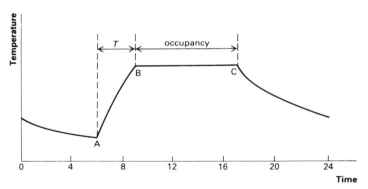

Figure G33
The temperature variation inside a building against time during a 24-hour period. The heating system is switched on at A and off at C.

b Compare the rate of loss of energy from the building at A and B.

c What physical factors concerning the construction of the building will affect the time lag, *T*?

d What is the disadvantage of constructing a building with a large time lag?

e We expect the temperature to continue to rise at B. This may be prevented by opening windows. Suggest a more economical way of achieving a constant temperature. Would your suggestion have any disadvantages?

f Redraw the graph to show how the temperature in the building would change if the heating system remained on at full power from A to C and no attempt was made to ventilate the building.

g Draw a graph to show how the temperature inside a lightly constructed hut, without ventilation, would change if a powerful electric heater were to be switched on.

h Suggest some of the factors an architect should consider when designing a building in order to heat it economically.

59(P) An occupied room requires ventilation. Calculate the power needed to warm air (from $0\,°C$ to $20\,°C$) in a room $5\,m \times 6\,m \times 3\,m$ which has two complete air changes per hour. Compare this with the loss of energy through a closed window as calculated in questions **50** and **52**. (Heating capacity of air per unit volume $= 1.3\,kJ\,m^{-3}\,K^{-1}$.)

60(R)a It is recommended that the air change in a school should be $0.014\,m^3\,s^{-1}$ for each person in the room. Estimate the power required to maintain the air temperature in winter in your classroom, with this rate of ventilation.

b If each person converts the energy from food at a rate of $100\,W$, what percentage of the power needed is supplied by the occupants?

SECTION G4
ENERGY OPTIONS

In this course you can only touch on the large amount of information available on energy sources. But while you cannot study it all you ought to know something about its range and be familiar with at least part of it. This is what this Section is designed to achieve.

You are expected to find information by reading, and later report on your findings in some suitable way. To help you get started, each major option (for example, solar energy) has been divided into a number of smaller topics (passive devices; active devices; photovoltaic devices; biomass; wind energy). Some study items are suggested for each topic; you should tackle these items at least.

Part of the skill of reading for learning is to generate new questions (Do I understand this? What is it I don't understand? What is X (or Y or Z) that is referred to? Is it important? Where can I find out more about it?). Another important skill is to recognize new topics and to appraise their importance for you. You should remember that this is not an exercise in copying out large sections of material from your sources into your report. Some suitable articles have been included in the background books (*Energy sources: data, references, and readings* and *Energy options: a reader*), but you should not limit your search to these. They are there, like the study items, to help you to get started.

Reporting back on your findings helps others to broaden their knowledge; it also shows you how well you understand your topic. Here are some suggested ways of reporting your findings.

Give a short verbal report: not a very original idea perhaps, but it is effective if well prepared.

Give out a short duplicated report. You will probably also need to speak about the report or be able to answer questions on it.

Both of these would effectively be articles, written by you, presenting the most important findings. Much of the detail could be organized into appendices which you could keep for reference purposes.

You might make slides of appropriate features (wind turbines, power stations, etc.) or collect illustrations and use these as the centrepiece of your talk. This might even be extended to the production of a tape–slide sequence suitable for use in general studies, or for a younger class.

Display your report on a noticeboard alongside all the others from your group. This way of communicating is often used at scientific conferences and is called a 'poster session'. It is especially important to present the report well and highlight the main points clearly.

STUDY ITEMS

To save repeating items in each option, here is a list of the characteristics of energy sources discussed in Section G1 and which should be

considered in connection with each option (except for the few wh[e]
they are specifically excluded).

1 *Intensity or energy density* – how intense is the energy source? Wh[a]
implications does this have for its use?

2 *Time* – is the option available at any time it might be required? If no[t]
what limitation does this place on its use?

3 *Transportability* – is the energy source available everywhere, is it easil[y]
transported, or is it limited to particular places? What implications doe[s]
this have for its use?

4 *Feasibility* – can the option be implemented at once? Alternatively doe[s]
it need new technology; or does it need a large input of money or energ[y]
before it can be implemented; or do people need to be persuaded abou[t]
its value? What are the implications for its use?

To help to remind you that these four factors should be considered fo[r]
most options, the other study items for each topic are numbered from [5]
onwards. You may not find it sensible to tackle the study items i[n]
numerical order. Try reading them all through to start with, but com[e]
back to them every so often.

Finally, since the topics vary considerably in length and difficult[y]
you may well need to attempt more than one.

NUCLEAR POWER

A Nuclear power stations

Figure G34
The UKAEA Prototype Fast Reactor at
Dounreay, Caithness, Scotland.
United Kingdom Atomic Energy Authority.

5 Choose one particular type of reactor and study it in detail. There ar[e]
many types to choose from, including: Prototype Fast (PFR[)]
(figure G34), Advanced Gas-cooled (AGR) (figure G35), Pressurize[d]
Water (PWR), and Fast Breeder (FBR). How is energy extracted from
the reactor you have chosen to study?

6 How are the principles described in Section G2 put into practice?

7 What is the moderator (if any), the coolant, the fuel? How are these arranged in the reactor?

8 What are the operating conditions and the efficiency?

9 What are the safety requirements?

10 What are the World's resources of suitable fuels?

11 What factors determine the cost of the energy produced?

12 What environmental implications does this option have?

B **Nuclear fusion**

5 Find out what the possible fusion reactions are. Which seems to be the most promising? Are there suitable fuel resources? How large are they?

6 The principal requirement for a fusion reactor is high temperature. Why? What is the Lawson criterion?

7 Choose one suggestion for a practical reactor and find out what you can about it. How is energy extracted from such a reactor?

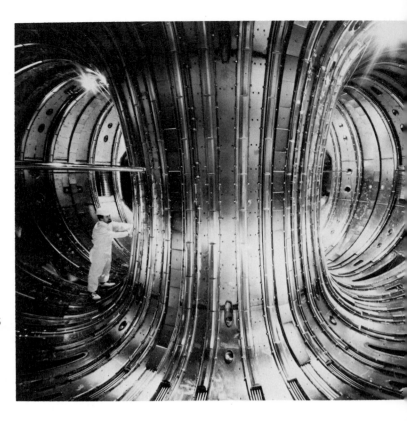

Figure G36
The interior of the vacuum vessel of the World's largest toroidal fusion experiment, the Joint European Torus. JET is staffed and funded by many European countries.
JET Joint Undertaking.

8 What safety precautions are necessary?

9 What factors will determine the cost of the energy produced?

10 What environmental implications does this option have?

C Nuclear safety

Since this is not an energy source, the four general study items are n⟨ appropriate to this topic.

1 What are the biological effects of various radiations? How are dos⟨ and dose rates measured, *i.e.*, what are the units and what instrumen⟨ are used?

2 Trace the path of nuclear fuel from mining the ore to its use. (Th nuclear fuel cycle.) What are the dangers at each stage?

3 What are the hazards in the operation of a nuclear power station? Ho are they minimized? How do they compare with other hazards – natur⟨ and Man-made?

4 The waste products from reprocessing the fuel present two main hazards: **a** radioactivity; **b** heating the environment. How does **b** arise from **a**? These are problems partly because of their intensity and partly because of their duration. Find out about them.

How are the hazards dealt with in the short, medium, and long term?

SOLAR ENERGY

D Passive solar devices

5 What is the common characteristic of passive systems? How large a contribution to our energy needs could this resource make?

6 How can buildings be designed to maximize the input of solar energy? (You need to consider at least the orientation of the building; the position and size of windows; the construction of walls.)

7 How can buildings be designed to conserve the energy which they gain? (You need to consider at least the position and size of windows; the construction of the building, particularly from the point of view of insulation and retention of internal energy.)

Figure G38
Roof panels to utilize solar energy.
United Kingdom Atomic Energy Authority.

8   What factors determine the cost of the energy?

9   What environmental implications does this option have?

E   Active solar devices

5   What are the common characteristics of these devices? How large contribution to our energy needs could such devices make?

6   How is the input of energy maximized? You will need to consider least the following: coating the collector; covering the collector with transparent cover; orientation of the collector; use of reflectors; contr systems.

7   What factors determine the cost of the energy?

8   What environmental implications does this option have?

F   Photovoltaic devices

5   What are solar cells? How are they made? How do they work? Ho large a contribution to our energy needs might be made by photovolta devices?

6   What are the characteristics of solar cells? You will need to consider least the following: e.m.f.; power output; impedance; spectral respons efficiency. It may be possible to investigate some of the experimentally.

7   What factors determine the cost of the energy?

8   What environmental implications does this option have?

G   Biomass

5   What is biomass? By what process is it produced? How much fuel coul be produced? What is the efficiency of the process?

6   Here are some possible systems: growing crops which are intended to l burned as fuel; growing crops which are processed to produce fue growing crops for food but using waste products for fuel; usir naturally growing vegetable material (rather than cultivated crops) fe fuel.
   Give examples of these. Are there other systems? How successful a they? What is the potential size of the resource?

7   What are the advantages and disadvantages of this option?

Figure G39
2 kW aero generator in Orkney.
Ilyn Satterley.

H Wind energy

5 Prove that the maximum power which could be developed by a wind turbine $P = \frac{1}{2}\rho Av^3$, where ρ is the density of air, A is the area swept out by the blades, and v is the wind speed. Not all of this power can be extracted. Why not?

6 What does the equation in item **5** tell you about designing a turbine to produce as much power as possible? What are the practical limitations on this?

7 How does wind speed vary with place, time, and height above the land surface? What contribution could this resource make to our energy needs?

8 What are the advantages and disadvantages of the various designs of turbine?

9 What factors determine the cost of the energy?

10 What environmental implications does this option have?

FOSSIL FUELS

I Power stations

5 Make yourself familiar with the components of a fossil-fuelled power station and their functions. You should consider at least the following: boiler; cooling system; high- and low-pressure turbines. How do fossil-fuel and nuclear power stations differ?

6 What is done to maximize efficiency? What places limits on efficiency?

7 Find out about 'Combined heat and power'. How does the system work? What are its advantages and disadvantages?

8 What are 'the carbon dioxide problem' and 'acid rain'? Find out about them. How big a contribution do power stations make to these problems? What other environmental implications do power stations have?

9 Remind yourself of the size of the resources which are available. What are the advantages and disadvantages of using fossil fuels to generate electricity?

G

Figure G40
The 2000 MW coal-fired Eggborough power station in Yorkshire. Coal is delivered by rail on the merry-go-round system.
Central Electricity Generating Board.

J   Alternative sources of fossil fuels

5   *Oil shale*. What is it? How big a resource of oil is it? How is the o
extracted? What factors determine the cost of the oil produced? Wha
environmental implications are there in its use?

6   *Coal liquefaction*. How is it done? How big a resource of oil is it? Wha
factors determine the cost of the oil? How would its use affect th
lifetime of coal as a resource? What are the environmental implication
of coal liquefaction?

7   *Coal gasification*. What methods are available? How big a source of ga
is it? What advantages does it have? Does it have any disadvantages?

OTHER SOURCES OF INTERNAL ENERGY

K   Geothermal

5   What is the origin of geothermal energy? What size is the resource
How does its availability vary with factors like geographical area
geological structure, and depth?

6   Is the energy used directly for heating or is it converted into electrica
energy? What factors determine this choice?

7 What techniques are used for extracting geothermal energy? What factors determine the cost of the energy?

8 What are the arguments for regarding geothermal energy as a renewable or a non-renewable energy source?

L Heat pumps

5 What is the principle of the heat pump? What does 'coefficient of performance' mean?

6 What are common applications of heat pumps? Why is the coefficient of performance different for different applications? What factors determine the cost of the energy?

7 What are the advantages and disadvantages of heat pumps?

M Combustion of refuse

5 What size is the resource?

6 Is it a viable proposition? Have any countries used it? Which? How do they use it?

WATER POWER

N Hydro-power

5 What size of resource is available? How is it distributed over the Earth's surface?

6 Sometimes the power is utilized directly as it has been in the past. What methods are used and what conditions are required? What limitations does this place on the usefulness of the energy source?

7 Hydro-power is sometimes used for generating electricity. What factors determine the cost of the electricity produced?

8 What environmental implications does this option have?

O Wave energy

5 What size of resource is available? How is it distributed over the Earth's surface?

6 What methods have been proposed for utilizing this resource? Describe one of these in detail.

G

7 What factors determine the cost of the energy produced?

8 What environmental implications does this option have?

Figure G41
Waves.
Picturepoint – London.

P Tidal energy

5 How are tides produced? Why does their size vary in different places and at different times?

6 How is tidal energy utilized? What conditions are necessary for this? Study the installation on the estuary of the river Rance in France.

7 What determines the cost of the energy produced?

8 What environmental implications does this option have?

CONSERVATION

Conservation is not, of course, a source of fuel or of energy. However, saving a quantity of fuel which would otherwise be consumed is equivalent to discovering a new source for that quantity of fuel. The effect of conservation is to enable limited resources of fuel to last longer than they would otherwise.

It is therefore important to study conservation together with energy sources. However, since the four basic study items relate more to the fuel which is saved than to the process of conservation, these items will not be included here.

Q Domestic

1 What methods are available to save fuel in the home?

2 What savings in fuel cost can each produce in a year? Are there other advantages?

3 What does each cost to install? Are there other disadvantages?

4 What is the payback time? Is each method viable?

5 How can better control of heating systems save fuel?

R Industrial

1 How much fuel is used in industry in the U.K.? What proportion is used for process heating compared with space heating?

2 Compare the amount of energy needed to produce different materials, such as steel, aluminium, cement, and paper. How can savings be made?

3 Prepare a case study on process heating (breweries are an example).

4 How can the redesign of products or the re-cycling of materials lead to reduced energy costs? (One example is packaging for food and drink.)

5 What can be done to reduce the energy required for space heating and lighting in industrial buildings, while maintaining or improving working conditions (for example 'heat wheels')?

6 How have improvements in the efficiency of combustion helped fuel conservation?

G

S Transport

1 How much fuel is used for transport? What are the energy costs of different systems of transport?

2 How can improvements in design of vehicles help to conserve fuel (for example reduction in mass; drag; improved carburation; use of diesel engines)?

3 What alternatives to oil-based fuels for transport are possible? What are their advantages and disadvantages?

4 What are the implications of increased use of the more fuel-efficient transport systems (such as public transport, increased numbers of passengers in cars)?

T Storage

This is not an energy source so the four general study items listed on page 460 are not relevant here.

1 Why and where is energy storage important?

2 Here are some possible energy storage systems: chemical, for example, rechargeable cells; potential, for example, pumped storage hydroelectric schemes; kinetic, for example, flywheels; internal, for example, store for solar energy; electrical, for example, capacitors; magnetic, for example, magnetic fields.

Find additional examples of the various categories given. Can you find other categories?

3 What is the energy density of a store? What values are obtained for the systems listed in **2**? What other criteria are important in choosing an energy store?

HOME EXPERIMENTS

GH1 The cost of a 'cuppa'

Use the gas or electricity meter in your home to measure the quantity used to boil 1 kg of water. Work out the efficiency of the process (specific heat capacity of water is $4200 \, J \, kg^{-1} \, K^{-1}$). Look at a recent gas bill for information about the total energy transformed when $1 \, m^3$ of gas is burned, and its cost; or look at an electricity bill to find the cost of one unit (1 kW h).

Compare the cost and time taken, as well as the efficiencies. Use the data given in figure G4 (between pages 430 and 431) to compare the quantities of primary fuel used in the two heating processes.

GH2 Surface layer

If there are surface layers near walls and windows, then the temperature at the surface will not be the same as that of the bulk of the air in the room. See whether you can detect any such temperature differences using your hands or face. How valid are your observations? For instance, are your hands reliable temperature sensors? (You probably know the simple experiment which involves placing a finger of one hand in cold water, a finger from the other hand in hot water, and, when they are acclimatized, putting both in tepid water, and seeing whether both fingers give the same temperature indication. Also, do you get the same sensation of hotness or coldness when you touch pieces of glass, metal, or expanded polystyrene which *are* at the same temperature? What causes these differences?) How might you make more reliable tests of the existence of a surface layer?

G

ANSWERS TO SELECTED QUESTIONS

UNIT A
Materials and mechanics

The graph is shown in figure Q1.

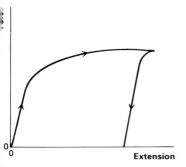

Figure Q1

1 Question 1 asked for a translation from words to a graph: this one asks you to reverse the process. Being able to 'read' a graph is an important skill to acquire. The four substances are: polythene strip, copper or aluminium wire, rubber cord, and steel or cast iron wire (this last one is too curved to be a thin glass rod)

3a $100\,\mathrm{N\,m^{-1}}$
b $0.02\,\mathrm{m}$; $50\,\mathrm{N\,m^{-1}}$
$0.005\,\mathrm{m}$; $200\,\mathrm{N\,m^{-1}}$
d Halved; doubled.

5a Glass is stiffer than Perspex; it requires more force per unit cross-sectional area to stretch it the same amount.
b It is easier to flex wood across the grain - in the low Young modulus direction. Wood.

7a 10^{11} to $10^{12}\,\mathrm{N\,m^{-2}}$
b 100 to 1000 times greater.
c $3.4\,\mathrm{mm}$

10a $\mathrm{M\,L\,T^{-2}}$
b $\mathrm{M\,T^{-2}}$
c $\mathrm{T^{-1}}$
d $\mathrm{T^{-1}}$, *i.e.*, frequency.
$1/2\pi$ is purely a number. It does not represent a mass, length, or any other physical quantity, as k does.
11 [stress] $= \mathrm{M\,L^{-1}\,T^{-2}}$; strain is simply the ratio of two lengths, so it has no dimensions; [the Young modulus] $= \mathrm{M\,L^{-1}\,T^{-2}}$.

12e,f If the force, F, were proportional to extension, x, so that $F = kx$, then the energy $\frac{1}{2}Fx$ would be $\frac{1}{2}kx^2$. The graph of potential energy against extension would be a parabola.

13a 1 N
bi 2 N to the left.
ii 2.5 m s^{-2} to the left.
ci 0.2 J
ii 0.25 J
d 0.05 J

17ai About 0.1 J.
ii About 0.07 J.
b The graphs are shown in figure Q2.

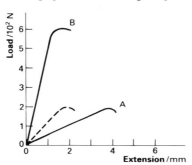

Figure Q2

c The longer cable can stretch more than a shorter one before breaking, and can store more energy. The longer one is therefore less likely to snap when there are sudden jerks.

18b $2.4 \times 10^{-9}\,\mathrm{m}$
c 10 %; 1 %; 30 %
d The layer might be more than one molecule thick.
e About 10.
f About 10^{19}.

23a 3
b Each bubble attracts other bubbles, and the result is that they cluster as closely as possible. The bubbles are round, and attract other bubbles equally in any direction, whereas, if the forces tended to act in certain special directions, some arrangement like **1** or **4** might result.
c By analogy, it might be true that the attractive forces between atoms in such metals were equal in all directions.

d Because the sizes of the Na$^+$ and Cl$^-$ ions are very different.

24a $7.12 \times 10^{-6}\,\mathrm{m^3}$
bi $8.68 \times 10^{-30}\,\mathrm{m^3}$
ii $8.20 \times 10^{23}\,\mathrm{mol^{-1}}$
iii This is an upper limit for L because the spherical copper atoms cannot fill all the space in a crystal of copper.
ci $16.6 \times 10^{-30}\,\mathrm{m^3}$
ii $4.30 \times 10^{23}\,\mathrm{mol^{-1}}$
iii This is a low value for L because the assumption of a 'square' array over estimates the volume taken up by each atom.
d $6.15 \times 10^{23}\,\mathrm{mol^{-1}}$

27a The potential energy of the two atoms decreases as they move together with a net attractive force. 0Q gives the mean equilibrium separation, r_0. As the repulsive force is predominant for closer distances than r_0 the potential energy rises. It becomes positive for separations less than 0P.
b 0Q on the potential energy–separation graph $= 0A$ on the force–separation graph.
c The sum of potential and kinetic energies must be constant if there is no external source or sink of energy. So as the kinetic energy increases the potential energy will decrease, and vice versa. The potential energy will vary by ΔE as the atoms oscillate between separations X and Y (see figure Q3).

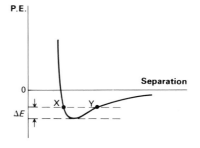

Figure Q3

d The depth of the potential well at Q gives the binding energy, E (see figure Q4).

28a A
b B
c B
d B

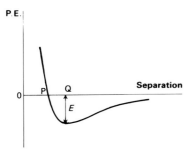

Figure Q4

29a 2 N
b $400\,\mathrm{N\,m^{-1}}$

34 '…in the next picture the dislocation has moved on by one atom (See figure Q5.) In each succeeding picture the dislocation moves on by one atom until it reaches the edge'.

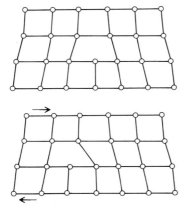

Figure Q5

37a The diagrams are shown in figure Q6.
b The tension $T = W/2\cos\theta$. As $\theta\to0$ so $T\to W/2$. The tension is less in a long cord than a short one.

38a $F = N\sin\theta$
b $W = N\cos\theta$
c $F = W\tan\theta$
d $F\cos\theta = W\sin\theta$

41 Forces are 9.7 kN and 12.3 kN at left- and righthand supports respectively.

42a mg
b Greater
i $T = mg/\cos\theta$
ii $F_e = mg\tan\theta$
c For small angles $\tan\theta\approx\theta$, and $\theta\propto d$.

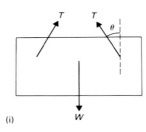

(i)

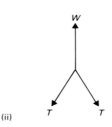

(ii)

Figure Q6
(i) Forces on the picture.
(ii) Forces on the hook.

43a*i* 400 N
ii 346 N
iii 346 N to the right.
b*i* 200 N
ii 173 N
iii 173 N to the right; 100 N upwards.
c*i* 600 N
ii 520 N
iii 520 N to the right; 100 N upwards.
d*i* 795 N
ii 521 N
iii 521 N to the right; 200 N *downwards*.

47a The weight of the earth above the pipe forces it into an oval shape. The breaks occur where the greatest (tension) stresses are.
b X is at a point of maximum compression rather than tension. Concrete is strong in compression and weak in tension.
c The concrete surface is under compression from the plastic. So when the pipe is buried the concrete remains in reduced compression or only low tension. The plastic cover also has to stretch.
d Concrete is porous, the plastic prevents leakage. Surface damage is also stopped.

50a $8\,\mathrm{m\,s^{-1}}$
b $6\,\mathrm{m\,s^{-1}}$
c 1.4 J
d $1.4\,\mathrm{kg\,m\,s^{-1}}$
e 35 N

52a 6000 N s
b $6000\,\mathrm{kg\,m\,s^{-1}}$
c About $10\,\mathrm{m\,s^{-1}}$.
d The force at any instant is proportional to the slope of the graph of velocity against time. The graph is shown in figure Q7.

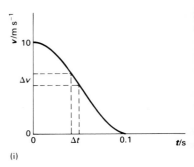

(i)

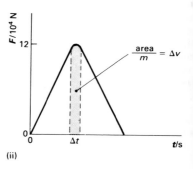

(ii)

Figure Q7

54a The Earth.
b The pull of the Earth on the book, the push of the table on the book. The force pairs of Newton's Third Law are the two gravitational forces (Earth on book, book on Earth), and the two contact forces (book on table, table on book). Newton's Third Law does not require that the two forces acting on the book be equal. If they were not the book would accelerate (Newton 1).
c The Earth moves up towards the body with equal and opposite momentum.
d The train pushes on the Earth which moves 'backwards' as the train moves 'forwards'. The Earth and train make the closed system.

The ratio of masses ensures that the effect on the Earth's motion is negligible in both cases.

1a 42 moles
b 7.0 kg
6.0 m^3

5a 5.65×10^{-21} J
b 4.71×10^{-26} kg
c 5.65×10^{-21} J
d 1300 m s^{-1}

7a 210 m s^{-1}
b 2.4 ms
c 110 km; 1.2×10^{12}.
d 9.2×10^{-8} m, *i.e.*, approximately 10^{-7} m.
e 1.5×10^{-10} m
Yes. The assumption for the theoretical calculation is that all molecules are of the same diameter. This is approximately true.

9a pA
b $pA\Delta l$
c $p\Delta V$
d $p\Delta V$
e No frictional losses at piston; change happens slowly.
f When air is compressed the work done on it increases its internal energy; when it expands it does work, its internal energy is reduced.

UNIT B
Currents, circuits, and charge

1a 0.16 C
b 10^{18} particles.
c 6×10^{22} particles per metre length.
d 1.67×10^{-5} m
e 1.0×10^{-6} m s^{-1}
f The algebraic answers to **a–e** become:
It, It/Q, nA, It/QnA, I/QnA
g The speed at which anything moves obviously does not depend on the length of time for which its motion has been observed.
Reducing n increases v; fewer particles have to move faster.
Reducing Q raises v; each carries less charge, so they must move faster to carry the same total charge in a given time.
A molar solution is strong, but can be made, and contains 6×10^{23} solute molecules in one litre, and so 6×10^{26} molecules in one cubic metre. If every molecule provides an ion, there will be 6×10^{26} ions in each cubic metre. Water molecules are not by any means all dissociated into ions in very pure water while, say, paraffin would contain almost no ions.

3a A possible estimate can be reached by considering the time for which the cell could light a torch bulb (say 5 hours) carrying a current of, say, 300 mA.
b Estimate mass and cruising velocity and calculate $\frac{1}{2}mv^2$. An allowance for inefficiency is necessary (typically the overall efficiency of conversion of internal energy ('heat') into mechanical energy is about 30 %). An all-electric train itself will have a much higher efficiency *by itself* but the overall inefficiency is transferred back to the power station.
c With no losses, about 110 seconds.
d *i* 1000 A *ii* 100 A *iii* $(10)^2 =$ 100 times greater *iv* 100 times longer

4 B

7 The resistance of (a) is larger than the resistance of (b) and both seem to be linear components, obeying Ohm's Law. (c) might be a diode, as its resistance changes when the p.d. is reversed. (d) could be a filament lamp, for its resistance increases as the current increases. Initially, (d) conducts better than any of the others. You could also compare the forward and reverse resistances of (c) with the resistances of (a) and (b).

9a Only graph (b) obeys Ohm's Law as stated. Graph (a) is linear, but there is current when there is zero applied p.d. (perhaps the component included a source of p.d.). (c) shows a constant current, not a constant rate of change of current with change of p.d. Graph (d) is non-linear, though it might be said to obey Ohm's Law for p.d.s in one direction only.
b Yes. A graph of current against p.d. is a straight line through the origin. Alternatively, equal increases in current go with equal increases in voltage. The resistance at 2.0 mA, and at all other currents given, is 14.5×10^3 Ω.
c The answers here are partly matters of opinion. Certainly Ohm's Law is not applicable to all objects that conduct, but it does compactly describe the behaviour of some. Suggestion 3 avoids trouble by defining it away; sometimes physicists do this, but in this particular case they usually speak of 'ohmic' or 'linear' materials. 4 is wrong, for Ohm's Law requires the current–voltage graph to pass through the origin and a small section of the curve, though nearly straight, will not project back to the origin. Whether Ohm's Law is important and general enough to

count as a law is a matter of opinion. It does not rank with those 'Laws of Physics' to which we know no exception (give an example).

11ai The resistance of A increases with applied p.d.
ii The resistance of B is constant.
b For 2 A, $V_A = 5$ V and $V_B = 10$ V, so the combined resistance is

$$\frac{(5+10) \text{ V}}{2 \text{ A}} = 7.5 \, \Omega.$$

c If they are in parallel, the p.d. is the same across each. A p.d. of 5 V produces $I_A = 2$ A and $I_B = 1$ A, *i.e.*, a total current of 3 A.
d The resistance of a parallel combination is less than the resistance of the smallest component. The parallel combination of A and B has a minimum resistance when that of A is as low as possible, *i.e.*, at a low value of p.d.

13a Their resistivities rise in constant ratio.
b The logarithm to base ten of the resistivity.
c More than 10^{20} km, over ten million light years; further from us than the nearest galaxy.

15a 1.6 A
b $2R$

17a $I/A = V/AR$
b $I/A = V/\rho l$
c 5.9×10^7 A m^{-2}; current in a 1-mm^2 wire is 59 A.
d $v = 3.7 \times 10^{26}/n$
e 8.5×10^{28} atoms, or electrons, per cubic metre.
f About 4 mm s^{-1}.

18a A is connected electrically to B, B is not connected to D (unless by a very high resistance). A may or may not be connected to C (or D).
b The effective resistance between A and B has 1.5 V across it when a current of (4.5/1000) A passes, and is thus just over 330 Ω.
c The effective resistance between A and B is just over 330 Ω, in agreement with **a** and **b**.
d If there were resistors connected between A and C or between B and D, the meter would read more than 18 mA (but **a** suggested that B and D were not connected).

e If A and C were connected, the meter would give a reading when the switch was open. So they are not. The reading of 18 mA when the switch is closed confirms earlier suggestions that the circuit is as in figure Q8.

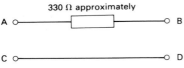

330 Ω approximately

A ○——————[ ]————————○ B

C ○—————————————————————○ D

Figure Q8

20a 11 000 Ω
b 0.27 mA
c Three-quarters of the length of CD.
d The lamp resistance (15 Ω working) is so low compared with the resistance of the part of the potentiometer across which it is connected, that it effectively short-circuits the output, producing very nearly zero p.d. across it.

22 1 Ω. You should find, and perhaps be able to prove, that the power delivered has a maximum value when R is 1 Ω, that is, the resistance of the external circuit is equal to the internal resistance of the cell.

27a From the sliding contact into the meter.
b $I_1 - I_2$
c Three; only two are needed.
d 83 μA

28a The current is zero at first, rises sharply to 2 A, remains steady for 10 hours, and then drops sharply to zero at 6 p.m.
b 2 C
c 7.2×10^4 C
d The charge passed is represented by the shaded area in figure Q9.

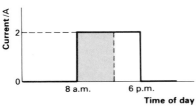

Figure Q9

e The charge passed is still shown by the area between the graph and the time axis.
f About 5×10^4 C.

29a The same as meter 1.
b 5 divisions to the left.

c None!
d Charging, anti-clockwise; discharging, clockwise.
e When the capacitor is being charged the equality of meter readings 1 and 2 shows that as much charge flows onto one plate as flows off the other.

30a As in figure Q10.

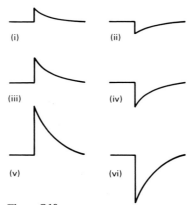

(i)   (ii)

(iii)   (iv)

(v)   (vi)

Figure Q10

b *i–iv* As for **a**i.   *v* As for **a**vi.
c Equal changes in p.d. produce equal flows of charge (identical traces).

32a 10^{-3} C
b 6.25×10^{15}
c The resistance would have to be decreased at a steady rate.
d 10^{-4} A
e See figure Q11.

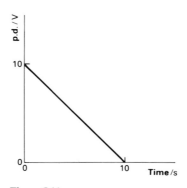

Figure Q11

34a 100 μC
b 200 μC
c 10 V
d 10 V
e 300 μC
f 30 μF
g Yes!

35a *i* No   *ii* No
b No, from **a**. (Or consider the isolated part of the circuit consisting of the righthand plate of C_1 and the lefthand plate of C_2.)
c Q/C_1
d Q/C_2
f $C = Q/V$

37a 0.01 F
b 0.02 C
c 0.38 J. The taller shaded strip in figure Q12.

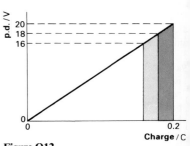

Figure Q12

d 0.34 J
e The two shaded strips in figure Q12.
f $0.38 J + 0.34 J + \ldots\ldots + 0.20 J = 2 J$
g The total area of the triangle under the graph ($\frac{1}{2} \times$ base $\times$ height) represents the energy transformed.

44 You could use a circuit with a capacitor discharging through a variable resistor, with the discharge current holding the relay on while it exceeds 0.1 mA. The time constant of this circuit determines how long the relay is closed and hence for how long the lamp lights. The variable resistor could be calibrated for the required lighting time. Sketch the curve of discharge current against time for these approximate values: resistance 10 kΩ, capacitance 700 μF, charging p.d. 10 V.

47 The current is the ratio of power to potential difference. An estimate of 10 W power gives a current of 5×10^{-4} A. This is the same as 3×10^{15} electrons per second on an area of 10^5 mm² which, gives a density of 3×10^{10} electrons per mm². Notice that the number of electrons per square millimetre of screen must be large otherwise it might be possible to see random fluctuations in the brightness as the electrons arrived randomly.

48a QV

b Fd

c $QV = Fd$. If $F = W$, then $Q = Wd/V$.

d See table Q1.

V p.d. to balance drop/V	Q Charge on drop/C	n Multiple
470	4.75×10^{-19}	3
820	3.18×10^{-19}	2
230	6.44×10^{-19}	4
770	1.64×10^{-19}	1
1030	1.57×10^{-19}	1
395	7.83×10^{-19}	5

Table Q1

e The basic charge, e, is about 1.6×10^{-19} C.

f The values of Q and n in the table lead to the following values for e:
1.58×10^{-19} C, 1.59×10^{-19} C, 1.61×10^{-19} C, 1.64×10^{-19} C, 1.57×10^{-19} C, 1.56×10^{-19} C. Given that the true value of e is 1.60×10^{-19} C, the uncertainty in these calculated values is about 3% suggesting that the *sum* of the percentage uncertainties in V, W, and d is also about 3%. From the results, V, which is usually given to two significant figures only, seems to have the largest uncertainty, and d the least.

UNIT C
Digital electronic systems

1a Suppose the numbers are 1342 and 5768. Select the end digits of each number (2 and 8). Add them. The sum is 10. Store the digit 0 as the first digit of the answer, and carry the digit 1 to the next step. Select the next two digits (4 and 6). Add them, and the 1 carried over, getting 11. Store 1 and carry 1. Then continue. A better system would have a stopping device that turned the adding process off when all of the incoming digits (carry as well) became zero. Without it a mindless computer would go on for ever, producing the answer

......0000000000000000000000000007110

b When the clock reaches a preset time it switches on the power to a heater immersed in cold water. When the water boils, the heater is switched off. The 'water boiling' signal also operates a device that pours the water onto the tea leaves. After a suitable delay for the tea to infuse, a 'tea ready' alarm is rung. The human part of the system then adds milk and sugar.

c Figures Q13 to Q15 show systems, or parts of systems, to do the suggested jobs.

5 There are only two digits in the binary system: 0 and 1. It is easy to define two voltage levels – high and low; supply voltage and zero volts – representing the digits 0 and 1. The binary system readily lends itself to electronic systems. Defining intermediate levels to represent more digits is much more complicated.

6i A simple on–off system, information easily transmitted by a digital signal.

ii Temperature is continuously variable, so it is difficult to use a simple digital code (although not impossible, as you may see later).

iii Depends on the number of directions. Two would be easy; more directions would require the use of more than one binary digit. For example, four directions could be coded using two 'bits'. If the directions are north, east, south, and west, then north could be 00, east 01, south 10, and west 11.

iv The magnitude of an electric current is like temperature: continuously variable.

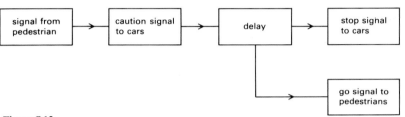

Figure Q13
Part of a system for controlling cars. Other parts should be added to operate go and stop signals for cars and pedestrians.

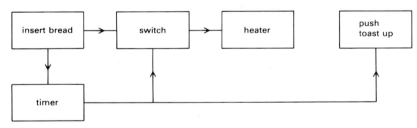

Figure Q14
A system for a 'pop-up' toaster.

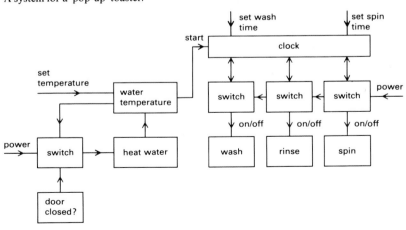

Figure Q15
A system for an automatic washing machine.

However, direction is likely in most cases to be easier to transmit with a binary code.
v The number of pages in this book can be expressed as a binary number which can be transmitted as a sequence of 1s and 0s. For example, if there are 76 pages: decimal 76 is binary 1001100.
vi The area of the page may also be expressed as a number and so can be expressed in binary form.
vii Again, this a number and can be transmitted in digital form.

8 Circuits (a) and (b) are both inverters.
(c) This circuit functions as an inverter.
(d) In this circuit the output is always low.
(e) In this circuit the output is always high.
(f) This circuit functions as an inverter.

9 The circuit is an AND gate.

| inputs | | | | output |
A	B	C	D	E
0	0	1	1	0
1	0	0	1	0
0	1	1	0	0
1	1	0	0	1

Figure Q16

11 The solution to this problem is called a Parity gate. It is the inverse of an Exclusive OR gate (see question **10b**).

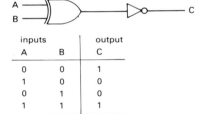

| inputs | | output |
A	B	C
0	0	1
1	0	0
0	1	0
1	1	1

Figure Q17
A Parity gate.

13 If the system is thought of as a box with four inputs and three outputs, then truth tables can be constructed to satisfy the conditions specified in the question. For example, the red L.E.D. is controlled by sensors A and B in the way that the truth table in figure Q18 shows.
There are two other truth tables for the green and the yellow lights. The circuits in figure Q19 satisfy the truth tables.

14 This circuit will have three inputs and one output. Without showing the whole truth table, the necessary combination of

A	B	output red
1	1	1
1	0	0
0	1	0
0	0	0

Figure Q18

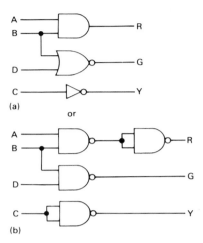

(a)

or

(b)

Figure Q19

inputs to make the output high is shown in figure Q20(a) The circuit in figure Q20(b) satisfies this condition. Any other combination of inputs gives a low output.

A	B	C	output
1	0	0	1

(a)

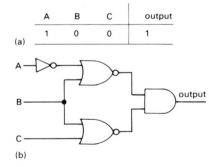

(b)

Figure Q20

17 The truth table is that of the Exclusive OR gate, although the second input is now labelled 'control'. Examination of this table shows that the 'input' is passed unchanged to the 'output' if the 'control' is 0. If 'control' is 1, then the 'output' is the inverse of the 'input'. (Figure Q21.)

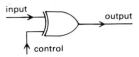

input	control	output
0	0	0
0	1	1
1	0	1
1	1	0

Figure Q21

19a 5 volts. All the kits will work from a 5 V supply.
b When the output of one gate is connected to the input of another, the 'output low' of the first gate must be less than or equal to the 'input low' of the second gate, if the logic is to be preserved.
For the same reason, 'output voltage "high" is higher than 'input voltage "high"'
c Basic unit and TTL gates have operating limits that are defined by fixed power supply limits. Examine the characteristic curves in figure C13 (page 167).
CMOS gates have a very sharp transition from high to low, but the level of this depends on the supply voltage. Since the supply voltage (V_s) can vary over a wide range, it is best to express the limits in terms of V_s.

21a Figure C72(b). As the resistance of the thermistor drops so does the potential difference across it and therefore the input to the inverter goes low. So the output of the inverter becomes high, lighting the warning lamp.
The circuit shown in figure C72(a) will warn against low temperatures.
b The circuit in figure C73 would be used to warn against decreases or increases in temperature. The indicator, which in this circuit is on when conditions are safe, will go out if either A or B goes high. The resistors are chosen so that the inputs are low when the temperature is within the safe range. If an OR gate were used instead, the indicator would be off when both inputs were low. It would then light when either input went high, that is, when the temperature went out of the safe range.

24a Figure Q22 shows a half adder using NOR gates.
b See figure C17 (page 169) for full adder from half adders.

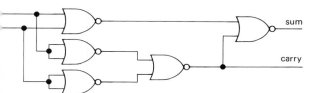

Figure Q22

See figure C17. Only if A = 1 and B = 1 is the carry from the first half adder 1. But then the sum from the first half adder is 0, so there can be no carry from the second half adder.

25a A pulse producer – see figure C19.
b Input 1 goes from low to high; output 1 goes low, stays there for a while, and then returns to high. When output 1 goes high this triggers circuit 2, whose output then goes low, stays there for a while, and then goes high again.
c The process would continue indefinitely – just like an astable circuit.

27 The output D is high when either input A **or** input B are high.

A	B	C	D
0	0	1	0
0	1	0	1
1	0	0	1
1	1	0	1

Figure Q23

29a The output of the second also goes from low to high.
b Nothing happens to it next. It stays high. The output from the second keeps it high.

UNIT D
Oscillations and waves

1a It is probable that the car body has initially been set into oscillation, and the energy of oscillation is gradually being reduced due to damping. (This particular car has rather worn shock absorbers – the energy should disappear in only one or two cycles.) The energy is being distributed among the many atoms making up the surroundings of the car body. The reverse process – concentration of energy from many individual atoms into oscillations of one object – never happens.

But the motion shown could be reversed if the car body were being given a properly timed series of pushes by an external agent – that is, being set into forced oscillation.

3 The story of Harrison and his chronometers is told in a booklet obtainable from the National Maritime Museum.

A ship carried a clock, telling the time at some definite longitude, which gradually came to be that of Greenwich. The ship's unknown longitude was found by comparing local time (at, say, noon, when the Sun is highest in the sky) with the time indicated by the clock. If, for instance, local noon is six hours behind Greenwich, the ship is one-quarter of the way around the Earth West of Greenwich. At the Equator, a time difference of one minute corresponds to a distance in longitude of nearly thirty kilometres, so an accurate clock is needed.

The way to test the clock is to determine, by making astronomical observations, the longitude of the place reached (Jamaica), and to calculate the expected time of noon compared with that at Greenwich. The clock, if it is accurate, will still show Greenwich time, and its reading can be compared with the calculated time.

5 This is discussed in the reading 'Quartz and atomic clocks' on pages 237–9.

8ai 0.5 %
ii 0.006 %
iii 0
bi $\cos\theta \approx 1$
ii $\tan\theta \approx \theta$

9 $6.7\,\mathrm{m\,s^{-1}}$

10a

	Velocity	Acceleration
Q	zero (but about to move rapidly down)	large, downward
R	large, upward	zero
S	zero	large, downward

Table Q2

bi Tension in the Slinky (T) – a large tension will mean that for a given displacement there will be a large restoring force, accelerating displaced coils back towards their mean position more quickly.
ii Mass per unit length of the Slinky (μ) – a given restoring force acting on a large mass, will accelerate displaced coils more slowly towards their mean position.

In fact

$$c = \sqrt{\frac{T}{\mu}}$$

14c The amplitude would only be zero at a minimum if the amplitudes of the disturbances arriving from S_1 and S_2 were equal. The wave energy is spreading out from the sources, so the amplitude of an individual wave decreases with distance travelled. Since in general waves from S_1 and S_2 will have travelled different distances to reach a minimum point, they will not have equal amplitudes.

15 The path difference has increased by an odd number of half-wavelengths. So λ could be 2 m, or $\frac{2}{3}$ m, or $\frac{2}{5}$ m, or … .

17 There are many answers. Wavelength measurements usually involve using the fact that waves superpose to give a large or small resultant effect, depending on whether the path difference is an even or an odd number of half-wavelengths.

Two-source experiments are possible in practice with waves of reasonable wavelength, for the sources need not then be very close together. But the sources must emit in phase. For light, two-source experiments are still possible, but the two sources must be imitated by splitting the light from one narrow source.

Diffraction gratings can be used, and are especially suitable for visible light and for X-rays. The grating spacing must not be very much larger than one wavelength if the diffraction angle is to be reasonably large. For X-rays, the grating is tilted so that the X-rays graze its surface, and the spacing looks small to the X-rays.

In the microwave and v.h.f. region (wavelength from 0.01 m to 1 m roughly), simple arrangements of reflectors which introduce a path difference can be used. A problem here is to have big enough reflectors, for a reflector must be bigger than one wavelength in linear dimensions to reflect an appreciable amount of wave energy.

18b As S_2 is moving further from T, it is reflecting a decreasing amount of wave energy; and as R is also now further away, a smaller proportion of this amount reaches R. Hence the two superposing signals will no longer have equal amplitudes, which they must have to cancel out exactly.

20c The light following paths A and C superposes destructively – thus reducing the intensity reflected. B and D, however, give constructive superposition, increasing the intensity transmitted.
d There will be relatively little light intensity reflected in the middle of the visible band, and relatively more at the ends – that is, blue and red. These combine to make the reflected light appear purple.

21 78 m

27 (b) will oscillate. (Do not confuse the 'weightlessness' of the mass with lack of inertia. Its mass is the same as on the Earth, and it behaves as an oscillator with the same period, neglecting any relativistic effects.) You can decide about the others.

28c From diagram, i $\omega = 1100° \, \text{s}^{-1}$
ii $\omega = 19 \, \text{rad s}^{-1}$.

29a $-1 \, \text{m s}^{-2}$
b $-0.05 \, \text{m s}^{-1}$
c $-0.05 \, \text{m s}^{-1}$
d $-0.005 \, \text{m}$
e $0.095 \, \text{m}$
f $-0.95 \, \text{m s}^{-2}$
g $-0.095 \, \text{m s}^{-1}$
h $-0.145 \, \text{m s}^{-1}$
Table Q3 shows the other values.

t/s	s/m	$a/\text{m s}^{-2}$	$v/\text{m s}^{-1}$	$\Delta s/\text{m}$
0	0.1	-1	0	
0.05			-0.05	-0.005
0.10	0.095	-0.95		
0.15			-0.145	-0.0145
0.20	0.0805	-0.805		
0.25			-0.226	-0.0226
0.30	0.0580	-0.580		
0.35			-0.283	-0.0283
0.40	0.0296	-0.296		
0.45			-0.313	-0.0313
0.50	-0.0017			

Table Q3

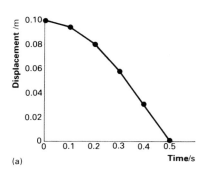

(a)

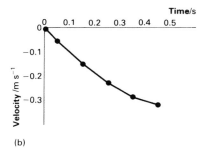

(b)

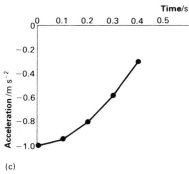

(c)

Figure Q24

i The graphs are shown in figure Q24.
j 2.0 s. It compares well (1.99 s).
k These answers would all be doubled; the period would be shorter.
l These answers would all be halved; the period would be longer.

31 The description is plausible; but the lack of definition in the photograph, and the associated uncertainty in plotting points on a graph, make it difficult to tell for certain.

32ci Acceleration $= -\dfrac{2xg}{l}$

ii Angular velocity $= \sqrt{\dfrac{2g}{l}}$

iii Periodic time $= \pi \sqrt{\dfrac{2l}{g}}$

33 $2\pi \sqrt{\dfrac{hd}{\rho g}} = 2\pi \sqrt{\dfrac{L}{g}}$

35c 1.14 p.m.

36b $400 \, \text{N m}^{-1}$
ci $2.8 \times 10^{-20} \, \text{J}$ $\qquad ii$ $0.18 \, \text{eV}$

37a 0.1 m
b $10 \, \text{rad s}^{-1}$
c 0.16 s
d $1 \, \text{m s}^{-1}$

41a $8 \times 10^{12} \, \text{m s}^{-1}$
b $5600 \, \text{m s}^{-1}$

42b $\sqrt{\dfrac{E}{\rho}}$

46b $3.8 \times 10^{-26} \, \text{kg}$
c $1.2 \times 10^{13} \, \text{Hz}$ (effective $k = 200 \, \text{N m}^{-1}$)
The assumption is not a very good one as the Cl^- ion is less than twice as heavy as the Na^+ ion (mass ratio 23:35)
di $2.6 \times 10^{-5} \, \text{m}$

54c $1.7 \times 10^{-8} \, \text{s}$
d 5 m

56c $60 \, \text{m s}^{-1}$
d The system responds with large amplitude only at special resonant frequencies. The mass-on-spring has only one resonant frequency, however, whereas the string has a series of resonant frequencies: its fundamental frequency, and multiples of this. The resonance of the string depends on waves arriving back from the far end in phase with the new waves being generated: and this can happen when the path of the waves (4 m) is any whole number of wavelengths.

57 192 Hz; 320 Hz.

UNIT E
Field and potential

1a To the right. The charges on the plates cause a force to be exerted on the positively-charged ball in this direction.
b The ball becomes negatively charged. It moves to the left.
c Negative.
d Negative.
e Anticlockwise.
f Clockwise.
g So that charge can flow on and off the surface.
h The p.d., V, of the supply, and the separation, d, of the plates. The frequency rises with increasing V or decreasing d.
i Charge is displaced over the conducting surface of the ball; the larger force on the near side of the ball causes it to be attracted to the plate, whereupon it acquires the same charge as the plate and the process continues as before.

3a Fd
b QV
c $F/Q = V/d$
d $V m^{-1}$
ei joules/coulomb
ii newtons × metres
iii $\dfrac{V}{m} = \dfrac{J}{Cm} = \dfrac{Nm}{Cm} = \dfrac{N}{C}$
fi $10^4 V m^{-1}$
ii $10^{-9} C$
iii $10^4 N$

4a Use $E = V/d$. $V \approx 2000 V$.
b Briefly, the higher the pressure the shorter the distance a particle moves before a collision with another. If in that distance the occasional ion or electron can acquire enough energy (about 30 electronvolts, or about 5×10^{-18} J) from the p.d. it moves through, it may ionize the molecule it hits, and the ion or electron so freed may make another in the same way, leading to an 'avalanche' of ions, and a spark. At higher pressures, then, a greater p.d. is needed to start a spark, for the ions have to acquire a fixed energy in a shorter distance. The question suggests that the sparking p.d. in a car engine exceeds 2 kV, for the pressure in the cylinder is several times atmospheric pressure.

7a A large charge would affect the charges on the plates, completely altering the field between them. A small charge,

however, would give such a small force that it could not be measured.
b With no flame, the probe may acquire an induced charge and thus affect the field around it, altering the potential at the tip. The flame contains ions which discharge the tip of the probe so that there is no potential difference between it and its surroundings. The electroscope gives little or no indication (unless a charged plate is touched) without the flame.
c Any source of ions would do, for instance a radioactive source giving out alpha particles. These are in fact used in probes in balloons in the upper atmosphere.
d Set up as described, it will acquire positive charge.
e Towards the electroscope.
f If the charge does not leak off the electroscope, then this current will drop to zero. In practice, it settles at a small steady value, enough to keep the electroscope 'topped up'.

8a Positive.
b To the left.
c Negative.
d Increase.
e The graph is shown in figure Q25.

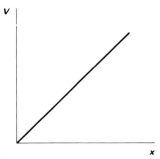

Figure Q25

f Positive.
g $E = -\dfrac{\Delta V}{\Delta x}$ or
'field strength = – potential gradient'.

9a The graph is shown in figure Q26.

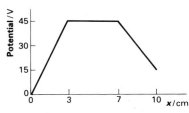

Figure Q26

b For CG_1 it is $-1500 V m^{-1}$;
for G_1G_2 it is $0 V m^{-1}$;
and for G_2A it is $1000 V m^{-1}$.
c Between C and G_1 there is constant acceleration up to maximum energy of 45 eV.
Between G_1 and G_2 there is constant speed.
Between G_2 and A there is constant deceleration down to energy of 15 eV.
d Since it has only 10 eV it cannot reach A, so will 'fall back' towards G_2.

12a $10^{-11} F$
b Approximately $10^{-8} F$.
c Approximately 0.1 %.
d The effective capacitance is now about $2 \times 10^{-11} F$. Only 50% of the charge would be transferred; nevertheless the p.d. across the voltmeter would be about 500 V, which could cause serious damage if an electrometer was being used as the high resistance voltmeter.

13a The statement is correct: $Q = 10 \mu C$ so $I = 0.5 mA$ at 50 Hz.
b $10^{-9} F$

15a When it is rolled up, otherwise B and D will touch and short circuit.
b About 100 m.
c 2
d It would be halved, *i.e.* about 50 m.
e The cross-sectional area can be considered as πr^2, or as length (50 m) × thickness of sandwich (0.2 mm). This gives a diameter of about 0.15 m, rather larger than one might want of a 1 μF capacitor. In practice paper of this thickness would not be used.

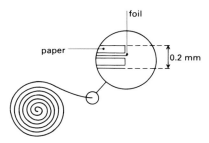

Figure Q27

16ai halved   **ii** same   **iii** halved
iv halved   **v** halved.
bi halved   **ii** same   **iii** doubled
iv doubled   **v** same.

c In **a** energy is transformed and heats the battery as half the charge flows 'the wrong way' through the battery. In **b** energy has to be supplied to move the plates apart.

19a $4\,\mathrm{m\,s^{-1}}$
b $4\,\mathrm{m\,s^{-1}}$
ci $8\,\mathrm{J}$ **ii** $8\,\mathrm{J}$ **iii** $640\,\mathrm{J}$
d Approximately 1 m apart for 10 J intervals of energy; 100 m apart for 1000 J intervals.
e Contour lines, as on a map, joining points at the same potential energy.

20ai $50\,000\,\mathrm{J}$ **ii** $200\,000\,\mathrm{J}$
iii $-50\,000\,\mathrm{J}$ (*i.e.* it *loses* potential energy).
bi $50\,\mathrm{J}$ **ii** $200\,\mathrm{J}$ **iii** $-50\,\mathrm{J}$
ci $50\,\mathrm{J\,kg^{-1}}$ **ii** $200\,\mathrm{J\,kg^{-1}}$
iii $-50\,\mathrm{J\,kg^{-1}}$
di $150\,\mathrm{J\,kg^{-1}}$ **ii** $-100\,\mathrm{J\,kg^{-1}}$
e $120\,000\,\mathrm{J}$

23a The Earth's gravitational pull slows it down since it acts almost in the opposite direction to the motion.
b $7.65 \times 10^{-3}\,\mathrm{m\,s^{-2}}$
c $7.65 \times 10^{-3}\,\mathrm{N\,kg^{-1}}$
d $7.9 \times 10^{-3}\,\mathrm{N\,kg^{-1}}$
e $2.2 \times 10^{-4}\,\mathrm{N\,kg^{-1}}$. Its contribution is beginning to become significant, although still small ($\approx 3\,\%$ of the Earth's field strength).

24a

Mean distance $r/10^6\,\mathrm{m}$	Mean acceleration $g/\mathrm{m\,s^{-2}}$
27.7	-0.453
55.4	-0.122
96.5	-0.042
170.4	-0.013

b Numerical values of gravitational field strength are identical to those of the acceleration; units are $\mathrm{N\,kg^{-1}}$.
c The graph is shown in figure Q28.

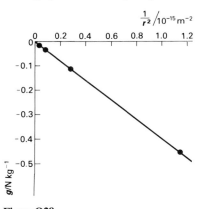

Figure Q28

25a One step. Assuming that g is constant over such a large interval leads to a poor estimate of the area under the graph.
b 1000 steps. This yields the best estimate of ΔV_g.
c One could use an even smaller step size.
d The program would take a very long time to run.
e $2\,\%$
f Nearest the Earth, since this is where g varies most.

26a The graph is shown in figure Q29.

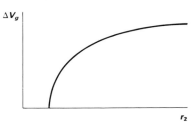

Figure Q29

bi $42.5 \times 10^9\,\mathrm{J}$
ii $10.6 \times 10^9\,\mathrm{J}$
iii $2 \times 10^6\,\mathrm{J}$
c $62.5\,\mathrm{MJ}$
di It will reach a distance of about $32 \times 10^6\,\mathrm{m}$ from the centre of the Earth before falling back towards it.
ii It will escape completely from the Earth, its kinetic energy tending towards $7.5\,\mathrm{MJ}$, when it is a long distance from the Earth.
iii It will just escape from the Earth (assuming no other forces act on it), its kinetic energy tending towards zero as its separation from Earth increases.

27ai The graph is shown in figure Q30.

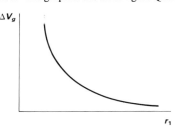

Figure Q30

ii The field strength is decreasing as r_1 increases, so the energy required to move one kilogram a given distance also decreases.
iii Suggestions might include $\Delta V_g \propto 1/r_1$; $\Delta V_g \propto 1/r_1^2$; $\Delta V_g \propto e^{-r_1}$. But note that

whereas $\Delta V_g = 0$ when $r_1 = 50 \times 10^6\,\mathrm{m}$, ΔV_g will never become zero in any of the above relationships, however large r_1 becomes.
bi The graph is shown in figure Q31.

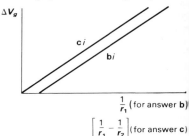

$\left[\dfrac{1}{r_1} - \dfrac{1}{r_2}\right]$ (for answer **c**)

Figure Q31

ii Gradient $\approx 4.0 \times 10^{14}\,\mathrm{N\,m^2\,kg^{-1}}$.
iii Intercept on $\dfrac{1}{r_1}$ axis is very close to the value of $\dfrac{1}{r_2}$, *i.e.*, $2 \times 10^{-8}\,\mathrm{m^{-1}}$.
iv $\Delta V_g = 4 \times 10^{14} \times \left(\dfrac{1}{r_1} - \dfrac{1}{r_2}\right)$
ci See figure Q31.
d It would have had the same gradient but a smaller intercept.

28a Zero.
b No.
c ΔV_g now represents the amount of energy required to remove 1 kilogram to an infinite distance from the Earth, or 'as far away as one would wish'.
d Gravitational potential energy is lost.
e Negative.
f $-62.5 \times 10^6\,\mathrm{J\,kg^{-1}}$
g The graph is shown in figure Q32.

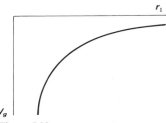

Figure Q32

29a $\Delta V_g \approx \dfrac{4 \times 10^{14}}{r_1}$

b $V_g \approx -\dfrac{4 \times 10^{14}}{r_1}$

c $GM = 3.98 \times 10^{14}\,\mathrm{N\,m^2\,kg^{-1}}$. The two expressions are equal.
d $63 \times 10^6\,\mathrm{J\,kg^{-1}}$, using the approximate value for GM.
$62.5 \times 10^6\,\mathrm{J\,kg^{-1}}$, using the more accurate value.

a They are equal in magnitude and opposite in sign, assuming that no other forces act.
$-34.41 \times 10^6 \, \text{J kg}^{-1}$
$34.41 \times 10^6 \, \text{J kg}^{-1}$
Zero.
$35.33 \times 10^6 \, \text{J kg}^{-1}$
The graph is shown in figure Q33.

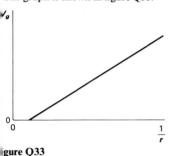

Figure Q33

The gradient is very close to $3.98 \times 10^{14} \, \text{N m}^2 \, \text{kg}^{-1}$, the value of GM_E. The intercept is approximately 2×10^{-9}, giving a value of r_0 of 5×10^8 m.
Zero.
$\Delta V_g = \dfrac{GM_E}{r}$

a 10^5 J.
10 N. This is approximately the force at the Earth's surface, though it decreases with height.
61.5×10^6 J
Because the force is much less than 10 N for most of the distance.
62.5×10^6 J
$11\,200 \, \text{m s}^{-1}$
The energy needed is $\dfrac{GMm}{r} = \frac{1}{2}mv^2$.
m cancels so a larger mass will require the same *velocity*, though of course it has greater *energy*.
The graph is shown in figure Q34.

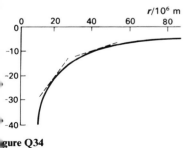

Figure Q34

$1.0 \, \text{N kg}^{-1}$; $0.25 \, \text{N kg}^{-1}$
They represent the magnitudes of the field strengths at those points.

k The ratio is 4:1. Since the ratio of the distances is 1:2, this is quite consistent with an inverse-square field.

32a $-62.5 \times 10^6 \, \text{J kg}^{-1}$; $-2.8 \times 10^6 \, \text{J kg}^{-1}$
b $59.7 \times 10^6 \, \text{J kg}^{-1}$
c 597×10^9 J
d i $-62.5 \times 10^6 \, \text{J kg}^{-1}$
ii $-4.02 \times 10^6 \, \text{J kg}^{-1}$
iii $-1.36 \times 10^6 \, \text{J kg}^{-1}$
iv $-1.32 \times 10^6 \, \text{J kg}^{-1}$
v $-1.35 \times 10^6 \, \text{J kg}^{-1}$
vi $-3.94 \times 10^6 \, \text{J kg}^{-1}$
e, f, g The graph is shown in figure Q35.

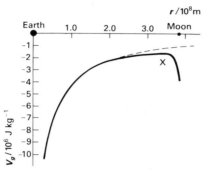

Figure Q35

h 612×10^9 J
i Extra energy will be gained in 'falling' from X towards the Moon. This must be transformed using reverse thrust. Also energy will have been 'lost' as the spacecraft is heated in passing through the Earth's atmosphere; and much of the fuel itself has been transported some distance from the Earth, which requires further energy.
j $2400 \, \text{m s}^{-1}$. It needs only sufficient energy to reach X, which is only about two-thirds of that required to escape completely. After X it will 'fall' towards Earth.
k Many molecules will have speeds in excess of the escape velocity (which is the same for all masses) and so will escape. However many molecules might have been present, a good proportion will always escape, so any atmosphere will soon dwindle to zero.

33a $r_1\theta$
b $\pi r_1^2\theta^2$
c $\sigma\pi r_1^2\theta^2$
d, e $G\dfrac{(\sigma\pi r_1^2\theta^2)}{r_1^2} = G\sigma\pi\theta^2$
f The result is of the same magnitude.
g Zero. The contributions are equal in

size but opposite in direction.
h Zero. There is no field anywhere inside the hollow sphere.
i No.

37a i $\Delta t = \frac{1}{2}T$ ii $\Delta v = 2v$
iii $a = \dfrac{2v}{\frac{1}{2}T} = \dfrac{4v}{T}$
iv $T = \dfrac{2\pi r}{v}$ $a = \dfrac{4}{2\pi}\dfrac{v^2}{r} \approx 0.64\dfrac{v^2}{r}$
b i $\Delta t = \frac{1}{4}T$ ii $\Delta v = \sqrt{2}v$
iii $a \approx 0.90\dfrac{v^2}{r}$
c i $\Delta t = \frac{1}{6}T$ ii $\Delta v = v$
iii $a \approx 0.95\dfrac{v^2}{r}$
d i $\Delta t = \dfrac{\theta}{2\pi}T$ ii $\Delta v \approx PQ = v\theta$
iii $a = \dfrac{v^2}{r}$
iv Acceleration is at right angles to motion.
e Centripetal acceleration $= \dfrac{v^2}{r}$
f Centripetal force, $F = \dfrac{mv^2}{r}$
g A negative sign, $F = -\dfrac{mv^2}{r}$

38a $\dfrac{2\pi r}{T}$
b $-\dfrac{v^2}{r}$
c $-\dfrac{4\pi^2}{T^2}r$
d $-\dfrac{GM}{r^2}$
e $\left(\dfrac{GMT^2}{4\pi^2}\right)^{\frac{1}{3}}$. Only one particular value of r gives the correct period.
f 36×10^6 m above the Earth's surface. $53 \times 10^6 \, \text{J kg}^{-1}$.
g Because nowhere else could the satellite stay directly above a point and move in a circular orbit *with the centre of the Earth as its centre*.
h Shorter.
i $v = \left(\dfrac{GM}{r}\right)^{\frac{1}{2}}$. Faster.
j Briefly, energy is transformed during the impact so the satellite loses energy and cannot remain in its present orbit. In dropping to a lower orbit, it speeds up, as potential energy is now transformed into kinetic energy.
k A satellite in 'geostationary' orbit above the Equator can 'see' nearly a third of the

Earth's surface and so can be used for telecommunications over a very extensive area. However, with nearly 200 communications satellites in this orbit, space is becoming a little crowded, especially in the popular region over the Atlantic. (Signals from nearby satellites must be modified to prevent interference.)

Satellites in low orbits lose energy quickly as they encounter the upper levels of the Earth's atmosphere. Sometimes their short life is an acceptable price to pay for the better 'view' they afford, especially if they are designed for monitoring or military purposes. The more stable orbits further out tend to accommodate satellites serving more civilian purposes such as weather forecasting. Many satellites engaged in scientific research have extremely eccentric elliptical orbits which enable them to penetrate wide regions of the magnetosphere. All in all the first 25 years of space exploration have seen over 3000 satellites go into orbit.

41a Twice the distance.
b Twice the height.
c The same.

d The same.
e Four times the area.
f One-quarter.
g One-ninth.
h It is a constant (called the 'flux' of the light).
i In a parallel beam of light both intensity and area remain constant with distance, and so does the flux.

Figure Q36

But wherever light converges to or diverges from a focal point, the change in intensity with distance must be inverse square if flux is to remain constant. The 'conservation of flux' is a key concept in more advanced physics and you will certainly meet the idea later in the context of another type of field – magnetism.

42a Approximately, F is proportional to d for small angles. The deflection is doubled.
b The points on a graph of d against $1/r^2$ lie near to a straight line.

c $0.01\,\mu\text{F} = 10^{-8}\,\text{F}$. At 0.5 V the charge on it is $0.5 \times 10^{-8}\,\text{C}$. This is nearly the charge that was on the ball. The fluctuations in charge could explain why the points plotted in **b** are scattered on either side of a straight line.
d, e The force constant is of the order $10^{10}\,\text{N m}^2\,\text{C}^{-2}$, and certainly lies between 5×10^9 and $5 \times 10^{10}\,\text{N m}^2\,\text{C}^{-2}$. You may think the limits are closer than this.

43a $10^{-4}\,\text{N}$
b $10^{-9}\,\text{C}$
c $10^{-11}\,\text{A}$
d $10^3\,\text{V}$
e $10^{14}\,\Omega$

45a $Q/4\pi\varepsilon_0 r^2$
b $2r$
c $4Q$
d $4Q/4\pi\varepsilon_0(2r)^2 = Q/4\pi\varepsilon_0 r^2$
ei Upwards.
ii Small.
iii Approximately uniform.
f Approximately inverse-square.

49a You may produce a pattern something like that shown in Q37.
b See figure Q38.

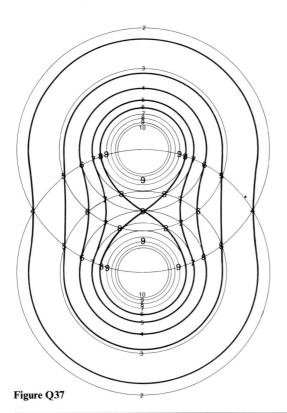

Figure Q37

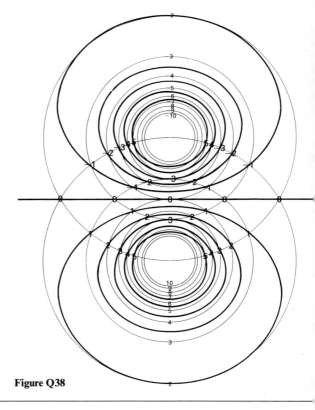

Figure Q38

The field patterns should be similar to those obtained between two point electrodes with semolina particles in experiment E3.

a 900 V

1200 V at X, 1080 V at Y; by simple addition of the potentials due to A and B.

Yes, because there is a potential gradient across it.

Charge would flow towards Y until the potential gradient fell to zero.

No.

The surface of a conductor must always be an equipotential, even if some parts of the surface are charged.

2a 10^{-47} N

10^{40}

Both the electric force and gravity obey inverse-square laws so one force falls off with distance at the same rate as the other.

Approximately 10^{36}.

No.

It falls off very sharply indeed and is quite insignificant outside the nucleus.

3a 1.4×10^{11} N C^{-1} $(= 1$ unit$)$

$\frac{1}{100}$ unit

$-\frac{1}{9}$ unit at C; $-\frac{1}{144}$ unit at D.

$\frac{8}{9}$ unit at C; $\frac{44}{14400}$ unit at D.

$\frac{11}{3200} \approx \frac{1}{290}$.

More rapidly (particularly if you consider that the distance of D from the middle of the dipole is only just over five times that of C). From a good distance the field strengths due to A and B almost cancel; indeed the dipole 'looks' almost like a neutral object.

The array is neutral when 'viewed' from a distance and it is only very near any individual charge that the local field dominates significantly.

5a Electrical/gravitational field strength the force acting on unit charge/mass.

By taking the negative of the gradient of the graph.

By computing the area under the graph between the two points.

By multiplying by the mass or charge.

The negative of the force.

The force is always attractive, so the force vector is directed *inwards*, in the opposite direction to the separation or displacement vector.

This is taken care of by the charges which may be positive or negative, giving both attractive (negative) and repulsive (positive) forces.

h Because the potential energy of a mass at all points is less than that at the agreed zero of potential (at 'infinity').

i A positive charge would have positive potential energy and a negative one negative potential energy.

UNIT F
Radioactivity and the nuclear atom

2a 10^6 V m^{-1}

b 1.6×10^{-13} N

c 1.6×10^{16} m s^{-2}

d 3.3×10^{-10} s

e 9×10^{-4} m ≈ 1 mm

f If $L = l$, the deflection y will be more than twice as big as s, because the straight path from the edge of the plates to the screen cuts the undeflected path through the plates (if projected back) somewhere between the plates, not at the far end of the plates. An estimate of 3 or 4 mm for y might be fair.

4a Towards the end of a track the particle has lost a lot of energy and is travelling more slowly. It is therefore spending more time in the vicinity of the atoms of the air; these atoms are more likely to be ionized.

b The number of ion-pairs is given by the area below the graph from the distance 50 mm back to the origin. One rough estimate put the number of ion pairs at about 200 000.

c The total energy of such an alpha particle is around 6 MeV.

6a 8.7×10^{-19} J

b 8×10^{-13} J

c 2.4×10^{12} J

d The energy released in the radioactive decay of radium is greater than that released on oxidation by a factor of over a million.

e Even with 1 mole of radium (226 g – a very large sample), the energy is dissipated over thousands of years. In the first 1600 years, 0.5 mole will have decayed.

7 The laboratory notes for this Unit are prefaced by a note about the precautions which you should observe when handling radioactive substances. You ought to read this.

Important points to make in explaining the precautions to be taken when using radioactive materials include:

The radiation from a radioactive substance can harm you in several ways if it is absorbed by part of your body. Because one kind of change it can produce is a change of the genetic information stored in the genes in cells, it is especially important to keep such radiation away from the reproductive organs. This does not mean that it is safe to allow the radiation to reach other parts of the body: it is not.

The simplest precaution, and perhaps the most effective, is to keep the source at a distance. Few alpha particles travel more than 50 mm or so in air, so an alpha source held in tongs is pretty safe, as long as the source itself is sealed and you cannot breathe or ingest any of the radioactive substance. Gamma rays spread out so that their intensity falls off as the inverse-square of the distance, so again distance helps, though it does not reduce the radiation received to zero. Lead blocks are used to absorb gamma rays from sources stronger than those in schools.

The only radioactive substances you are allowed to handle which are not in a sealed form are the naturally occurring salts of uranium, thorium, and potassium. These have long half-lives, and the radiation from them is small. The strength of the sealed sources you may use is also limited.

Everybody is continually exposed to radiation from both cosmic rays and the radioactive minerals in the Earth. The various regulations aim to make sure that you and your teacher cannot receive in a year more than a very small extra fraction of the dose you will in any case receive from this natural background.

It is very important that rules and precautions to protect people handling radioactive sources should be framed with care, revised from time to time, and, above all, obeyed.

8 197 g of gold contain 6×10^{23} atoms and occupy $\dfrac{197 \times 10^{-3}}{19.3 \times 10^3}$ m^3. 1 atom of gold occupies $1/(6 \times 10^{23})$ of this space, which is 17×10^{-30} m^3. The cube root of this gives an estimate of the diameter of a gold atom ($\approx 2.6 \times 10^{-10}$ m).

9 If the gold atom is about 2.6×10^{-10} m across, the number of layers, n, in a foil 6×10^{-7} m thick is 2.3×10^3 approximately.

If d is the diameter of a nucleus, then

$$\frac{d^2}{(2.6 \times 10^{-10})^2} \approx \frac{1}{8000n}$$

$$d \approx 6 \times 10^{-14} \, \text{m}$$

The roughness of the data does not justify an accurate calculation, for example allowing for the way gold atoms pack together. The important thing is the order of magnitude of the answer, some 10^4 times smaller than the diameter of the atom.

11a $8 \times 10^{-13} \, \text{J}$
b As the alpha particle approaches the nucleus, this kinetic energy is converted to electrical potential energy. When the speed of the particle is zero, this conversion is complete and the energy stored in the system is $8 \times 10^{-13} \, \text{J}$.

The potential energy is given by

$$E_p = \frac{(2e)(79e)}{4\pi\varepsilon_0 r}$$

on the assumption that the gold nucleus acquires very little kinetic energy.
c $r = 4.6 \times 10^{-14} \, \text{m}$
d The electric potential at this distance is
$$V = \frac{79e}{4\pi\varepsilon_0 r} = 2.5 \times 10^6 \, \text{V}.$$
e Since the alpha particle carries charge $+2e$, its original kinetic energy in eV will be double this number, i.e., $5.0 \times 10^6 \, \text{MeV}$.
f The electric field at this distance
$$E = \frac{V}{r} = \frac{2.5 \times 10^6 \, \text{V}}{4.6 \times 10^{-14} \, \text{m}} = 5.4 \times 10^{19} \, \text{V m}^{-1}$$
The force on a charge of $2e$ is about 17 N.

This is about the same as the weight of a mass of 1.5 kg – an enormous force to find exerted on a single atomic particle. The book was the best guess.

12a An attractive force which decreases with distance from A.
b A repulsive force which decreases with distance from A.
c No significant force at the distances shown.
d A repulsive force which suddenly affects P at a small distance from A.

14ai Not very good. It describes roughly what happens, but may give the wrong impression about why. The particles come in by chance, and if the chance of arrival stays the same, over a long time there will be a steady average rate.

ii We tried to make this a correct description.
iii Wrong. The time between arrivals may have any value, but not all intervals are equally likely. The average interval is most likely; much longer or shorter intervals are less likely.
b To say that in the kinetic model the molecules of a gas have random motion, implies that a molecule may be moving in any direction at any instant (not one of which is preferred). Molecular speeds are distributed randomly about a mean value; this mean value is the most likely to be found whilst far higher or far smaller speeds are unlikely.

16 A common answer is that a radioactive sample has an infinite lifetime 'in theory but not in practice'. Although not absolutely wrong, this seems to us to be a bit feeble.

The smooth mathematical model of exponential decay does not exactly fit the behaviour of radioactive atoms. As you may have found by experiment, the rate of decay of a sample fluctuates considerably. The average rate of decay is close to the value to be expected from the mathematical model, but need not be equal to it.

When the sample is reduced to only a few atoms, the smooth exponential model is a poor fit. Yet it is not enough to say that 'the theory breaks down'. The smooth exponential decay is a consequence of supposing that there is a constant chance of decay for each atom in each time interval, and also that there are very many atoms. When there are not many atoms left, the smooth decay is not to be expected. But so far as is known, the chance of decay remains constant. When one atom remains, that atom may decay in the next second, or in the next hundred years. It is possible to say how long it will last on average: that is, how long one atom will last in many trials. But it is not possible to say how long one particular atom will last.

19a Both are exponential. The lengths of the straws decrease in a fixed proportion.
b Plot the natural logarithm of y against x; or check whether y changes by a constant factor as x increases in equal steps.

21a Although the graph of ln (capacity in GW) against time does not follow a

straight line, there are periods when it does so with reasonable accuracy. From 1951 to 1958 it was very nearly straight (and the growth exponential), and also from 1964 to 1970.
b The growth constant diminished in about 1962. During the 1970s growth ceased and the capacity remained approximately constant.
c The 'doubling times' were
i about 10 years, and
ii about 11 years.

24a The change in the number of families with video recorders.
b The rate at which the number of such families changes.
c The model $\Delta N/\Delta t = kN$ suggests that the rate at which the number of families with video recorders grows is directly proportional to the number of families already having video recorders. This would give an exponential growth. In real life this could be the result of 'keeping up with the Joneses'.
d There could be a limit on the supply of money available, costs might rise unexpectedly, or the supply of the equipment might become limited. Such changes would reduce the value of k after a time.
e The existence of television could increase the value of k.
f See figure Q39.

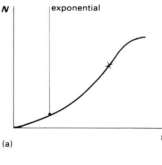

(a)

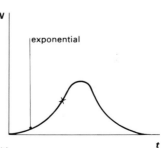

(b)

Figure Q39

25a Press, in order, the keys 2, x^y, 7, = (or the keys 2, log, ×, 7, =, 10^x; or multiply the logarithm of 2 by 7 and find the number whose logarithm is this product).

b $N = 1.414$

c,d The graphs start at $N = 1$ because at $t = 0$, $N = a^0 = 1$.

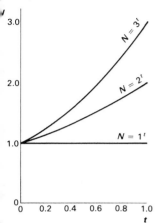

Figure Q40

Plots of $N = 1^t$, $N = 2^t$, $N = 3^t$.

e At each step of 0.2 in t, the value of N is multiplied by the same factor, which is $2^{0.2}$, or 1.15.

f $2^{0.2} = 1.15$; $2^{3.2} = 1.15 \times 8 = 9.20$.

g In the series a^t, a^{2t}, a^{3t}, etc., each exponent is equal to the previous one with the addition of t. Adding t to the exponent of a means multiplying by a^t; similarly, adding x to the exponent of a means multiplying by a^x.

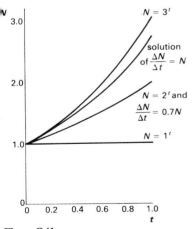

Figure Q41

Plots of $N = 1^t$, $N = 2^t$, $N = 3^t$, and solution of $\dfrac{\Delta N}{\Delta t} = +(1.0)\, N$.

26a, b (See figure Q41.)

a ΔN is a little larger in the second 0.1 s interval because the starting value, N, is now larger (1.1 instead of 1.0).

b Growth is slower because the constant ratio is now 0.7, not 1.0.

c If $N = 2^t$, then $N = 2$ at $t = 1$.

d $N = a$ at $t = 1$.

e For $\Delta N/\Delta t = (1.0)N$, $N \approx 2.7$ at $t = 1$, so $a \approx 2.7$.
For $\Delta N/\Delta t = (0.7)N$, $N \approx 2.0$ at $t = 1$, so $a \approx 2.0$.

27a $e^{0.7} = 2.01 \approx 2.0$.

b Since $(e^{0.7})^t = e^{0.7t}$, the best speculation is $a = e^k$.

29a $1.993 \times 10^{-26}\,\text{kg}$

b $1.661 \times 10^{-27}\,\text{kg}$

c 1.007 u; 1.008 u

d 0.000 548 u

32a One gram of carbon contains 0.5×10^{23} atoms, since 12 g contain 6×10^{23} atoms. Of these, about 0.5×10^{13} atoms are the isotope ^{14}C. In one second, 4×10^{-12} of the ^{14}C atoms will decay, so in 1 g of carbon one may expect around 20 decays per second, an activity of 20 Bq.

b 11 000 years is close to two half-lives, so the activity might be around 5 Bq from one gram.

c The rate of decay is small, and the smaller it is, the harder it will be to measure the rate accurately unless one is prepared to measure it over a very long period. Even with counting times of about a day, the ^{14}C method is subject to pretty large errors for times above about two half-lives, and is not of great assistance much beyond three half-lives, say 20 000 years.

34a Mass of alpha particle $\approx 7 \times 10^{-27}$ kg.

b Speed $\approx 7 \times 10^6\,\text{m s}^{-1}$. Note that to use the equation, kinetic energy $= \frac{1}{2}mv^2$, with m in kilograms, the energy must be in joules, not electronvolts.

36a These are the inert gases, which rarely form compounds with other elements. They have high ionization energies. It is not easy to remove an electron from one of these atoms.

b These are the alkali metals, all very reactive, readily forming singly charged positive ions. In a crystal of sodium chloride, the sodium atoms are all ionized, and the crystal is a vast assembly of Na^+ and Cl^- ions. All the alkali metals have low ionization energies, indicating the ease with which an atom loses an electron.

c Neon has a large ionization energy; sodium, with one more electron, has a low ionization energy; it seems that the 'last' electron in a sodium atom is rather loosely held. (In fact, it takes 47 eV to remove a second electron from sodium, so it is only the 'last' one which is nearly free.)

d Yes, each comes first on the rise following a trough. They are members of another family with chemical similarities.

e If francium is the element one before radium, it must be an alkali metal, like Li, Na, K, Rb, Cs. Their ionization energies are all seen to be low, from 3 to 5 eV. The value drops slowly as one goes along the list, so francium will probably have an ionization energy nearer the lower end of this range.

38a -2.18×10^{-18} J

b $-e^2/4\pi\varepsilon_0 r$

c 1.0×10^{-10} m

d The *total* energy of the electron is -2.18×10^{-18} J. Since it has some kinetic energy, which must be positive, the potential energy is more negative than -2.18×10^{-18}, *i.e.*, the electron is closer to the proton than this calculation suggests. (In fact, as will be seen in Unit L, 'Waves, particles, and atoms', the electron is not precisely located at one distance from the proton. But this calculation gives the correct order of magnitude for the proton–electron distance.)

42a F_e is repulsive if Q_1 and Q_2 are both positive or both negative; F_g is always attractive.

b $F_e/F_g = Q_1 Q_2/4\pi\varepsilon_0\, Gm_1 m_2$

c Both forces vary in the same way with distance (inverse-square law), so their ratio is the same at all distances.

d 1.2×10^{36}

e No; it is much smaller than the repulsive electrical force.

f Nuclei of neighbouring atoms would be attracted to each other: they would not stay about 10^{-10} m apart.

43a ^{12}C is the more stable.

b ^{2}H, ^{6}Li, ^{4}He, ^{235}U, ^{12}C, ^{208}Pb, ^{56}Fe.

c Helium is at a minimum in the curve. The average binding energy per nucleon is lower for some heavier nuclei (*e.g.*, ^{12}C, ^{16}O), but to form nuclei like ^{6}Li which are likely to be involved in intermediate steps in the build-up of heavier nuclei requires an input of energy.

d C, O, Fe.

You can read more about the evolution of the elements in the article 'Our nuclear history' in the Reader *Particles, imaging, and nuclei*.

45a 0.001 kg (1 gram).
b 0.5 J (taking $k = 100\,\mathrm{N\,m^{-1}}$, and extension $= 0.1$ m).
c 6×10^{-18} kg
d 6×10^{-13} per cent.

46a $2m_\mathrm{p} + 2m_\mathrm{n} + 2m_\mathrm{e} = 4.033\,0$ u. The mass of an atom of $^4_2$He is 0.030 4 u *less* than the mass of its parts.
bi -4.54×10^{-12} J   ii -28.4 MeV
c -7.1 MeV
(Note that because the information refers to a neutral helium atom – nucleus plus electrons – the binding energy calculated includes the binding energy of the electrons as well as that of the nucleons. However, since electrons are so much more weakly bound than nucleons, the value obtained is still a good estimate for the average binding energy per nucleon.)

UNIT G
Energy sources

1ai The potential energy of the mass is the work which would be done by the gravitational field if the mass were allowed to fall to some reference point (often the surface of the Earth). The mass needs the presence of the Earth in order to have potential energy, which is therefore a property of the Earth–mass system.
ii The gravitational field acting on the mass transfers energy to the generator by working. The mass loses energy and the generator gains energy (to an equal extent in an ideal system).
iii This means that a quantity of energy (15 J in this example) would be transformed if the generator were brought to rest.
b Boxes should be inserted at every stage of conversion. They all represent internal energy of the surroundings.
c The boxes should have solid boundaries since they represent continuing 'losses' to the system.
d It depends what you mean by the system. If this is taken to be the 'Universe' then the energy is constant, though clearly it is not so if the mass–generator–lamp

arrangement is regarded as the system. In this latter case energy is lost from the system. Because the 'losses' are only losses in this limited sense quotation marks are used.

2 The Sankey diagram is shown in figure Q42.

4a 3.6 MJ
b About 35 MJ.
c About 29×10^9 J.
d 3.6×10^{12} J
e 10^3 J
f About 250 kJ.
g 2.9×10^{20} J
h 1.6×10^{-11} J
i 1060 J

6a About 32 %.
b About 93 %.
c About 29 %.
d About 92 %.
e About 91 %.
f Roughly $3\frac{1}{2}$ times as much as the useful output.

7a For equal heating costs electricity can be $1\frac{1}{3}$ times as expensive as gas.
b 1 kW h of heating by electricity requires about $2\frac{1}{2}$ times as much primary fuel as using gas for the same purpose.

8a World population has been growing approximately exponentially since about 1950 with a doubling-time of about 37 years.
b Between 1925 and 1986 world fuel consumption increased by a factor of about 7, whereas the population grew by factor of about $2\frac{1}{2}$. Clearly the fuel consumption *per capita* increased during this period.

9a The range of values is such that the graphs can only be plotted over about 12 hours, from 0–12 or from 12–24 hours.
b One hour before it is full, *i.e.*, after 23 hours.
c Probably not until $\frac{1}{8}$ or $\frac{1}{4}$ of the nutrient is used up. This means at 21 or 22 hours, *i.e.*, with only 2 or 3 hours to go before all the resource is used.
d 1 jar; 2 jars.
e It would only give one more hour's life.
This question emphasizes in various ways the very rapid use of resources which occurs when growth of consumption is exponential.

10 Doubling-time is 35 months, to the nearest month.

11 $t_\mathrm{D} = 0.693/g$

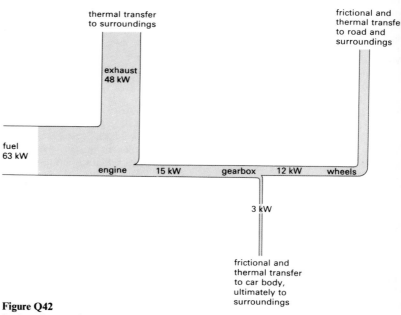

Figure Q42
Sankey diagram for a car.

2 Coal 199 years; oil, including shale, tarsands, etc., 56 years or 28 years for oil only; natural gas 46 years.

It is not clear in table G10 whether the reserves of oil and gas are proved reserves or recoverable reserves. If they represent proved reserves, the lifetimes would be less or the smaller proportion which can be recovered. On the other hand, oil and gas discoveries continue to be made which will increase the lifetimes. The main point is, however, that the lifetimes are very short, even assuming zero growth.

3a Gas consumption increased by 41.5% over ten years from 1976 to 1986. Assuming a growth rate of 3.15% *per annum* gives a doubling-time of 0.693/0.0315 = 22 years.

Similarly, the doubling-time for coal consumption is 24 years. "Although oil consumption was almost the same in 1986 as in 1976, there were fluctuations in the decade. Consumption reached a peak in 1979, fell until 1983, and then rose again from 1983 to 1986."

b Taking the growth rate as an average of ten years' growth smooths out the variations which occur over a shorter period. If, however, the growth is very rapid (as in the case of nuclear power from 1976 to 1986) the average value over ten years is not the same as the year-on-year value which applies throughout the period.

In fact, an annual growth rate of 3.15% leads to growth in ten years by a factor $e^{(0.0315 \times 10)} = 1.370$, rather than 1.315, *i.e.*, using the 'average' growth rate overestimates the growth.

c Using 1986 figures the lifetimes of the fuels are: oil – 28 years; coal – 66 years; natural gas – 28 years.

d The additional resources increase the lifetimes as follows: oil – 56 years (assuming the oil shale and tarsands reserves are all recoverable); coal – 131 years; natural gas – no change since no additional resources are listed.

e If the price of fuel rises, the figure for recoverable reserves also rise since reserves which could not previously be recovered economically now become viable.

19a The U.K. and U.S.S.R. have a slight overall surplus of production over supply, North America has a slight deficit, and Japan has a large deficit.

b For water power and nuclear energy the figures for production and consumption are identical. Only one set of figures is published, presumably because storage of electrical energy on any reasonable scale is not possible. The figures for coal, natural gas, and oil can be processed as in part **a**. The calculations are simple but the real point of the question is dealt with in parts **c** and **d**.

c Short-term shortfalls can be dealt with by importing from nations which have surplus resources. The importing nations must have some means of earning the currency to buy the fuel and the fuel-exporting nations may have some degree of control over the economies of the importers (for example, the Organization of Petroleum Exporting Countries (OPEC) in the late 1970s).

d Nations like Japan, with continuing shortfalls, are likely to wish to reduce their dependence on imported fuels. They may have to develop alternatives much more purposefully than do nations with fossil fuel resources. To some extent, what happens to Japan today happens to the World tomorrow.

21a*i* The area needed is 0.13 km^2.
ii 0.007 miner.
iii Area of panel needed = 70 m^2.
b Roughly three times as much resource would be needed.

23a 0.04%
b 4%
c 5.5%
d 7.5%
e 0.25%

24a A base load capacity of about 39 GW is required. About 9 GW of hydroelectric power will also be needed.
b 9000 m^3 of water would be needed per second to provide the peak power output. The total energy required is about 98 GW h. This would be provided by the contents of a lake of area about 12 km^2 and 30 m depth falling through 100 m.

25a Volume of the ring main = 56 000 m^3.
b The pressure falls.
c About 1.4×10^6 m^3.
d About 3.9×10^6 m^3.
e The increase in pressure which would result from storing all this gas in the main would be about 70×10^5 Pa (*i.e.*, from 25×10^5 to 95×10^5 Pa) which takes the main pressure above its limit. Over half ($\frac{40}{70}$) would have to be stored elsewhere.

28 About 2.3%.

29 From question **27** increasing expectations seem to account for a rate of growth of 2.1%. Table G11 shows much larger increases for less developed countries than for developed ones.

31a*i* No. The number of electrons is the same on each side.
ii Change in mass is 0.192 u or 3.18×10^{-28} kg.
iii 2.86×10^{-11} J
b*i* $\dfrac{Q_1 Q_2}{4\pi\varepsilon_0 d}$
ii 4.61×10^{-11} J
c*i* 9×10^{13} J
ii About the same.
iii 143 kg
iv 143 tonnes
v About 30 times as much material needs to be used to provide the same amount of energy from coal as from uranium.

32a a/A
b $\dfrac{nAva}{A} = nav$
c NAd
d $nav \times NAd$
e $R = \dfrac{nav \times NAd}{A \times d} = nNav$
f $a = R/nNv$

33 0.1 eV. This is a factor of about 10^7 less than the energy of a neutron released in fission of ^{238}U.

37a $\dfrac{\Delta E_k}{E_k} = \dfrac{4_m M}{(M + m)^2}$
b M must be 1 u.
d Hydrogen.

38 54

40a 3.1×10^{19}
b 1.2×10^{-5} kg
c Total mass of ^{235}U = 1150 kg.
d The mass of the core is 52×10^3 kg.

42a 0.63 K W^{-1}
b 1.3 kW

43a 2.7×10^{-3} K W^{-1}
b 16×10^{-3} K W^{-1}
c*i* 1.3 kW
ii 7.5 kW
iii 8.8 kW
d*i* $I_T = I_1 + I_2$
ii The two calculations are analogous.

44a 19×10^{-3} K W^{-1}
b 570 W

46a $X/2$, $X/3$, X/A.
b $\mathscr{R} = X/A$. If, in spite of being led to the answer, you have ended up with the relation $X = \mathscr{R}/A$, think again why this cannot be correct.
c $m^2 K W^{-1}$
d $X = l/k$

47a $0.17 \, m^2 \, K \, W^{-1}$
b $0.13 \, m^2 \, K \, W^{-1}$

48a $2 \times 10^{-3} \, K \, W^{-1}$
b $10 \, kW$

50a Series
b $0.194 \, m^2 \, K \, W^{-1}$. The power loss is reduced to about one-fiftieth of its previous value.
c Outer surface of glass is at $6.2 \,^{\circ}C$. Inner surface of glass is at $6.6 \,^{\circ}C$.

52b Power loss through double-glazed window is about 120 W, *i.e.*, reduced to 0.6 of the single-glazed value.
c, d See figure Q43.

56a See figure Q44
b $0.114 \, K \, W^{-1}$
c 140 W
d $15.6 \,^{\circ}C$
e $0.074 \, K \, W^{-1}$
f 217 W
g $11.7 \,^{\circ}C$

57 Area of window $= 1.25 \, m^2$.

59 Power needed to heat the air $= 1.3 \, kW$. This is substantially greater than the loss through either single- or double-glazed windows in questions **50** and **52**.

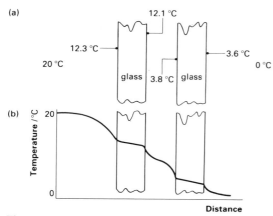

Figure Q43
(a) Temperatures at surfaces in a double-glazed window.
(b) Temperature variation with distance near a double-glazed window.

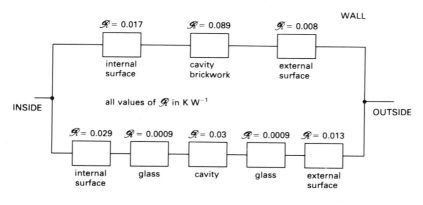

Figure Q44
Thermal resistances for wall with window.

REFERENCE MATERIAL

Textbooks and further reading

Textbooks that are useful throughout the course are listed here. Other books and references that are particularly relevant to specific Units are listed individually.

AKRILL, T. B., BENNET, G. A. G., and MILLAR, C. J. *Physics*. Arnold, 1979.
BOLTON, W. *Patterns in physics*. McGraw-Hill, 1974.
DUNCAN, T. *Physics: a textbook for advanced level students*. Murray, 1982.
OR
DUNCAN, T. *Advanced physics: materials and mechanics*. 2nd edn. Murray, 1981.
DUNCAN, T. *Advanced physics: fields, waves, and atoms*. 2nd edn. Murray, 1981.
WENHAM, E. J., DORLING, G. W., SNELL, J. A. N., and TAYLOR, B. *Physics: concepts and models*. 2nd edn. Addison-Wesley, 1984.

Unit A Materials and mechanics

In addition to the general list of textbooks above, these books are particularly useful for Unit A.

Strongly recommended for background reading
CRAC Hobsons Science Support Series, *Stress, strain and strength*. Hobsons Press, 1983.
CRAC Hobsons Science Support Series, *Gases and gas laws*. Hobsons Press, 1983.
GORDON, J. E. *The new science of strong materials*. Pitman, 1979. (First published by Penguin, 1968.)
GORDON, J. E. *Structures*. Pitman, 1979. (First published by Penguin, 1978.)
OGBORN, J. *Molecules and motion*. Longman, 1973. (Out of print.)
Materials. W. H. Freeman, 1967. (A *Scientific American* book.)

Textbooks for reference
AKRILL, T. B. and MILLAR, C. J. *Mechanics, vibrations, and waves*. Murray, 1974.
BOLTON, W. *Study Topics in Physics Volume 1: Motion and force*. Butterworths, 1980.
BOLTON, W. *Study Topics in Physics Volume 2: Materials*. Butterworths, 1980.
COLLIEU, A. MCB. and POWNEY, D. J. *The mechanical and thermal properties of materials*. Arnold, 1973.

MARTIN, J. W. and HULL, R. A. *Elementary science of metals*. Wykeham, 1969.
NUFFIELD REVISED ADVANCED CHEMISTRY *Students' book I*. Longman, 1984. (Topics 1, 3, 4, and 7.)
NUFFIELD REVISED PHYSICS *Pupils' texts*. Longman, 1978. (Especially Year 4.)

Further reading
BRONOWSKI, J. *The ascent of Man*. Futura, 1981. (First published by B.B.C., 1973.)
CLARKE, A. C. *The fountains of paradise*. Gollancz, 1979.
SCHOOLS COUNCIL Engineering Science Project, *Structures*. Macmillan, 1982.
SCHOOLS COUNCIL Engineering Science Project, *Dynamics*. Macmillan, 1982.
SCHOOLS COUNCIL Engineering Science Project, *The use of materials*. Macmillan, 1975. (Out of print.)
FARRER, R. A. Methuen Studies in Science, *The mechanical properties of materials*. Methuen, 1971. (Out of print.)
BLUNDELL, A., HAWKINS, R., and LUDDINGTON, D. Schools Council Modular Courses in Technology, *Structures*. Oliver and Boyd, 1981.
MCSHEA, J. Schools Council Modular Courses in Technology, *Materials technology*. Oliver and Boyd, 1981.
NUFFIELD ADVANCED PHYSICS *Physics and the engineer*. Longman, 1973. (Out of print.)
NUFFIELD CHEMISTRY Chemistry Background Book, *The structure of substances*. Longman, 1967. (Out of print.)
NUFFIELD CHEMISTRY Chemistry Background Book, *Plastics*. Longman, 1967. (Out of print.)
NUFFIELD REVISED CHEMISTRY *Handbook for pupils*. Longman, 1978.
NUFFIELD REVISED ADVANCED CHEMISTRY Special Studies, *Metals as materials*. Longman, 1985.

Unit B Currents, circuits, and charge

Textbook for reference
BENNET, G. A. G. *Electricity and modern physics*. 2nd edn. Arnold, 1974.

Background reading
CRAC Hobsons Science Support Series, *Instrumentation systems*. Hobsons Press, 1982.

MILLIKAN, R. A. Phoenix Science Series, *The electron*. University of Chicago Press, 1963.

Unit C Digital electronic systems

A large and rapidly growing number of books is available covering the work of this Unit. Some that you may find useful are listed below; try other books and magazines too.

General
ELECTRONIC SYSTEMS TEACHING PROGRAMME ESP 700 Book 3, *Processing systems*. 2nd edn. Feedback instruments, 1978.

The following books can be obtained in the U.K. from R.S. Components Ltd, but they must be ordered on school or college headed notepaper.
TEXAS INSTRUMENTS 'Understanding' series:
CANNON, D. L. and LUECKE, G. *Understanding communications systems*. Texas Instruments Inc., 1980.
CANNON, D. L. and LUECKE, G. *Understanding microprocessors*. Texas Instruments Inc., 1979.
MCWHORTER, G. *Understanding digital electronics*. Texas Instruments Inc., 1978.
TEXAS INSTRUMENTS LEARNING CENTER *Understanding solid state electronics*. 3rd edn. Texas Instruments Inc., 1978.

Computers
THOMPSON, D. L. *Inside the micro*. Unilab, 1982.

Practical circuits
MARSTON, R. M. *110 C-MOS digital I.C. projects for the home constructor*. Newnes, 1976.

The development and impact of microelectronics
BRAUN, E. and MACDONALD, S. *Revolution in miniature: history and impact of semiconductor electronics*. 2nd edn. Cambridge University Press, 1982.
Microelectronics. W. H. Freeman, 1977. (A *Scientific American* book of the September 1977 issue.)

Unit D Oscillations and waves

Textbooks for reference
As well as the general list of textbooks on page 491, these books are particularly useful for Unit D.

AKRILL, T. B. and MILLAR, C. J. *Mechanics, vibrations and waves.* Murray, 1974.

FEYNMAN, R. P., LEIGHTON, R. B., and SANDS, M. The Feynman lectures on physics, *Volume 1: Mainly mechanics, radiation, and heat.* Addison-Wesley, 1963.

PSSC *Physics.* 5th edn. Heath, 1981.

ROGERS, E. M. *Physics for the inquiring mind.* Oxford University Press, 1960.

Books for further reading
BISHOP, R. E. D. *Vibration.* 2nd edn. Cambridge University Press, 1979.

CHAUNDY, D. C. F. Longman Physics Topics, *Waves.* Longman, 1972.

DORLING, G. W. Longman Physics Topics, *Time.* Longman, 1973.

GRIFFIN, D. R. Science Study Series, No. 4, *Echoes of bats and men.* Heinemann, 1960. (Out of print.)

HOWSE, D. *Greenwich time and the discovery of the longitude.* Oxford University Press, 1980.

HUTCHINS, C. M. *The physics of music.* Freeman, 1978. (A *Scientific American* book.)

PROJECT PHYSICS Reader Unit 3. *The triumph of mechanics.* Holt, Rinehart, and Winston, 1971.

TRICKER, R. A. R. *Bores, breakers, waves, and wakes.* Mills and Boon, 1965. (Out of print.)

BASCOM, W. 'Ocean waves'. *Scientific American.* Volume **201**(2), Aug. 1959. (Offprint No. 828.)

BERNSTEIN, J. 'Tsunamis'. *Scientific American.* Volume **191**(2), Aug. 1954. (Offprint No. 829.)

BULLEN, K. E. 'The interior of the Earth'. *Scientific American.* Volume **193**(3), Sept. 1955. (Offprint No. 804.)

CRAC Hobsons Science Support Series, *Vibrations.* Hobsons Press, 1983.

CRAC Hobsons Science Support Series, *Waves and sound.* Hobsons Press, 1982.

GOULD, R. T. *John Harrison and his timekeepers.* 4th edn. National Maritime Museum, 1978.

GRIFFIN, D. R. 'More about bat radar'. *Scientific American.* Volume **199**(1), July 1958. (Offprint No. 1121.)

LYONS, H. 'Atomic clocks'. *Scientific American.* Volume **196**(2), Feb. 1967. (Offprint No. 225.)

OLIVER, J. 'Long earthquake waves'. *Scientific American.* Volume **200**(3), Mar. 1959. (Offprint No. 827.)

Note: Although *Scientific American* Offprints can no longer be purchased in Europe, the titles listed here may still be available in many schools and colleges.

Unit E Field and potential

Textbook for reference
ROGERS, E. M. *Physics for the inquiring mind.* Oxford University Press, 1960.

Further reading
ASIMOV, I. *The collapsing Universe: the story of black holes.* Corgi, 1978.

BRONOWSKI, J. *The ascent of Man.* Futura, 1981. (First published by B.B.C., 1973.)

CALDER, N. *The key to the Universe.* B.B.C., 1977. (Out of print.)

DAVIES, P. C. W. *The forces of nature.* Cambridge University Press, 1979.

FEYNMAN, R. P., LEIGHTON, R. B., and SANDS, M. The Feynman lectures on physics, *Volume 1: Mainly mechanics, radiation, and heat.* Addison-Wesley, 1963.

GAMOW, G. *Gravity.* Heinemann, 1962. (Out of print.)

KOESTLER, A. *The sleepwalkers.* Penguin, 1970. (First published by Hutchinson, 1968.)

Articles
COHEN, I. B. 'Newton's discovery of gravity'. *Scientific American.* Volume **244**(3), March 1981, page 122.

HUGHES, J. E.. 'Industrial hazards of electrostatics'. *Phys. Technol.* Volume **12**, January 1981, page 10.

MOORE, A. D. 'Electrostatics'. *Scientific American.* Volume **226**(3), March 1972, page 46.

NUFFIELD ADVANCED PHYSICS *Physics and the engineer.* Longman, 1972. (Out of print.) (The reference is to the article, 'Electrostatic engineering' by N. J. Felici.)

ROSE-INNES, A. 'Static electricity'. *New Scientist.* Volume **94**, 6 May 1982, page 333.

Unit F Radioactivity and the nuclear atom

Textbooks for reference
BENNET, G. A. G. *Electricity and modern physics.* 2nd edn. Arnold, 1974.

CARO, D. E., MCDONNELL, J. A., and

SPICER, B. M. *Modern physics.* 3rd edn. Arnold, 1978.

LEWIS, J. L. Longman Physics Topics, *Electrons and atoms.* Longman, 1972.

LEWIS, J. L. and WENHAM, E. J. Longman Physics Topics, *Radioactivity.* Longman, 1970.

ROGERS, E. M. *Physics for the inquiring mind.* Oxford University Press, 1960.

WRIGHT, S. (Ed.) *Classical scientific papers – physics.* Mills and Boon, 1964. (Out of print.)

Further reading
CLARK, D. H. *The universe and Man.* Rutherford Appleton Laboratory Monograph, 1981.

CLOSE, F. E. *Atoms, particles, leptons and quarks.* Rutherford Appleton Laboratory Monographs, 1980.

CROW, J. F. 'Ionizing radiation and evolution'. *Scientific American.* Volume **201**(3), September 1959, p138.

DAMERELL, C. J. S. *Experimental particle physics.* Rutherford Appleton Laboratory Monographs, 1981.

DAVIES, P. C. W. *The forces of nature.* Cambridge University Press, 1979.

HUGHES, D. J. Science Study Series No 1, *The neutron story.* Heinemann, 1960. (Out of print.)

HURLEY, P. M. Science Study Series No. 5, *How old is the Earth?* Heinemann, 1960. (Out of print.)

PROJECT PHYSICS Text Unit 6, *The nucleus.* Holt, Rinehart, and Winston, 1981.

The monographs are available free from:
The Library
Rutherford Appleton Laboratory
Chitton
Didcot
Oxfordshire
OX11 0QX

Unit G Energy sources

CHAPMAN, P. *Fuel's paradise.* Penguin, 1980.

ELKINGTON, J. *Sun traps.* Penguin, 1984.

FOLEY, G. *The energy question.* 2nd edn. Penguin, 1981.

LEWIS, J. L. (Ed.) *Science in society.* Heinemann, 1981.

MCMULLEN, J. T., MORGAN, R., and MURRAY, R. B. *Energy resources.* 2nd edn. Arnold, 1983.

RAMAGE, J. *Energy: a guide book.* Oxford University Press, 1983.

SOLOMON, J. Science In a Social Context Series. Blackwell/ASE, 1983.

DATA; FORMULAE AND RELATIONSHIPS; SYMBOLS

Data

(Values are given to three significant figures, except where more – or less – are useful)

Physical constants

speed of light	c	$3.00 \times 10^8 \, \text{m s}^{-1}$
permittivity of free space	ε_0	$8.85 \times 10^{-12} \, \text{C}^2 \, \text{N}^{-1} \, \text{m}^{-2}$ (or F m^{-1})
electric force constant	$\dfrac{1}{4\pi\varepsilon_0}$	$8.98 \times 10^9 \, \text{N m}^2 \, \text{C}^{-2} \, (\approx 9 \times 10^9 \, \text{N m}^2 \, \text{C}^{-2})$
permeability of free space	μ_0	$4\pi \times 10^{-7} \, \text{N A}^{-2}$ (or H m^{-1})
charge on electron	e	$-1.60 \times 10^{-19} \, \text{C}$
mass of electron	m_e	$9.11 \times 10^{-31} \, \text{kg} = 0.000\,55 \, \text{u}$
mass of proton	m_p	$1.673 \times 10^{-27} \, \text{kg} = 1.007\,3 \, \text{u}$
mass of neutron	m_n	$1.675 \times 10^{-27} \, \text{kg} = 1.008\,7 \, \text{u}$
mass of alpha particle	m_α	$6.646 \times 10^{-27} \, \text{kg} = 4.001\,5 \, \text{u}$
Avogadro constant	L, N_A	$6.02 \times 10^{23} \, \text{mol}^{-1}$
Planck constant	h	$6.63 \times 10^{-34} \, \text{J s}$
Boltzmann constant	k	$1.38 \times 10^{-23} \, \text{J K}^{-1}$
molar gas constant	R	$8.31 \, \text{J mol}^{-1} \, \text{K}^{-1}$
gravitational force constant	G	$6.67 \times 10^{-11} \, \text{N m}^2 \, \text{kg}^{-2}$

Other data

molar volume of a gas at s.t.p.	V_m	$2.24 \times 10^{-2} \, \text{m}^3$
standard temperature and pressure		$273 \, \text{K} \, (0°\text{C})$, $1.01 \times 10^5 \, \text{Pa}$ (1 atmosphere)
gravitational field strength at Earth's surface	g	$9.81 \, \text{N kg}^{-1}$
mass of Earth		$5.98 \times 10^{24} \, \text{kg}$
GM for Earth		$\approx 4 \times 10^{14} \, \text{N m}^2 \, \text{kg}^{-1}$
mass of Moon		$7.35 \times 10^{22} \, \text{kg}$
average separation of Earth and Moon		$3.82 \times 10^8 \, \text{m}$
mean radius of Earth		$6.37 \times 10^6 \, \text{m}$

| mean radius of Moon | | 1.74×10^6 m |
| the number e, the base of natural logarithms | e | 2.718... |

Conversion factor

| unified atomic mass unit | 1 u | $= 1.661 \times 10^{-27}$ kg |

Formulae and relationships

Motion and forces

$$\text{linear momentum} = mv \qquad (\text{mass } m, \text{ velocity } v)$$

$$\text{force} = \text{rate of change of momentum}$$

$$F = ma \text{ if mass is constant (force } F, \text{ acceleration } a)$$

$$\text{impulse} = F\Delta t$$

$$\text{translational kinetic energy} = \tfrac{1}{2}mv^2$$

$$\text{gravitational potential energy difference} = mgh \qquad (\text{uniform field strength } g, \text{ height } h)$$

$$\text{energy transformed (work)} = \text{component of force} \times \text{displacement}$$

components of force in two perpendicular directions:

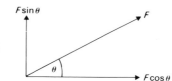

$$\text{moment of force about a point} = \text{force} \times \text{perpendicular distance from point to line of action of force}$$

static equilibrium conditions:

$$\Sigma F = 0$$

$$\Sigma \text{ moments} = 0$$

limiting friction

$$F = \mu N \qquad (\text{coefficient of friction } \mu, \text{ normal force } N)$$

circular motion

$$a = v^2/r \qquad (\text{acceleration } a, \text{ speed } v, \text{ radius } r)$$

$$F = mv^2/r \qquad (\text{centripetal force } F, \text{ mass } m)$$

Solids

For a material in tension

Hooke's Law:

$$F = kx \qquad (\text{tension } F, \text{ spring constant } k, \text{ extension } x)$$

$$\text{stress} = \text{tension/cross-sectional area}$$

$$\text{strain} = \text{extension/original length}$$

$$\text{Young modulus} = \text{stress/strain}$$

$$\text{elastic strain energy} = \tfrac{1}{2}kx^2$$

$$\text{elastic strain energy per unit volume} = \tfrac{1}{2} \text{ stress} \times \text{strain}$$

Gases

Ideal gas equation

for n moles

$$pV = nRT$$

(pressure p, volume V, molar gas constant R, temperature T)

for one mole

$$pV_m = RT$$

(molar volume V_m)

Kinetic theory of gases

$$pV = \tfrac{1}{3}Nm\overline{c^2}$$

(number of molecules N, mass of molecule m, mean square speed $\overline{c^2}$)

$$p = \tfrac{1}{3}\rho\overline{c^2}$$

(density ρ)

mean kinetic energy of translation of one mole of an ideal gas $= \tfrac{3}{2}RT$

Electricity

Flow

$$I = AvnQ$$

(current I, area A, velocity of carriers v, carrier density n, charge Q)

Resistance

$$R = V/I$$

(resistance R, potential difference V)

$$R = \rho l/A$$

(resistivity ρ, length l, area A)

$$R = R_1 + R_2 + \ldots$$

(resistors in series)

$$1/R = 1/R_1 + 1/R_2 + \ldots$$

(resistors in parallel)

Charge

$$\Delta Q = I\Delta t$$

(charge Q, time t)

Capacitance

$$C = Q/V$$

(capacitance C)

$$\text{energy stored} = \tfrac{1}{2}QV$$

$$1/C = 1/C_1 + 1/C_2 + \ldots$$

(capacitors in series)

$$C = C_1 + C_2 + \ldots$$

(capacitors in parallel)

Oscillations

Simple harmonic motion

equation of motion

$$a = -(k/m)s$$

(acceleration a, force per unit displacement k, mass m, displacement s)

displacement–time relation

$$s = A\cos\omega t$$

(amplitude A, angular frequency ω, time t)

$$\omega^2 = k/m$$

$$T = 2\pi/\omega$$

(periodic time T)

$$= 2\pi\sqrt{\frac{m}{k}}$$

$$f = 1/T = \omega/2\pi$$

(frequency f)

$$f = \frac{1}{2\pi}\sqrt{\frac{k}{m}}$$

$$v_{max} = \omega A$$

(maximum velocity v_{max})

$$a_{max} = \omega v_{max} = \omega^2 A$$

(maximum acceleration a_{max})

$$\text{kinetic energy} = \tfrac{1}{2}mv^2$$

$$\text{potential energy} = \tfrac{1}{2}ks^2$$

$$\text{total energy} = \tfrac{1}{2}kA^2$$

Quality factor

$$Q = = 2\pi\frac{\text{energy stored in oscillator}}{\text{energy lost per cycle}}$$

Waves

Wave speeds

for all waves	$c = f\lambda$	(wave speed c, frequency f, wavelength λ)
compression wave in mass-spring system	$c = x\sqrt{\dfrac{k}{m}}$	
sound in a solid	$c = \sqrt{\dfrac{E}{\rho}}$	(Young modulus E, density ρ)
transverse wave on string	$c = \sqrt{\dfrac{T}{\mu}}$	(tension T, mass per unit length μ)

Field and potential

All fields

$$E = -dV/dr$$
$$(\approx -\Delta V/\Delta r)$$

(field strength E, potential gradient dV/dr)

electric field	$E = F/Q$	(electric field strength E, force F, charge Q)
uniform field between parallel plates	$E = V/d$	(potential difference V, separation d)
	$\sigma = \varepsilon_0 E$	(charge density σ, permittivity of free space ε_0)
parallel plate capacitor	$C = \varepsilon_0 \varepsilon_r A/d$	(capacitance C, relative permittivity ε_r, area A, separation d)
point charges	$F = \dfrac{1}{4\pi\varepsilon_0}\dfrac{Q_1 Q_2}{r^2}$	(charges Q_1, Q_2, separation r)
	$E = \dfrac{1}{4\pi\varepsilon_0}\dfrac{Q}{r^2}$	(electric field strength E)
	$V = \dfrac{1}{4\pi\varepsilon_0}\dfrac{Q}{r}$	(electric potential V)
gravitational field	$g = F/m$	(gravitational field strength g, force F, mass m)
	$F = -Gm_1 m_2/r^2$	(universal gravitational constant G, masses m_1, m_2, separation of centres r)
	$g = -GM/r^2$	(mass of Earth, or other body, M)
	$V_g = -GM/r$	(gravitational potential V_g)
	$\Delta V_g = GM(1/r_1 - 1/r_2)$	(gravitational potential difference ΔV_g)
uniform gravitational field	$\Delta V_g = g\Delta h$	(height h)

Atomic and nuclear physics

Radioactive decay

	$dN/dt = -\lambda N$	(number N, decay constant λ)
	$N = N_0 e^{-\lambda t}$	(initial number N_0)
	$T_{\frac{1}{2}} = \dfrac{\ln 2}{\lambda}$	
	$= \dfrac{0.693}{\lambda}$	(half-life $T_{\frac{1}{2}}$)
mass–energy relationship	$\Delta E = c^2 \Delta m$	(energy E, mass m, speed of light c)

nergy transfer

$$\text{efficiency} = \frac{\text{useful energy output}}{\text{total energy input}}$$

$$\text{efficiency of heat engine} = \frac{T_1 - T_2}{T_1} \qquad \text{(temperature of source } T_1,$$
$$\text{temperature of sink } T_2)$$

$$\Delta T = \phi \mathscr{R} \qquad \text{(temperature difference } \Delta T, \text{ rate of}$$
thermal transfer of energy ϕ,
thermal resistance $\mathscr{R}$)

$$\mathscr{R} = l/kA \qquad \text{(area } A, \text{ length } l, \text{ thermal}$$
conductivity k)

$$X = \mathscr{R}A \qquad \text{(thermal resistance coefficient } X)$$

lectrical circuit symbols

ome of the symbols used for circuit diagrams are shown below:

Vires, junctions, terminals
rossing of wires,
o electrical contact

unction

louble junction

erminal

erial

arth

rame or chassis connection

_amps
ignal lamp

amp for illumination

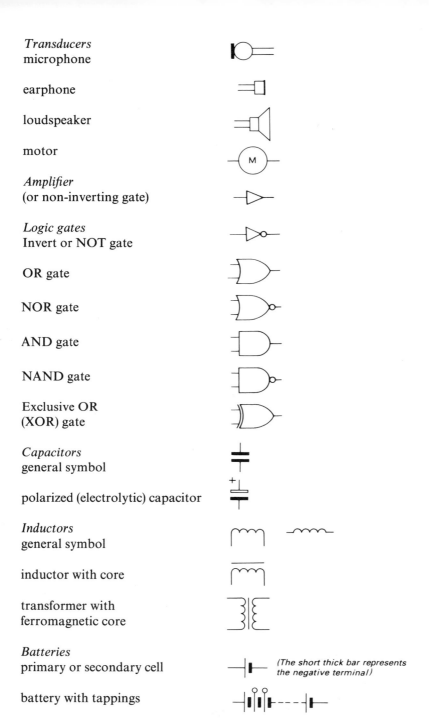

Transducers
microphone

earphone

loudspeaker

motor

Amplifier
(or non-inverting gate)

Logic gates
Invert or NOT gate

OR gate

NOR gate

AND gate

NAND gate

Exclusive OR
(XOR) gate

Capacitors
general symbol

polarized (electrolytic) capacitor

Inductors
general symbol

inductor with core

transformer with
ferromagnetic core

Batteries
primary or secondary cell   *(The short thick bar represents the negative terminal)*

battery with tappings

Diodes
diode/rectifier

light sensitive diode

light emitting diode
(L.E.D.)

Measuring instruments
voltmeter

ammeter

galvanometer

Switches, relays
normally open switch

normally closed switch

relay coil

relay contact

Resistors
general symbol

variable resistor

resistor with preset adjustment

potentiometer

resistor with inherent
variability (e.g. thermistor)

light-sensitive resistor

INDEX

waves, 213–18, 244
  longitudinal, 216; on Slinky spring, 246;
    on trolley-and-spring model, 253–4
  standing, 229–31, 258–61, 263
  superposition, 216–18, 246, 247–51
  transverse, 216; on springs, 244–5, 246;
    on trolley-and-spring model, 245
Wheatstone bridge circuit, 98
  applications, 108–13
wig-wag, *see* inertia balance

wind energy, 465
windows, energy flow through, 450,
  452–3, 454–6, 457
wires, *see under* metals
wood, as structural material, 34
wool, 39
work hardening, 12
  effect on metal structure, 27
'wrap round', 309